TURBULENCE STRUCTURE AND
VORTEX DYNAMICS

The Isaac Newton Institute of Mathematical Sciences of the University of Cambridge exists to stimulate research in all branches of the mathematical sciences, including pure mathematics, statistics, applied mathematics, theoretical physics, theoretical computer science, mathematical biology and economics. The research programmes it runs each year bring together leading mathematical scientists from all over the world to exchange ideas through seminars, teaching and informal interaction.

TURBULENCE STRUCTURE AND VORTEX DYNAMICS

edited by

J.C.R. Hunt

University College London

and

J.C. Vassilicos

University of Cambridge

CAMBRIDGE
UNIVERSITY PRESS

CAMBRIDGE UNIVERSITY PRESS
Cambridge, New York, Melbourne, Madrid, Cape Town,
Singapore, São Paulo, Delhi, Tokyo, Mexico City

Cambridge University Press
The Edinburgh Building, Cambridge CB2 8RU, UK

Published in the United States of America by Cambridge University Press, New York

www.cambridge.org
Information on this title: www.cambridge.org/9780521175128

First published 2000
First paperback edition 2011

A catalogue record for this publication is available from the British Library

ISBN 978-0-521-78131-2 Hardback
ISBN 978-0-521-17512-8 Paperback

Additional resources for this publication at www.cambridge.org/9780521175128

CONTENTS

Contributors

C.F. Barenghi, Mathematics Department, University of Newcastle, Newcastle upon Tyne NE1 7RU, UK
*C.F.Barenghi@newcastle.ac.uk

C. Cambon, Laboratoire de Mécanique de Fluides et d'Acoustique, UMR 5509, Ecole Centrale de Lyon, BP 163, 69131 Ecully Cedex France
cambon@mecaflu.ec-lyon.fr

Pierre Comte, Institut de Mécanique des Fluides, 2 rue Boussingault, 67000 Strasbourg Cedex, France

B. Galanti, Department of Chemical Physics, Weizmann Institute of Science, Rehovot, Israel 76100
galanti@chemphys.weizmann.ac.il

J.D. Gibbon, Department of Mathematics, Imperial College of Science, Technology & Medicine, 180 Queen's Gate, London SW7 2BZ, UK
j.gibbon@ic.ac.uk

Y. Fukumoto, Graduate School of Mathematics, Kyushu University 33, Fukuoka 812-8581, Japan
yasuhide@math.kyushu-u.ac.jp

J.C.R. Hunt, Department of Space and Climate Physics, University College, Gower St., London WC1, UK
jcrh@mssl.ucl.ac.uk

F. Hussain, Department of Mechanical Engineering, University of Houston, Houston TX 77204-4792, USA
fhussain@UH.edu

Craig Johnston, Department of Physics & Astronomy, The University of Edinburgh, The King's Buildings, Mayfield Road, Edinburgh EH9 3JZ, UK
C.Johnston@ed.ac.uk

R.M. Kerr, National Center for Atmospheric Research, P.O. Box 3000, Boulder, Colorado 80307, USA
kerrrobt@ncar.ucar.edu

S. le Dizès, Institut de Recherche sur les Phénomes Hors Equilibre, Université Aix-Marseille, 12 avenue Général Leclerc, F 13003 Marseille, France
ledizes@marius.univ-mrs.fr

A. Leonard, Graduate Aeronautical Laboratories, California Institute of Technology, Pasadena, CA 91125, USA
tony@galcit.caltech.edu

M. Lesieur, Laboratoires de Ecoulements Géophysiques et Industriels, BP 53, 38041 Grenoble Cedex, France
Marcel.Lesieur@hmg.inpg.fr

T. Leweke, Institut de Recherche sur les Phénomes Hors Equilibre, Université Aix-Marseille, 12 avenue Général Leclerc, F 13003 Marseille, France
leweke@marius.univ-mrs.fr

W.D. McComb, Department of Physics & Astronomy, The University of Edinburgh, The King's Buildings, Mayfield Road, Edinburgh EH9 3JZ., UK
wilm@holyrood.ed.ac.uk

Olivier Metais, LEGI, BP 53, 38041 Grenoble Cedex 9, France

G.D. Miller, Boeing Airplane Group, Seattle WA 98124, USA

H.K. Moffatt, Isaac Newton Institute for Mathematical Sciences, University of Cambridge, 20 Clarkson Road, Cambridge CB3 0EH, UK
hkm2@newton.cam.ac.uk

E.A. Novikov, Institute for Nonlinear Science, University of California, San Diego, La Jolla CA 92093-0402, USA
enovikov@ucsd.edu

D.S. Pradeep, Department of Mechanical Engineering, University of Houston, Houston TX 77204–4792, USA

A. Tsinober, Department of Fluid Mechanics and Heat Transfer, Tel Aviv University, Ramat Aviv 69978, Israel
tsinober@eng.tau.ac.il

J.C. Vassilicos, DAMTP, University of Cambridge, Cambridge, CB3 9EW, UK
J.C.Vassilicos@damtp.cam.ac.uk

C.H.K. Williamson, Dept of Mechanical & Aerospace Engineering, Upson Hall, Cornell University, Ithaca NY 14853, USA
cw26@cornell.edu

Z. Warhaft, Department of Mechanical & Aerospace Engineering, Upson Hall, Cornell University, Ithaca NY 14853, USA
zw16@cornell.edu

Introduction

Leonardo da Vinci's drawings of eddies below waterfalls, John Constable's paintings of swirling and disintegrating cloud shapes and L.F. Richardson's Swiftian rhyme all show different aspects of the essential nature of turbulence. When expressed in prosaic scientific language the modern understanding of turbulence is that it is a collection of weakly correlated vortical motions, which, despite their intermittent and chaotic distribution over a wide range of space and time scales, actually consist of local characteristic 'eddy' patterns that persist as they move around under the influences of their own and other eddies' vorticity fields. Numerical simulations and experimental observations have now identified basic forms and even the 'life-cycles' of some of these structures. Some of them, for example, seem to appear as local shear layers, then evolve into vortex tubes and finally break up. In some cases quite extreme distortion and interaction between vortices lead to very large local velocities. These universal features occur in all highly turbulent flows. However, because the largest scale eddies extend across the whole flow and are strongly influenced by the boundary conditions they are not universal; nevertheless they tend to have the same characteristic forms in each type of turbulent flow.

In the Isaac Newton Institute (INI) programme on turbulence held between January and July 1999 there were several workshops and conferences on different aspects of the subject. All of them succeeded in bringing together physicists, engineers, mathematicians and experimentalists, as can be seen in this and other volumes and review articles describing the programme (Voke, Sandham & Kleiser 1999; Launder & Sandham 2000; Vassilicos 2000; Hunt, Sandham, Vassilicos, Launder, Monkewitz & Hewitt 2000).

In the Symposium on Vortex Dynamics and Turbulence Structure there were lectures and discussions on a number of key questions that have engaged turbulence researchers for many years. What is the overall significance for turbulent flows of vortical structures? How should one study their persistence and characteristic structure; do they correspond to some kind of eigensolutions of the basic equations or of some reduced form of these equations; what are their geometrical statistics and their stability, given that they exist in a chaotic environment with many other structures surrounding them? How do they interact or not interact with each other and with surrounding turbulence, and what are their dissipative properties? Are the near-singularities of the turbulence or the conjectured finite-time singularities related to the vortical or other (e.g. straining) structures, and if so what kind? What are the Eulerian and Lagrangian properties of such structures, and how do their conditional statistics relate to the well-established unconditional Eulerian and

Lagrangian statistics (e.g. spectra, energy cascades up- and down-scale, relative motions of particles) and the scaling properties of the entire flow? To what extent can turbulence be represented in terms of space-filling functions such as Fourier or Chebychev basis functions or is it necessary to work in terms of localised functions such as wavelets.

The articles in this volume address all these questions. Most involve mathematical analysis, but some describe numerical simulations and experimental results that focus on these questions. Some of the papers focused on the deterministic kinds of vortical motion that characterise eddy motions, while others also relate these studies to the overall statistics of the turbulent flows which can be measured more readily than the details of individual eddies. Only one paper is exclusively concerned with the statistical dynamics of turbulence.

Deterministic analyses were applied to isolated vortices, to their response when subject to large scale rotational and irrotational straining, and to their interaction with each other. In some situations large scale straining is a reasonable 'mean-field' approximation for the average effects of all other vortices. But in other situations it is necessary to consider specific interactions between small numbers of vortices. **Fukumoto & Moffatt** analyse the effect of viscosity on the motion of a vortex ring, and how the diffusion of vorticity changes its motion. The straining of vortices are considered in three papers; **Gibbon, Galanti & Kerr** consider the general mathematical properties of the stretching and compression of vorticity, including the surprising fact that its tendency to become a singularity at any point in the flow is related to the overall properties of the flow.

There are many different ways that finite amplitude vortices can be stretched and distorted, and **Le Dizes** presents an analysis of a new family of stretched non-axisymmetric vortices. As elongated vortices are stretched and distorted by external straining fields, oscillations and waves can develop and lead to the formation of new structures and ultimately to the total breakdown into small scale chaotic motion. The basic mechanisms of these 'core dynamics' are reviewed by **Pradeep & Hussain**. In some cases the external motions are caused by adjacent vortices and then the instability and transformations are coupled in a global sense, as shown in the experimental paper of **Williamson, Leweke & Miller**. In 'classical' fluids such as air and water at ambient temperature, the vorticity in a vortex diffuses out of vortices or is exchanged when vortices interact as a result of molecular diffusion. In superfluids at very low temperatures these diffusion processes do not occur and therefore vortices move and interact with each other according to the theory of ideal inviscid flow. However certain quantum effects also lead to dissipative phenomena such as reconnection. This is the motivation of **Barenghi's** paper on ideal fluid turbulence and its relation to normal fluid turbulence.

Other papers here show how a combination of deterministic and statistical analyses of turbulent velocity fields is leading to a better understanding of the qualitative characteristics of the eddy motion in turbulence as well as to quantitative predictions. Much research is based on the assumption that this is the key to improving the approximate models of turbulence (such as Large Eddy Simulation and spectral models) and to assessing their accuracy and range of application. **Leonard** analyses, following the earlier ideas of Synge & Lin (1943), the dynamics and kinematics of small individual eddies or packets of vorticity, strained by eddy motions with larger length scales. He explores the limits when the lengths of the strained eddies become comparable with the larger ones, and tend to form elongated and randomly twisted 'ribbons'. The consequences for the spectra are worked out.

Novikov explains why this dynamical interaction implies that small scale turbulence may not be as statistically independent of the large scales as is assumed in Kolmogorov's theory; there may be fewer degrees of freedom and some aspects of their motion may be 'slaved' to the larger scales on some 'slow' manifold. He derives some statistical conditions based on this concept. However the eddy motions do need to be considered because they determine the intermittency of turbulence which he explains as being crucial to the interpretation of the overall turbulence statistics.

Warhaft's discussion of experimental measurements of small scale turbulence also takes up this theme. The higher the order of the statistical moments the more they are anisotropic. These are associated with small scale organised structures, in which there are strong local gradients in both the velocity and scalar fields,. He demonstrates that the structures can be defined more precisely if measurements are made at three rather than two points simultaneously, which has been usual up to now.

Flow visualisation and experiments have indicated that these structures are quite geometrically complex, often approximating to sheets of vorticity and scalars wound up into spiral forms which correspond to a type of ideal mathematical singularity. **Vassilicos** analyses such velocity fields and their effects on the diffusion of scalars; he also shows how these types of eddy can be detected when they occur at random positions in numerical simulations of turbulent flows. He demonstrates how such structures are consistent with the 'anomalous' scaling laws found in statistical correlations in fully developed turbulence.

Tsinober analyses the dynamical equations governing correlations between the straining and vorticity fields of small scale turbulence, in order to clarify the relative roles of vortex stretching and straining, or relative advection, in producing even smaller scales and thence dissipation. His results suggest that it may be necessary to consider a cycle of stretching and straining of eddy motions to understand the full dynamics; indeed the simple, rather static concept of vortex stretching is quite inconsistent with the production of smaller scales. Like Betchov in 1956 he has the temerity to propose an amendment to L.F. Richardson's rhyme about the roles of great whirls and lesser whirls in the cascade process!

Hunt's paper is similar to Leonard's in assuming that the analysis of the non-linear interactions in turbulence can be usefully idealised as a sequence of events when small scale vortices are strained by large scale motions. He discusses how the weakly non-linear effects cause the vortex sheets to roll up, or become unstable. Curiously there is a geometrical problem to be solved: how to define the changes in these shapes, which are associated with the cycle of growth, transformation and breakup of small scale eddies, that Tsinober analyses using statistical data in his paper. Hunt also reviews an outstanding kinematical question about turbulence as to when and to what extent spectra reflect on the one hand the forms of the eddies themselves, especially their singularities, and on the other, the distribution of their amplitudes with wavenumber (or frequency).

Cambon takes up the question, touched on by Hunt, that when vortices are formed in turbulence, for example as a result of straining by larger scales, various kinds of waves and instabilities tend to grow. He reviews and relates a number of current mathematical techniques used for analysing these perturbations. He points out how some are local and some global; some are based on eigen solutions, while others are based on general linear solutions more dependent on initial conditions. Many interesting special cases are described in detail, and reasons are given why cyclonic eddies are more stable than anticyclonic.

In numerical simulations the resolution is now fine enough for even small scale flow structure to be described for high Reynolds number turbulence. **Lesieur, Comte and Metais** use Large Eddy Simulation techniques to examine the structure of the vortices that form in shear flows and rotating flows. They explain how the vortices contribute to the statistical distribution of kinetic energy in the turbulence, as well as describing in some detail how the different scales and orientations of vortices are related in these chaotic flows, which have a high degree of local organisation.

On long enough timescales it is likely that the internal eddy structure is unimportant, in which case turbulence can be analysed rather like a viscoelastic fluid, based on the concepts and methods of statistical physics. **McComb & Johnston** use methods involving the Renormalisation Group. In conjunction with novel assumptions about the statistical independence of the small eddy scales, they derive quantitatively the energy spectrum of turbulence and new results about the internal 'eddy viscosity' that controls the energy transfer between eddy scales. These methods may well have wider applications to more complex flows in future.

We, and we believe all the speakers at the workshop, are extremely grateful to Geoff Hewitt, Peter Monkewitz and Neil Sandham for their invaluable contribution to the organisation of the Isaac Newton Institute's Turbulence Research Programme: to Keith Moffatt and the wonderful staff of the Isaac Newton Institute, ERCOFTAC, the European Commission, the Royal Academy of Engineering and the Industrial Working Group under the chairmanship of Michael Reeks, for their support; and to the Isaac Newton Institute for sponsoring and hosting the workshops.

References

Hunt, J.C.R., Sandham, N., Vassilicos, J.C., Launder, B.E., Monkewitz, P.A. & Hewitt, G.F., 2000, 'Developments in turbulence research: a review based on the 1999 Programme of the Isaac Newton Institute, Cambridge', submitted to *J. Fluid Mech.*

Launder, B.E. & Sandham, N.D. (eds.), 2000, *Closure Strategies for Turbulent and Transitional Flows*, Cambridge University Press.

Synge, J.L. & Lin, C.C., 1943, 'On a statistical model of isotropic turbulence', *Trans. R. Soc. Canada*, **37**, 45–79.

Vassilicos, J.C., (ed.), 2000, *Intermittency in Turbulent Flows*, Cambridge University Press.

Voke, P.R., Sandham, N., & Kleiser, L. (eds.), 1999, *Direct Large-Eddy Simulation III*, Proceedings of the Isaac Newton Institute Symposium/ERCOFTAC Workshop, Cambridge UK, 12–14 May 1999, ERCOFTAC Series, Vol. 7, Kluwer Academic Publishers.

Julian Hunt

Christos Vassilicos

Motion and Expansion of a Viscous Vortex Ring: Elliptical Slowing Down and Diffusive Expansion

Yasuhide Fukumoto and H.K. Moffatt

1 Introduction

The motion of a vortex ring is a venerable problem, and, since the attempts of Helmholtz and Kelvin in the last century, extensive study has been made on various dynamical aspects, such as formation, traveling speed, waves, instability, interactions and so on. Concerning the steady solution for inviscid dynamics, analytical technique has been matured enough to make a highly nonlinear regime tractable. In contrast, the effect of viscosity on the nonlinear dynamics is poorly understood even for an isolated vortex ring.

In this article, we present a large-Reynolds-number asymptotic theory of the Navier–Stokes equations for the motion of an axisymmetric vortex ring of small cross-section. Our intention is to make the nonlinear effect amenable to analysis by constructing a framework for calculating higher-order asymptotics. The nonlinearity is featured by deformation of the core cross-section. We build a general formula for the translation speed incorporating the slowing-down effect caused by the elliptical deformation of the core. Moreover we show that viscosity has the action of expanding the ring radius, simultaneously with swelling the core; starting from an infinitely thin circular loop of radius R_0, the radii $R_s(t)$ of the loop of stagnation points relative to a comoving frame, $R_p(t)$ of the loop of peak vorticity, $R_c(t)$ of the centroid of vorticity all grow linearly in time t as $R_s \approx R_0 + 2.5902739\nu t/R_0$, $R_p \approx R_0 + 4.5902739\nu t/R_0$, and $R_c \approx R_0 + 3\nu t/R_0$. It is pointed out that the asymptotic values of R_p and R_c exhibit a discrepancy, at a finite Reynolds number, from the numerical result of Wang, Chu & Chang (1994).

To begin with, we briefly survey known results. Dyson (1893) (see also Fraenkel 1972) extended Kelvin's formula for the speed U of a thin axisymmetric vortex ring, steadily translating in an inviscid incompressible fluid of infinite extent, to third (virtually fourth) order in a small parameter $\varepsilon = \sigma/R_0$, the ratio of core radius σ to the ring radius R_0, as

$$U = \frac{\Gamma}{4\pi R_0}\left\{\log\left(\frac{8}{\varepsilon}\right) - \frac{1}{4} - \frac{3\varepsilon^2}{8}\left[\log\left(\frac{8}{\varepsilon}\right) - \frac{5}{4}\right] + O(\varepsilon^4\log\varepsilon)\right\}, \quad (1.1)$$

1

where Γ is the circulation carried by the ring. The vorticity is assumed to be in proportion to distance from the axis of symmetry. We consider Kelvin's formula (the first two terms) as the first-order and the $O(\varepsilon^2)$-terms as the third. The local self-induced flow consists not only of a uniform flow but also of a straining field. The latter manifests itself at $O(\varepsilon^2)$ and deforms the core into an ellipse, elongated in the propagating direction:

$$r = \sigma \left\{ 1 - \frac{3\varepsilon^2}{8} \left[\log \left(\frac{8}{\varepsilon} \right) - \frac{17}{12} \right] \cos 2\theta + \cdots \right\}, \qquad (1.2)$$

where (r, θ) are local moving cylindrical coordinates about the core center which will be introduced in §2. The inclusion of the third-order term in the propagating velocity gives a remarkable improvement in approximation; (1.1) compares well even with the exact value for the 'fat' limit of Hill's spherical vortex (Fraenkel 1972). In this limit, the parameter ε is as large as $\sqrt{2}$ under a suitable normalisation. This surprising agreement encourages us to explore a higher-order approximation in more general circumstances.

Viscosity acts to diffuse vorticity, and the motion ceases to be steady. Its influence on the traveling speed, at large Reynolds number, was first addressed by Tung & Ting (1967), using the matched asymptotic expansions, for the case where the the vorticity is, at a virtual instant $t = 0$, a 'δ-function' concentrated on a circle of radius R_0. By a different method, Saffman (1970) succeeded in deriving an explicit formula, valid up to first order in $\epsilon \equiv (\nu/\Gamma)^{1/2}$, as

$$U = \frac{\Gamma}{4\pi R_0} \left[\log \left(\frac{8R_0}{2\sqrt{\nu t}} \right) - \frac{1}{2}(1 - \gamma + \log 2) + \cdots \right], \qquad (1.3)$$

where ν is the viscosity, t is the time, and $\gamma = 0.57721566\cdots$ is Euler's constant (see also Callegari & Ting 1978). Wang, Chu & Chang (1994) employed a similar method to Tung & Ting (1967), but with a different choice $\sqrt{t}$ as small parameter, and gained a correction to (1.3) originating from the viscous diffusive effect. This correction vanishes in the limit of $\nu \to 0$. Unfortunately, the existing asymptotic theories all assume a circular symmetric core with a Gaussian distribution of vorticity. It implies that our knowledge of the non-linear effect is restricted to $O(\epsilon)$. For comprehensive lists of theories of vortex rings, the article of Shariff & Leonard (1992) should be referred to.

Motivated by intriguing pattern variation of the dissipation field visualised from numerical data of simulations of fully developed turbulence, Moffatt, Kida & Ohkitani (1994) developed a large-Reynolds-number asymptotic theory for a steady stretched vortex tube subjected to uniform non-axisymmetric irrotational strain. They demonstrated that the higher-order asymptotics satisfactorily account for the fine structure of the dissipation field previously obtained by numerical computation (Kida & Ohkitani 1992). The corresponding planar problem, though unsteady, is dealt with in a similar manner, and an extension of the result of Ting & Tung (1965) to a higher order was achieved by Jiménez, Moffatt & Vasco (1996). The structure of the solutions have much in common; at leading order, a columnar vortex with circular cores, an exact solution of the Navier–Stokes equations, is obtained. A quadrupole component enters at $O(\nu/\Gamma)$, which is realised as the deformation of the core cross-section into an ellipse. The distinguishing feature is that the major axis of the ellipse is aligned at $45°$ to the principal axis of the external stain. This result leads us to expectat that the strained cross-section of a vortex ring, observed in nature, is established as an equilibrium between self-induced strain and viscous diffusion. Along the line of this scenario, we elucidate the structure of this strained core and its influence on the traveling speed of an axisymmetric vortex ring.

A powerful technique for our purpose is the method of matched asymptotic expansions. It has been previously developed to derive the velocity of a slender curved vortex tube (see, for example, Ting & Klein 1991). However this method is limited to $O(\epsilon^2)$ (Moore & Saffman 1972; Fukumoto & Miyazaki 1991). In the viscous case also, the self-induced strain, with the resulting elliptical deformation of the core, makes its appearance at $O(\epsilon^2)$, and its influences on the translation speed come up at $O(\epsilon^3)$. We are thus requested to extend asymptotic expansions to a higher order.

In §2, we state the general problem. The existing asymptotic formula for the potential flow associated with a circular vortex loop is not sufficient to carry through our program. In order to work out the correct inner limit of the outer solution, we devise, in §3, a technique to produce a systematic asymptotic expression of the Biot–Savart integral accommodating an arbitrary vorticity distribution. In §4, the inner expansions are scrutinised to $O(\epsilon)$ and are extended to $O(\epsilon^2)$. Based on these, we demonstrate in §5.1 that the radii of the loops of the stagnation points, maximum vorticity and vorticity centroid all grow linearly in time owing to the action of viscosity. Thereafter, we establish in §5.2 a general formula for the translating velocity of a vortex ring. In §6, an equation governing the temporal evolution of the axisymmetric vorticity at $O(\epsilon^2)$ is derived, and an integral representation of the exact solution is given, by which the formula of the preceding section can be closed.

A few ambiguous steps lying in previous theories stand as obstacles to proceeding to higher orders. These highlight the significance of the dipoles distributed along the core centerline and oriented in the propagating direction. It turns out that their strength needs to be prescribed at an initial instant, which solves the problem of undetermined constants at $O(\epsilon)$. As a by-product, a clear interpretation is provided of the general mechanism of the self-induced motion of a curved vortex tube. Because of the limitation of space, we must omit the technical details. A comprehensive account of our theory will be available in the paper of Fukumoto & Moffatt (2000).

2   Formulation

Consider an axisymmetric vortex ring of circulation Γ moving in an infinite expanse of viscous fluid with kinematic viscosity ν. We suppose that the circulation Reynolds number Re_Γ is very large:

$$Re_\Gamma = \Gamma/\nu \gg 1 \,. \tag{2.1}$$

Two length scales are available, namely, measures of the core radius σ and the ring radius R_0. Suppose that their ratio σ/R_0 is very small. We focus attention on the translational motion of a 'quasi-steady' core. This means that we exclude stable or unstable wavy motion and fast core-area waves. Then, according to (1.1), the time-scale under question is of order $R_0/(\Gamma/R_0) = R_0^2/\Gamma$. The core spreads over this time to be of order $\sigma \sim (\nu t)^{1/2} \sim (\nu/\Gamma)^{1/2} R_0$. Our assumption of slenderness requires that the relevant small parameter $\epsilon (\ll 1)$ is

$$\epsilon = \sqrt{\nu/\Gamma} \,. \tag{2.2}$$

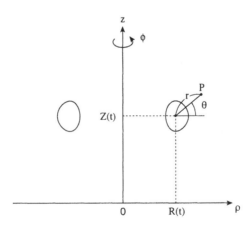

Figure 1

Choose cylindrical coordinates (ρ, ϕ, z) with the z-axis along the axis of symmetry and ϕ along the vortex lines as shown in Figure 1. We consider an axisymmetric distribution of vorticity $\boldsymbol{\omega} = \zeta(\rho, z)\boldsymbol{e}_\phi$ localised about the circle

$(\rho, z) = (R(t), Z(t))$, where e_ϕ is the unit vector in the azimuthal direction. The Stokes streamfunction ψ is given by

$$\psi(\rho, z) = -\frac{\rho}{4\pi} \int_{-\infty}^{\infty} \int_0^{2\pi} \int_0^{\infty} \frac{\zeta(\rho', z')\rho' \cos\phi' d\rho' d\phi' dz'}{\sqrt{\rho^2 - 2\rho\rho' \cos\phi' + \rho'^2 + (z - z')^2}}. \qquad (2.3)$$

The theorems of Kelvin and Helmholtz imply that determination of the ring motion necessitates a knowledge of the flow velocity in the vicinity of the core.

As is well known, the irrotational flow-velocity calculated from (2.3) for an infinitely thin core increases without limit primarily in inverse proportion to the distance from the core. In addition, it entails a logarithmic infinity originating from the curvature effect. These singularities may be resolved by matching the outer flow to an inner vortical flow which decays rapidly as the core center is approached. Thus we are led to inner and outer expansions (Ting & Tung 1965; Tung & Ting 1967). The inner region consists of the core itself and the surrounding toroidal region with thickness of the order of the core radius σ. There we develop an inner asymptotic expansion which matches at each level to the outer solution (2.3).

To this end, it is advantageous to introduce, in the axial plane, local polar coordinates (r, θ) moving with the core center[1] $(R(t), Z(t))$ with $\theta = 0$ in the ρ-direction (Figure 1):

$$\rho = R(t) + r\cos\theta, \qquad z = Z(t) + r\sin\theta. \qquad (2.4)$$

Let us make the inner variables dimensionless. The radial coordinate is normalised by the core radius $\epsilon R_0 (= \sigma)$ and the local velocity $v = (u, v)$, relative to the moving frame, by the maximum velocity $\Gamma/(\epsilon R_0)$. In view of (1.1), the normalisation parameter for the ring speed $(\dot{R}(t), \dot{Z}(t))$, the slow dynamics, should be Γ/R_0. The suitable dimensionless inner variables are thus defined as

$$r^* = r/\epsilon R_0, \quad t^* = t/\frac{R_0}{\Gamma}, \quad \psi^* = \frac{\psi}{\Gamma R_0}, \quad \zeta^* = \zeta/\frac{\Gamma}{R_0^2 \epsilon^2},$$
$$v^* = v/\frac{\Gamma}{R_0 \epsilon}, \quad (\dot{R}^*, \dot{Z}^*) = (\dot{R}, \dot{Z})/\frac{\Gamma}{R_0}. \qquad (2.5)$$

The difference in normalisation between the last two of (2.5) should be kept in mind.

The equations handled in the inner region are the coupled system of the vorticity equation and the subsidiary relation between ζ and ψ. Dropping the asterisks, they take the following form:

$$\frac{\partial \zeta}{\partial t} + \frac{1}{\epsilon^2}\left(u\frac{\partial \zeta}{\partial r} + \frac{v}{r}\frac{\partial \zeta}{\partial \theta}\right) - \frac{1}{\epsilon\rho^2}\left(\frac{\partial \psi}{\partial r}\sin\theta + \frac{1}{r}\frac{\partial \psi}{\partial \theta}\cos\theta\right)\zeta$$

---

[1] The definition of the 'core center' will be discussed at some length in §4.2

$$= \hat{\nu} \left[\Delta\zeta + \frac{\epsilon}{\rho} \left(\cos\theta \frac{\partial}{\partial r} - \frac{\sin\theta}{r} \frac{\partial}{\partial \theta} \right) \zeta - \frac{\epsilon^2}{\rho^2} \zeta \right], \qquad (2.6)$$

$$\zeta = \frac{1}{\rho} \Delta\psi - \frac{\epsilon}{\rho^2} \left(\cos\theta \frac{\partial}{\partial r} - \frac{\sin\theta}{r} \frac{\partial}{\partial \theta} \right) \psi, \qquad (2.7)$$

where $\hat{\nu} = 1$, $\rho = R + \epsilon r \cos\theta$, and Δ is the two-dimensional Laplacian,

$$\Delta = \frac{\partial^2}{\partial r^2} + \frac{1}{r} \frac{\partial}{\partial r} + \frac{1}{r^2} \frac{\partial^2}{\partial \theta^2}, \qquad (2.8)$$

and u and v are the r- and θ-components of the relative velocity v:

$$u = \frac{1}{r\rho} \frac{\partial\psi}{\partial\theta} - \epsilon(\dot{Z}\sin\theta + \dot{R}\cos\theta), \qquad (2.9a)$$

$$v = -\frac{1}{\rho} \frac{\partial\psi}{\partial r} - \epsilon(\dot{Z}\cos\theta - \dot{R}\sin\theta). \qquad (2.9b)$$

We now postulate the following series expansions of the solution:

$$\zeta = \zeta^{(0)} + \epsilon\zeta^{(1)} + \epsilon^2\zeta^{(2)} + \epsilon^3\zeta^{(3)} + \cdots, \qquad (2.10a)$$

$$\psi = \psi^{(0)} + \epsilon\psi^{(1)} + \epsilon^2\psi^{(2)} + \epsilon^3\psi^{(3)} + \cdots, \qquad (2.10b)$$

$$R = R^{(0)} + \epsilon R^{(1)} + \epsilon^2 R^{(2)} + \cdots, \qquad (2.10c)$$

$$Z = Z^{(0)} + \epsilon Z^{(1)} + \epsilon^2 Z^{(2)} + \cdots, \qquad (2.10d)$$

where $\zeta^{(i)}$ and $\psi^{(i)}$ ($i = 0, 1, 2, 3, \cdots$) are functions of r, θ and sometimes t. There arises $\log\epsilon$ as well, but we may conveniently take it to be of order unity, since multiples of $\log\epsilon$ happen to be ruled out at least to the above order. Inserting these expansions into (2.6) and (2.7), supplemented by (2.8)–(2.9b), and collecting terms with like powers of ϵ, we obtain the equations to be solved in the inner region.

The permissible solution must satisfy the condition:

$$u \text{ and } v \text{ are finite at } r = 0. \qquad (2.11)$$

We emphasise that this condition is better than the restrictive one that $u = v = 0$ at $r = 0$. The requirement that it smoothly match the asymptotic form, valid in the vicinity of the core, of the outer solution will determine the values of $\dot{R}^{(i)}$ and $\dot{Z}^{(i)}$ ($i = 0, 1, 2, \cdots$). This procedure was already performed by Tung & Ting (1967) and Callegari & Ting (1978) and others, up to first order. Our aim is to explore the second and third orders. Before that, we reconsider the earlier results.

3 Outer solution

For a circular vortex loop of unit strength placed at $(\rho, z) = (R, Z)$, $\zeta = \delta(\rho - R)\delta(z - Z)$ and the Stokes streamfunction (2.3) simplifies to

$$\psi_m(\rho, z; R) = -\frac{\rho}{4\pi} \int_0^{2\pi} \frac{R\cos\phi' d\phi'}{\sqrt{\rho^2 - 2\rho R\cos\phi' + R^2 + (z - Z)^2}}. \tag{3.1}$$

Use of the complete elliptic integrals K and E of the first and second kinds converts (3.1) into Maxwell's well-known formula. We call ψ_m the 'monopole field'. With the aid of the asymptotic behaviour of K and E for modulus close to unity, the asymptotic form of ψ_m for $r \ll R$ is obtainable at once (Dyson 1893; Tung & Ting 1967):

$$\begin{aligned}
\psi_m &= -\frac{\Gamma R}{2\pi} \bigg\{ \log\left(\frac{8R}{r}\right) + \frac{r}{2R}\left[\log\left(\frac{8R}{r}\right) - 1\right]\cos\theta \\
&\quad + \frac{r^2}{2^4 R^2}\left(\left[2\log\left(\frac{8R}{r}\right) + 1\right] + \left[-\log\left(\frac{8R}{r}\right) + 2\right]\cos 2\theta\right) \\
&\quad + \frac{r^3}{2^6 R^3}\left(\left[-3\log\left(\frac{8R}{r}\right) + 1\right]\cos\theta + \left[\log\left(\frac{8R}{r}\right) - \frac{7}{3}\right]\cos 3\theta\right) \bigg\} \\
&\quad + \cdots .
\end{aligned} \tag{3.2}$$

It turns out however that, when going to higher orders, (3.1) is not enough to qualify as the outer solution. Investigation of the detailed structure of (2.3) is unavoidable.

For this purpose, it is expedient to adapt Dyson's 'shift operator' technique to an arbitrary distribution of vorticity, and to cast (2.3) in the following form:

$$\psi = -\frac{\rho}{4\pi} \iint_{-\infty}^{\infty} dx' d\hat{z}' \zeta(x', \hat{z}') e^{x'\frac{\partial}{\partial R} - \hat{z}'\frac{\partial}{\partial z}} \int_0^{2\pi} \frac{R\cos\phi' d\phi'}{\sqrt{\rho^2 - 2\rho R\cos\phi' + R^2 + \hat{z}^2}}, \tag{3.3}$$

where $(x, \hat{z}) = (\rho - R, z - Z)$ are local Cartesian coordinates attached to the moving frame, and ζ is rewritten in terms of them. Hereafter, we use z for $\hat{z}$. Supposing rapid decrease of vorticity with distance from the local origin $r = 0$, the exponential function of the operators is formally expanded in Taylor series as

$$\begin{aligned}
\psi(\rho, z) &= \iint_{-\infty}^{\infty} dx' dz' \zeta(x', z') \bigg\{ 1 + \left(x'\frac{\partial}{\partial R} - z'\frac{\partial}{\partial z}\right) + \frac{1}{2!}\left(x'\frac{\partial}{\partial R} - z'\frac{\partial}{\partial z}\right)^2 \\
&\quad + \frac{1}{3!}\left(x'\frac{\partial}{\partial R} - z'\frac{\partial}{\partial z}\right)^3 + \frac{1}{4!}\left(x'\frac{\partial}{\partial R} - z'\frac{\partial}{\partial z}\right)^4 + \frac{1}{5!}\left(x'\frac{\partial}{\partial R} - z'\frac{\partial}{\partial z}\right)^5 \\
&\quad + \frac{1}{6!}\left(x'\frac{\partial}{\partial R} - z'\frac{\partial}{\partial z}\right)^6 + \cdots \bigg\} \psi_m(\rho, z; R).
\end{aligned} \tag{3.4}$$

We shall find in §4 that, up to $O(\epsilon^3)$, the vorticity distribution has the following dependence on the local polar coordinate θ:

$$\zeta(x,z) = \zeta_0 + \epsilon\,\zeta_{11}^{(1)}\cos\theta + \epsilon^2(\zeta_0^{(2)} + \zeta_{21}^{(2)}\cos 2\theta)$$
$$+\epsilon^3(\zeta_{11}^{(3)}\cos\theta + \zeta_{12}^{(3)}\sin\theta + \zeta_{31}^{(3)}\cos 3\theta) + \cdots, \qquad (3.5)$$

where $\zeta_{ij}^{(k)}$ are functions of r and t, and k stands for the order of perturbation, i labels the Fourier mode with $j = 1$ and 2 corresponding to $\cos i\theta$ and $\sin i\theta$ respectively.

With this form, (3.2) and (3.5), along with (2.10c), are substituted into (3.4) and the resulting expression is made dimensionless by use of the normalization (2.5). Using, in advance, $R^{(1)} = 0$ and (6.5), we eventually arrive at the asymptotic development of the Biot–Savart law, valid to $O(\epsilon^3)$, in a region $r \ll R$ surrounding the core:

$$\psi = -\frac{R^{(0)}\Gamma}{2\pi}\log\left(\frac{8R^{(0)}}{\epsilon r}\right) + \epsilon\left\{-\frac{\Gamma}{4\pi}\left[\log\left(\frac{8R^{(0)}}{\epsilon r}\right) - 1\right]r\cos\theta + d^{(1)}\frac{\cos\theta}{r}\right\}$$

$$+\epsilon^2\left\{-\frac{\Gamma}{2^5\pi R^{(0)}}\left(\left[2\log\left(\frac{8R^{(0)}}{\epsilon r}\right) + 1\right]r^2 - \left[\log\left(\frac{8R^{(0)}}{\epsilon r}\right) - 2\right]r^2\cos 2\theta\right)\right.$$

$$\left.+\frac{d^{(1)}}{2R^{(0)}}\left[\log\left(\frac{8R^{(0)}}{\epsilon r}\right) + \frac{\cos 2\theta}{2}\right] - \frac{\Gamma R^{(2)}}{2\pi}\log\left(\frac{8R^{(0)}}{\epsilon r}\right) + q^{(2)}\frac{\cos 2\theta}{r^2}\right\}$$

$$+\epsilon^3\left\{\frac{\Gamma}{2^7\pi(R^{(0)})^2}\left(\left[3\log\left(\frac{8R^{(0)}}{\epsilon r}\right) - 1\right]r^3\cos\theta - \left[\log\left(\frac{8R^{(0)}}{\epsilon r}\right) - \frac{7}{3}\right]r^3\cos 3\theta\right)\right.$$

$$-\frac{d^{(1)}}{8(R^{(0)})^2}\left(\left[\log\left(\frac{8R^{(0)}}{\epsilon r}\right) - \frac{7}{4}\right]r\cos\theta + \frac{r\cos 3\theta}{4}\right) - \frac{\Gamma R^{(2)}}{4\pi R^{(0)}}r\cos\theta$$

$$-\frac{1}{2\pi}\left(\frac{1}{4}\left[2\pi\int_0^\infty r^3\zeta_0^{(2)}dr\right] + R^{(0)}\left[\pi\int_0^\infty r^2\zeta_{11}^{(3)}dr\right] + \frac{1}{4}\left[\pi\int_0^\infty r^3\zeta_{21}^{(2)}dr\right]\right)\frac{\cos\theta}{r}$$

$$+\frac{q^{(2)}}{4R^{(0)}r}(\cos\theta + \cos 3\theta) - \frac{1}{\pi R^{(0)}}\left(\frac{1}{3\cdot 2^8}\left[2\pi\int_0^\infty r^7\zeta^{(0)}dr\right]\right.$$

$$-\frac{R^{(0)}}{8\cdot 4!}\left[\pi\int_0^\infty r^6\zeta_{11}^{(1)}dr\right] + \frac{(R^{(0)})^2}{4!}\left[\pi\int_0^\infty r^5\zeta_{21}^{(2)}dr\right]$$

$$\left.+\frac{(R^{(0)})^3}{6}\left[\pi\int_0^\infty r^4\zeta_{31}^{(3)}dr\right]\right)\frac{\cos 3\theta}{r^3} - \frac{R^{(0)}}{2\pi}\left[\pi\int_0^\infty r^2\zeta_{12}^{(3)}dr\right]\frac{\sin\theta}{r}\right\}$$

$$+\cdots, \qquad (3.6)$$

where

$$\Gamma = 2\pi\int_0^\infty r\zeta^{(0)}dr, \qquad (3.7a)$$

($\Gamma = 1$ when dimensionless), and $d^{(1)}$ and $q^{(2)}$ are the strength of the dipole at $O(\epsilon)$ and quadrupole at $O(\epsilon^2)$:

$$d^{(1)} = -\frac{1}{2\pi}\left\{\frac{1}{4}\left[2\pi\int_0^\infty r^3\zeta^{(0)}dr\right] + R^{(0)}\left[\pi\int_0^\infty r^2\zeta_{11}^{(1)}dr\right]\right\}, \qquad (3.7b)$$

$$q^{(2)} = -\frac{1}{2\pi R^{(0)}}\left\{-\frac{1}{2^6}\left[2\pi \int_0^\infty r^5 \zeta^{(0)} dr\right] + \frac{R^{(0)}}{8}\left[\pi \int_0^\infty r^4 \zeta_{11}^{(1)} dr\right]\right.$$
$$\left. +\frac{(R^{(0)})^2}{2}\left[\pi \int_0^\infty r^3 \zeta_{21}^{(2)} dr\right]\right\}. \tag{3.7c}$$

The terms multiplied by Γ stem from $\Gamma\psi_m$, and only these have been previously employed as the outer solution. We now recognize that, at higher orders, the monopole field needs to be corrected by the induction velocity associated with the di-, quadru-, hexa-poles ... distributed along the centerline $r = 0$ of the core. In the light of (3.7b) and (3.7c), the detailed profile of vorticity in the core is necessary to evaluate these multi-pole induction terms. Parts of (3.6) supply the matching conditions on the inner solution. The distributions of $\zeta_{11}^{(1)}$, $\zeta_0^{(2)}$, $\zeta_{21}^{(2)}$, $\zeta_{11}^{(3)}$, $\zeta_{12}^{(3)}$ and $\zeta_{31}^{(3)}$ are as yet unknown, but will be determined successively by the inner expansions and the matching procedure.

4 Inner expansions up to second order

In this section, we recall the inner expansions at leading and first orders, developed by Tung & Ting (1967), Widnall, Bliss & Zalay (1971) and Callegari & Ting (1978), and extend them to second order.

4.1 Zeroth order

At $O(\epsilon^0)$, the Navier–Stokes equations reduce to the Jacobian form of the steady Euler equations:

$$[\zeta^{(0)}, \psi^{(0)}] \equiv \frac{1}{r}\frac{\partial(\zeta^{(0)}, \psi^{(0)})}{\partial(r, \theta)} = 0, \tag{4.1}$$

resulting in $\zeta^{(0)} = \mathcal{F}(\psi^{(0)})$, for some function $\mathcal{F}$.

Suppose that the flow $\psi^{(0)}$ has a single stagnation point at $r = 0$, the streamlines being all closed around that point. Then it is probable that the solution of (4.1), coupled with the ζ-ψ relation

$$\zeta^{(0)} = \frac{1}{R^{(0)}}\Delta\psi^{(0)}, \tag{4.2}$$

must be 'radial'; the streamlines are necessarily all circles (Moffatt *et al.* 1994). This statement may stand as a corollary of the theorem proved by Caffarelli & Friedman (1980) and Fraenkel (1999). In any event, we may certainly *assume* that $\psi^{(0)} = \psi^{(0)}(r)$.

The functional form of $\psi^{(0)}$ and $\zeta^{(0)}$ remains undetermined at this level of approximation, but is determined through the axisymmetric (or θ-averaged) part of the vorticity equation at $O(\epsilon^2)$:

$$\frac{\partial \zeta^{(0)}}{\partial t} = \left(\zeta^{(0)} + \frac{r}{2}\frac{\partial \zeta^{(0)}}{\partial r}\right)\frac{\dot{R}^{(0)}}{R^{(0)}} + \hat{\nu}\left(\frac{\partial^2 \zeta^{(0)}}{\partial r^2} + \frac{1}{r}\frac{\partial \zeta^{(0)}}{\partial r}\right), \qquad (4.3)$$

(Tung & Ting 1967). It follows that viscosity plays the role of selecting the distribution. For instance, we restrict our attention to a specific initial distribution of a 'δ-function' vorticity concentrated on the circle of radius $R^{(0)}$:

$$\zeta^{(0)} = \delta(\rho - R^{(0)})\delta(z - Z^{(0)}) \quad \text{at} \ \ t = 0. \qquad (4.4)$$

When $R^{(0)}$ is constant, to be shown in the next subsection, we obtain the Oseen diffusing vortex:

$$\zeta^{(0)} = \frac{1}{4\pi\hat{\nu}t}e^{-r^2/4\hat{\nu}t}. \qquad (4.5)$$

In view of (2.9a), (2.9b) and (4.2), the leading-order variables are related to each other through

$$u^{(0)} = \frac{1}{R^{(0)}r}\frac{\partial \psi^{(0)}}{\partial \theta}, \quad v^{(0)} = -\frac{1}{R^{(0)}}\frac{\partial \psi^{(0)}}{\partial r}, \quad \zeta^{(0)} = -\frac{1}{r}\frac{\partial}{\partial r}\left(rv^{(0)}\right). \quad (4.6)$$

These are integrated to provide $u^{(0)} = 0$ and, in the case of the Oseen vortex,

$$v^{(0)} = -\frac{1}{2\pi r}\left(1 - e^{-r^2/4\hat{\nu}t}\right), \quad \psi^{(0)} = \frac{R^{(0)}}{2\pi}\int_0^r \frac{1}{r'}\left(1 - e^{-r'^2/4\hat{\nu}t}\right)dr'. \quad (4.7)$$

This solution automatically fulfills the matching condition, the leading-order part of (3.6).

4.2   First order

Combining the vorticity equation with ζ-ψ relation at $O(\epsilon)$, we see that the first-order perturbation $\psi^{(1)}$ satisfies

$$(\Delta - a)\psi^{(1)} = -\cos\theta v^{(0)} + R^{(0)}ra(\dot{Z}^{(0)}\cos\theta - \dot{R}^{(0)}\sin\theta) + 2r\zeta^{(0)}\cos\theta, \quad (4.8)$$

where

$$a(r, t) = -\frac{1}{v^{(0)}}\frac{\partial \zeta^{(0)}}{\partial r}. \qquad (4.9)$$

Here we have anticipated that $\zeta_0^{(1)} = 0$, which follows from an analysis of the vorticity equation at $O(\epsilon^3)$.

The solution satisfying the condition (2.11) at $O(\epsilon^3)$ is explicitly written in the following way. The θ-dependence is

$$\psi^{(1)} = \psi_{11}^{(1)}\cos\theta + \psi_{12}^{(1)}\sin\theta. \qquad (4.10)$$

The Fourier coefficients are conveniently decomposed into two parts:

$$\psi_{11}^{(1)} = \tilde{\psi}_{11}^{(1)} - R^{(0)} r \dot{Z}^{(0)}, \quad \psi_{12}^{(1)} = \tilde{\psi}_{12}^{(1)} + R^{(0)} r \dot{R}^{(0)}. \tag{4.11}$$

The equation for $\tilde{\psi}_{11}^{(1)}$ then becomes

$$\left[\frac{\partial^2}{\partial r^2} + \frac{1}{r} \frac{\partial}{\partial r} - \left(\frac{1}{r^2} + a \right) \right] \tilde{\psi}_{11}^{(1)} = -v^{(0)} + 2r\zeta^{(0)}, \tag{4.12}$$

and $\tilde{\psi}_{12}^{(1)}$ is governed simply by the homogeneous part of (4.12). By inspection, we find that $v^{(0)}$ is a solution of (4.12) (Widnall, *et al.* 1971; Callegari & Ting 1978). The general solution is then immediately obtainable as

$$\tilde{\psi}_{11}^{(1)} = \Psi_{11}^{(1)} + c_{11}^{(1)} v^{(0)}, \quad \tilde{\psi}_{12}^{(1)} = c_{12}^{(1)} v^{(0)}, \tag{4.13}$$

where $\Psi_{11}^{(1)}$ is a particular integral of (4.12):

$$\Psi_{11}^{(1)} = -v^{(0)} \left\{ \frac{r^2}{2} + \int_0^r \frac{dr'}{r'[v^{(0)}(r')]^2} \int_0^{r'} r'' \left[v^{(0)}(r'') \right]^2 dr'' \right\}, \tag{4.14}$$

and $c_{11}^{(1)}$ and $c_{12}^{(1)}$ are constants (which may depend on t).

To have an idea of the meaning of these constants, observe that, with spatial translation of the origin of the moving frame by $\epsilon\alpha$ in the ρ-direction and $\epsilon\beta$ in the z-direction, the streamfunction, when redefined as a function of the relative coordinates, is altered to

$$\psi(\rho - (R_0 + \epsilon\alpha), z - (Z_0 + \epsilon\beta))$$
$$= \psi^{(0)} + \epsilon \left(\psi^{(1)} + \alpha R_0 v^{(0)} \cos\theta + \beta R_0 v^{(0)} \sin\theta \right) + O(\epsilon^2). \tag{4.15}$$

Comparison of (4.15) with (4.11) and (4.13) suggests that $c_{11}^{(1)}$ is tied in with shift of the moving frame radially outward by $\epsilon c_{11}^{(1)}/R_0$ and $c_{12}^{(1)}$ with shift in the axial direction by $\epsilon c_{12}^{(1)}/R_0$. Alternatively, $c_{11}^{(1)}$ and $c_{12}^{(1)}$ may be reckoned upon as the parameters placing the circular core in a given moving frame, to an accuracy of $O(\epsilon)$, in terms of the inner spatial scale. Without loss of generality, we may put

$$c_{12}^{(1)} = 0. \tag{4.16}$$

It follows from (4.13), along with (4.7), that a proper formulation of the initial-value problem is completed with the specification of $d^{(1)}(0)$, and the matching condition includes the specification of the strength of dipole as well:

$$\psi^{(1)} \sim \left\{ -\frac{1}{4\pi} \left[\log \left(\frac{8R_0}{\epsilon r} \right) - 1 \right] r + \frac{d^{(1)}(t)}{r} \right\} \cos\theta \quad \text{as } r \to \infty, \tag{4.17}$$

This condition then gives rise to

$$\dot{R}^{(0)} = 0, \tag{4.18}$$

and

$$\dot{Z}^{(0)} = \frac{1}{4\pi R_0}\left[\log\left(\frac{8R_0}{\epsilon}\right) - \frac{1}{2} + A\right], \tag{4.19a}$$

where

$$A = \lim_{r\to\infty}\left\{4\pi^2\int_0^r r'[v^{(0)}(r')]^2 dr' - \log r\right\}. \tag{4.19b}$$

In (4.19a) and henceforth, we use, with some abuse of notation, R_0 in place of $R^{(0)}$. In the case of the Oseen vortex (4.5), the translation velocity (4.19a) and (4.19b), is identical with Saffman's formula (1.3). The parameter $c_{11}^{(1)}$ is related to $d^{(1)}$. Once the streamfunction is known, the distribution of vorticity is calculable through

$$\zeta^{(1)} = \frac{1}{R_0}\left(a\tilde{\psi}_{11}^{(1)} + r\zeta^{(0)}\right)\cos\theta + \frac{a}{R_0}\tilde{\psi}_{12}^{(1)}\sin\theta. \tag{4.20}$$

As yet, we have no way of finding the temporal evolution of $d^{(1)}(t)$ for $t > 0$. The explanation is that it may be arbitrary. We can verify that whatever the evolution of $d^{(1)}(t)$ or $c_{11}^{(1)}(t)$ for $t > 0$ may be, this arbitrariness is consistently absorbed at $O(\epsilon^3)$, producing the same radial velocity $\dot{R}^{(2)} + \dot{c}_{11}^{(1)}/R_0$ of the ring. This implies a redundant representation of the asymptotic solution of the Navier–Stokes equations which itself is unique. The speed of the ring is expressed, in an infinite variety of ways, as the sum of the speed of the moving coordinates and that of the ring in this frame.

It is informative to revisit the discrete model in an inviscid flow initially studied by Kelvin (1867) and Dyson (1893). The leading-order flow is 'the Rankine vortex', that is, a straight circular vortex tube of unit radius surrounded by an irrotational flow:

$$\zeta^{(0)} = \begin{cases} \frac{1}{\pi}, \\ 0, \end{cases} \qquad v^{(0)} = \begin{cases} -\frac{r}{2\pi}, & (r \le 1) \\ -\frac{1}{2\pi r}. & (r > 1) \end{cases} \tag{4.21}$$

The choice of $c_{11}^{(1)} = 5/8$ ensures continuity of the relative velocity across the core boundary ($r = 1$) to $O(\epsilon)$ (Widnall *et al.* 1971), and we have the streamfunctions to first order:

$$\psi^{(0)} = \begin{cases} \frac{R_0}{4\pi}\left[r^2 - 1 - 2\log\left(\frac{8R_0}{\epsilon}\right)\right], & (r \le 1) \\ -\frac{R_0}{2\pi}\log\left(\frac{8R_0}{\epsilon r}\right), & (r > 1) \end{cases} \tag{4.22}$$

$$\psi_{11}^{(1)} = \begin{cases} -\frac{1}{4\pi}\left[\log\left(\frac{8R_0}{\epsilon}\right) - 1\right] + \frac{5}{16\pi}(r^3 - r), & (r \le 1) \\ \frac{1}{4\pi}r\log r - \frac{r}{4\pi}\left[\log\left(\frac{8R_0}{\epsilon}\right) - 1\right] - \frac{3}{16\pi r}, & (r > 1) \end{cases} \tag{4.23}$$

where we have used

$$a = -2\delta(r - 1). \tag{4.24}$$

The first log-term of (4.23) for $r > 1$ is a particular solution to absorb the inhomogeneous term of (4.12), the curvature effect. Putting aside this term, $\psi_{11}^{(1)}$ signifies, outside the core ($r > 1$), the flow field around a circular cylinder of unit radius, moving in the z-direction with velocity $3/16\pi$, a dipole field, in an imposed uniform flow with positive velocity $[\log(8R_0/\epsilon) - 1]/4\pi$. The summation of these values is equal to Kelvin's velocity. This observation implies that a vortex ring is more than a passive entity convected by the self-induced flow.

Figure 2(a) displays the streamlines $\psi^{(0)} = $ const., Figure 2(b) $\tilde{\psi}_{11}^{(1)} \cos\theta = $ const., and Figure 2(c) $\psi^{(0)} + \epsilon\tilde{\psi}_{11}^{(1)} \cos\theta = $ const., stream patterns viewed from the moving frame. For clarity, the rather large value $\epsilon = 0.5$ is chosen. We confirm that the origin $r = 0$ coincides with the center of the circular core. As expected, Figure 2(b) exhibits a dipole flow associated with a pair of antiparallel vortices. Its source is twofold. One is the 'apparent' dipole due to displacement of the center as discussed above. The other has a kinematic origin. When a columnar vortex is bent to form a torus, vortex lines are stretched on the outer side and contracted on the inner side. As a consequence, the vorticity is enhanced on the outer side but is diminished on the inner side, implying the creation of a vortex pair at $O(\epsilon)$.

We speculate that the dipole is a key ingredient of a curved vortex tube. A vortex ring may be locally considered as a line of dipoles based at the core centerline embedded in the flow field induced by the circular line vortex. The driving mechanism of the self-propulsion is not only convection due to the self-induced flow but also the thrust provided by the dipoles. The dipole strength depends upon the distribution of vorticity in the core, and this is one of the reasons why we are concerned with the inner field.

4.3 Second order

The second-order equations reveal that the second-order perturbation $\psi^{(2)}$ comprises monopole and quadrupole terms:

$$\psi^{(2)} = \psi_0^{(2)} + \psi_{21}^{(2)} \cos 2\theta. \tag{4.25}$$

The latter reflects an elliptical core deformation. The governing equation for the monopole is

$$\left(\frac{\partial^2}{\partial r^2} + \frac{1}{r}\frac{\partial}{\partial r}\right)\psi_0^{(2)} = R_0\zeta_0^{(2)} + \frac{ra}{2R_0}\tilde{\psi}_{11}^{(1)}$$

$$+\frac{1}{2R_0}\left[rv^{(0)} + r^2\zeta^{(0)} + \frac{\partial\psi_{11}^{(1)}}{\partial r} + \frac{\psi_{11}^{(1)}}{r}\right] + R^{(2)}\zeta^{(0)}. \tag{4.26}$$

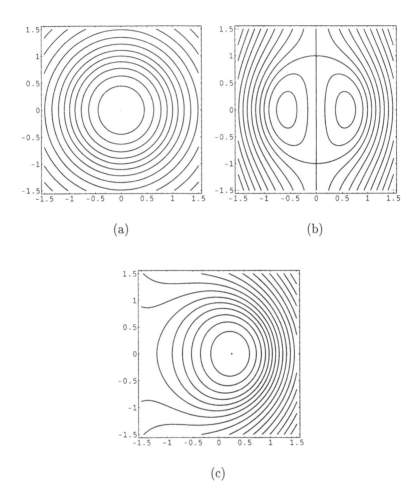

Figure 2.   Streamline patterns of the inviscid vortex ring with
uniform ζ/ρ, relative to the frame moving at the speed U given
by (1.1), up to $O(\epsilon)$ as obtained by (4.22) and (4.23). The small
parameter $\epsilon = 0.5$.   (a) $\psi^{(0)} = $ const.,   (b) $\tilde{\psi}_{11}^{(1)} \cos\theta = $ const.,
(c) $\psi^{(0)} + \epsilon\tilde{\psi}_{11}^{(1)} \cos\theta = $ const.

The functional form of $\zeta_0^{(2)}$, the axisymmetric vorticity component at $O(\epsilon^2)$,
will be found by solving the convection–diffusion equation obtained from the
solvability condition for the fourth-order equation (§6). The matching condi-

tion, a part of (3.6) at $O(\epsilon^2)$, is

$$\psi_0^{(2)} \sim -\frac{1}{2^4 \pi R_0}\left[\log\left(\frac{8R_0}{\epsilon r}\right) + \frac{1}{2}\right]r^2 + \left(\frac{d^{(1)}}{2R_0} - \frac{R^{(2)}}{2\pi}\right)\log\left(\frac{8R_0}{\epsilon r}\right) \qquad \text{as } r \to \infty.$$

(4.27)

Under the conditions of (4.27) and finiteness of velocity at $r = 0$, (4.26) is readily integrated once to give

$$\frac{\partial \psi_0^{(2)}}{\partial r} = \frac{R_0}{r}\int_0^r r'\zeta_0^{(2)}\,dr' + \frac{r}{2R_0}\frac{\partial \tilde{\psi}_{11}^{(1)}}{\partial r} + \left(\frac{r^2}{2R_0} - R^{(2)}\right)v^{(0)} - \frac{r}{2}\dot{Z}^{(0)}. \quad (4.28)$$

Next we turn to the quadrupole $\psi_{21}^{(2)}\cos 2\theta$. For convenience, we define $\tilde{\psi}_{21}^{(2)}$ through

$$\psi_{21}^{(2)} = \tilde{\psi}_{21}^{(2)} - \frac{1}{4}r^2\dot{Z}^{(0)}. \qquad (4.29)$$

Then the equation for $\tilde{\psi}_{21}^{(2)}$ takes the following form:

$$\left(\frac{\partial^2}{\partial r^2} + \frac{1}{r}\frac{\partial}{\partial r} - \frac{4}{r^2} - a\right)\tilde{\psi}_{21}^{(2)} = \frac{b}{4R_0}\left(\tilde{\psi}_{11}^{(1)}\right)^2 + \frac{ra}{R_0}\tilde{\psi}_{11}^{(1)}$$

$$+ \frac{1}{2R_0}\left(rv^{(0)} + r^2\zeta^{(0)} + \frac{\partial \tilde{\psi}_{11}^{(1)}}{\partial r} - \frac{\tilde{\psi}_{11}^{(1)}}{r}\right),$$

(4.30)

where

$$b(r,t) = -\frac{1}{v^{(0)}}\frac{\partial a}{\partial r}. \qquad (4.31)$$

The boundary conditions demand that

$$\tilde{\psi}_{21}^{(2)} \propto r^2 \quad \text{as } r \to 0, \qquad (4.32a)$$

$$\tilde{\psi}_{21}^{(2)} \sim \frac{r^2}{4}\left\{\dot{Z}^{(0)} + \frac{1}{8\pi R_0}\left[\log\left(\frac{8R_0}{\epsilon r}\right) - 2\right]r^2\right\} + \frac{d^{(1)}}{4R_0} \quad \text{as } r \to \infty. \quad (4.32b)$$

The vorticity distribution is calculable from

$$\zeta^{(2)} = \zeta_0^{(2)} + \left[\frac{a}{R_0}\tilde{\psi}_{21}^{(2)} + \frac{b}{4R_0^2}\left(\tilde{\psi}_{11}^{(1)}\right)^2 + \frac{ra}{2R_0^2}\tilde{\psi}_{11}^{(1)}\right]\cos 2\theta. \qquad (4.33)$$

In the general case, it is unlikely that (4.30) can be further integrated analytically. The numerical computation is postponed to a subsequent paper, but we content ourselves with an explicit solution for Dyson's discrete model:

$$\tilde{\psi}_{21}^{(2)} = \begin{cases} \dfrac{3}{64\pi R_0}r^4 + \dfrac{1}{16\pi R_0}\left[3\log\left(\dfrac{8R_0}{\epsilon}\right) - 5\right]r^2, & (r < 1) \\[2ex] \dfrac{1}{32\pi R_0}\left[3\log\left(\dfrac{8R_0}{\epsilon}\right) - \log r - \dfrac{5}{2}\right]r^2 - \dfrac{3}{64\pi R_0} \\[2ex] \qquad + \dfrac{3}{32\pi R_0}\left[\log\left(\dfrac{8R_0}{\epsilon}\right) - \dfrac{3}{2}\right]\dfrac{1}{r^2}. & (r > 1) \end{cases}$$

(4.34)

The axisymmetric part of vorticity is shown to be suppressed: $\zeta_0^{(2)} = 0$.

5   Third-order velocity of a vortex ring

At $O(\epsilon^3)$, a dipole field again shows up as the result of nonlinear interactions among the mono-, di- and quadru-poles of lower orders, and the streamfunction $\psi^{(3)}$ is written as

$$\psi^{(3)} = \psi_{11}^{(3)} \cos\theta + \psi_{12}^{(3)} \sin\theta + \psi_{31}^{(3)} \cos 3\theta \,. \tag{5.1}$$

The $\cos\theta$ and $\sin\theta$ components, the dipole field, are responsible for the axial and radial velocities respectively.

5.1   Radial expansion

It is not difficult to get $\dot{R}^{(2)}$ from the equation for $\psi_{12}^{(3)}$ in much the same way as getting $\dot{Z}^{(0)}$. Instead, by appeal to a fundamental conservation law, we can skip this procedure. Recall that the hydrodynamic impulse $\boldsymbol{P}$ is conserved, regardless of the inviscid or viscous character of the flow:

$$\boldsymbol{P} = \frac{1}{2} \iiint \boldsymbol{x} \times \boldsymbol{\omega} \, dV = \text{const.} \tag{5.2}$$

In the present axisymmetric problem, only the axial component P_z is nonzero, and upon substitution from $\boldsymbol{\omega} = \zeta \boldsymbol{e}_\phi$, (2.10a) and (2.10c), it becomes, after normalisation by ΓR_0^2,

$$P_z = \pi R_0^2 + \epsilon^2 P^{(2)} + \cdots \,, \tag{5.3}$$

where

$$P^{(2)} \equiv \pi \left(2R_0 R^{(2)}(t) - 4\pi d^{(1)}(t) \right) \,, \tag{5.4}$$

which should be constant throughout the time evolution. Since $R^{(2)} = 0$ at $t = 0$, the radial motion is completely ruled by the evolution of the dipole strength $d^{(1)}(t)$:

$$R^{(2)}(t) = \frac{2\pi}{R_0} \left[d^{(1)}(t) - d^{(1)}(0) \right] \,. \tag{5.5}$$

We concentrate on the vorticity distribution (4.5) starting from a δ-function core. In this case, $P_z = \pi R_0^2$ identically and hence the $O(\epsilon^2)$ correction term is absent: $P^{(2)}(t) = 0$ for $t \geq 0$. The numerical evaluation of the behaviour of $\Psi_{11}^{(1)}$, for large r, is carried out with ease to yield

$$\Psi_{11}^{(1)} = \frac{r}{4\pi} \left[\log r + \lim_{r \to \infty} \left(4\pi^2 \int_0^r r'[v^{(0)}(r')]^2 dr' - \log r \right) + \frac{1}{2} \right] + \frac{D^{(1)}}{r} + \cdots \,, \tag{5.6}$$

where

$$D^{(1)} \approx 0.41225489 \, t \,. \tag{5.7}$$

The first-order streamfunction with $c_{11}^{(1)} = 0$ corresponds to a dipole field whose stagnation point is permanently located at $r = 0$ (Klein & Knio 1995). By identifying $D^{(1)}(t)$ with $d^{(1)}(t)$ in (5.5) and restoring dimensional variables, we conclude that, given initially a circular line vortex of radius R_0, the location $\rho = R_s(t)$ of the stagnation point in the core, viewed from the comoving coordinates, drifts outwards linearly in time owing to the action of viscosity as

$$R_s \approx R_0 + 2.5902739 \, \frac{\nu t}{R_0} \, . \tag{5.8}$$

The temporal evolution of the radius $R_p(t)$ of the loop of peak vorticity is deducible by choosing $c_{11}^{(1)}(t)$ in (4.13) so that the local origin $r = 0$ is maintained at the maximum of $\zeta^{(0)} + \epsilon \zeta^{(1)}$. Inserting the Gaussian distribution (4.5) and (4.6) into (4.9) and (4.14), we manipulate the behaviour of (4.20) near $r = 0$ as

$$\zeta^{(1)} = \frac{1}{R_0} \left[\left(\frac{c_{11}^{(1)}}{8\pi t^2} + \frac{1}{4\pi t} \right) x + O(r^2 x) \right] , \tag{5.9}$$

where $x = r \cos \theta$. The condition of maximum vorticity $\zeta^{(0)} + \epsilon \zeta^{(1)}$ at $r = 0$ brings in

$$c_{11}^{(1)} = -2t \, . \tag{5.10}$$

In the light of (4.13) and (5.6) along with (5.7), the dipole strength becomes

$$d^{(1)}(t) \approx 4.5902739 \times \frac{t}{2\pi} \, , \tag{5.11}$$

when the origin of the moving frame is kept sitting at the location of the maximum vorticity. Relying upon the formula (5.5), we gain the expansion law of the circle of peak vorticity, which is expressed, in terms of the dimensional variables as

$$R_p \approx R_0 + 4.5902739 \, \frac{\nu t}{R_0} \, . \tag{5.12}$$

We point out that the asymptotic value $4.5902739 \cdots$ at large Reynolds number exhibits a marked difference from $1.65 \cdots$ at $Re_\Gamma = 10^4$ calculated numerically by Wang *et al.* (1994). In spite of this, at the same value of Re_Γ, their numerical value $0.5779 \cdots$ in the formula of the translation velocity is in conformity with Saffman's asymptotic value $0.57796576 \cdots$.

For completeness, we derive the expansion law of the circle of radial centroid $\rho = R_c(t)$ of vorticity defined by

$$R_c \equiv \frac{1}{2} \iiint \frac{(\boldsymbol{x} \times \boldsymbol{\omega}) \cdot \boldsymbol{P}}{P^2} \rho \, dV = \frac{\pi}{P_z} \iint \rho^3 \zeta r \, dr \, d\theta \, . \tag{5.13}$$

This is conveniently decomposed as

$$
\begin{aligned}
R_c &= \frac{\pi}{P_z} \iint (R + \epsilon r \cos \theta) \rho^2 \zeta r \, dr \, d\theta \\
&\approx R_0 + \epsilon^2 \left\{ R^{(2)} + \frac{\pi^2}{P_z} \int_0^\infty (2 R_0 r^3 \zeta^{(0)} + R_0^2 \zeta_{11}^{(1)}) r^2 \, dr \right\} .
\end{aligned} \tag{5.14}
$$

With the help of (3.7b) and (5.5), (5.14) is reduced to

$$R_c(t) \approx R_0 - \frac{2\pi\epsilon^2}{R_0}d^{(1)}(0) + \frac{3\epsilon^2}{4R_0}\left[2\pi\int_0^\infty r^3\zeta^{(0)}dr\right].$$ (5.15)

For the initial 'δ-function' core, the radial centroid evolve as

$$R_c(t) \approx R_0 + \frac{3\nu t}{R_0},$$ (5.16)

again being in contradiction with Wang *et al.*'s claim that R_c is a constant.

It is to be noted that the expansion laws of R_s, R_p and R_c do not include the parameter Γ. These laws are all attributable to the effect of viscous diffusion of curved vortex lines, which is linear in ν.

5.2 The third-order correction to the translation velocity

The equations to be manipulated to obtain $\dot{Z}^{(2)}$ reduce, after numerous cancellations, to

$$\frac{1}{r}\left\{v^{(0)}\frac{\partial}{\partial r}\left[\frac{1}{r}\frac{\partial}{\partial r}\left(r\psi_{11}^{(3)}\right)\right] + \frac{\partial\zeta^{(0)}}{\partial r}\psi_{11}^{(3)}\right\}$$

$$+R^{(2)}\dot{Z}^{(0)}\frac{\partial\zeta^{(0)}}{\partial r} + R_0\left(\dot{Z}^{(2)}\frac{\partial\zeta^{(0)}}{\partial r} + \dot{Z}^{(0)}\frac{\partial\zeta_0^{(2)}}{\partial r}\right)$$

$$-\frac{1}{r}\left\{\frac{1}{r}\left(\int_0^r r'\zeta_0^{(2)}dr'\right)\frac{\partial}{\partial r}\left[\frac{1}{r}\frac{\partial}{\partial r}\left(r\psi_{11}^{(1)}\right)\right] - \frac{\partial\zeta_0^{(2)}}{\partial r}\psi_{11}^{(1)}\right\} - \frac{2v^{(0)}}{r^2}\left(\int_0^r r'\zeta_0^{(2)}dr'\right)$$

$$+\frac{2}{R_0}\left(\frac{1}{R_0}\frac{\partial\psi^{(0)}}{\partial r}\Delta\psi_0^{(2)} + \frac{\partial\psi_0^{(2)}}{\partial r}\zeta^{(0)} + 2R^{(2)}v^{(0)}\zeta^{(0)}\right)$$

$$-\frac{v^{(0)}}{R_0 r}\left(\frac{1}{2}\frac{\partial\tilde{\psi}_{21}^{(2)}}{\partial r} + \frac{\tilde{\psi}_{21}^{(2)}}{r}\right)$$

$$+\frac{1}{r}\left\{\frac{1}{2}\left(-\frac{\partial\zeta_{21}^{(2)}}{\partial r}\tilde{\psi}_{11}^{(1)} + \frac{1}{R_0}a\tilde{\psi}_{11}^{(1)}\frac{\partial\tilde{\psi}_{21}^{(2)}}{\partial r}\right) - \zeta_{21}^{(2)}\frac{\partial\tilde{\psi}_{11}^{(1)}}{\partial r}\right.$$

$$\left.+\frac{1}{R_0}\left[\frac{\partial}{\partial r}\left(a\tilde{\psi}_{11}^{(1)}\right) + r\frac{\partial\zeta^{(0)}}{\partial r}\right]\tilde{\psi}_{21}^{(2)}\right\} - \frac{1}{4R_0^2}\left[a\tilde{\psi}_{11}^{(1)}\frac{\partial\tilde{\psi}_{11}^{(1)}}{\partial r} + \frac{a}{r}\left(\tilde{\psi}_{11}^{(1)}\right)^2\right]$$

$$+\frac{5v^{(0)}}{4R_0^2}\left(\frac{\partial\tilde{\psi}_{11}^{(1)}}{\partial r} + \frac{\tilde{\psi}_{11}^{(1)}}{r}\right) - \frac{r\zeta^{(0)}}{R_0^2}\frac{\partial\tilde{\psi}_{11}^{(1)}}{\partial r}$$

$$+\frac{3rv^{(0)}}{4R_0^2}a\tilde{\psi}_{11}^{(1)} + \frac{5}{4R_0^2}r\left(v^{(0)}\right)^2 + \frac{\dot{Z}^{(0)}}{R_0}\left(r\zeta^{(0)} - 2v^{(0)}\right) = 0.$$ (5.17)

The matching condition (3.6) reads

$$\psi_{11}^{(3)} \sim \frac{3}{2^7 \pi R_0^2}\left[\log\left(\frac{8R_0}{\epsilon r}\right) - \frac{1}{3}\right]r^3 - \frac{d^{(1)}}{8R_0^2}\left[\log\left(\frac{8R_0}{\epsilon r}\right) - \frac{7}{4}\right]r - \frac{R^{(2)}}{4\pi R_0}r$$

$$+\frac{d_1^{(3)}}{r} \quad \text{as } r \to \infty. \tag{5.18}$$

The coefficient $d_1^{(3)}$ is irrelevant to the determination of $\dot{Z}^{(2)}$. It is worth emphasising that the terms multiplied by $d^{(1)}$ represent the flow induced by a line of dipoles arranged on the core centerline, which has been overlooked in previous studies.

The remaining task is to multiply (5.17) by r^2, integrate with respect to r, and seek the limiting form as $r \to \infty$. Using (6.5) and the non-singular condition at $r = 0$, and taking the limit $r \to \infty$, we eventually arrive at the desired formula:

$$\dot{Z}^{(2)} = -\frac{3d^{(1)}}{8R_0^3}\left[\log\left(\frac{8R_0}{\epsilon}\right) + \frac{4}{3}A - \frac{7}{6}\right] - \frac{P^{(2)}}{8\pi^2 R_0^3}\left[\log\left(\frac{8R_0}{\epsilon}\right) + A - \frac{3}{2}\right]$$

$$-\frac{\pi}{4R_0^3}\left(B - \frac{13}{8}\int_0^\infty r^4 \zeta^{(0)}v^{(0)}dr\right)$$

$$-\frac{2\pi}{R_0}\int_0^\infty \left[\int_0^r r'\zeta_0^{(2)}(r')dr'\right]v^{(0)}(r)dr - \frac{\pi}{2R_0^2}\int_0^\infty \left(2ra + b\tilde{\psi}_{11}^{(1)}\right)rv^{(0)}\tilde{\psi}_{21}^{(2)}dr$$

$$+\frac{\pi}{8R_0^3}\int_0^\infty \left[ra\big(\tilde{\psi}_{11}^{(1)} - 3r\frac{\partial\tilde{\psi}_{11}^{(1)}}{\partial r}\big)\tilde{\psi}_{11}^{(1)} + b\big(\tilde{\psi}_{11}^{(1)} - r\frac{\partial\tilde{\psi}_{11}^{(1)}}{\partial r}\big)\big(\tilde{\psi}_{11}^{(1)}\big)^2\right]dr, \tag{5.19a}$$

where A and $P^{(2)}$ are given by (4.19b) and (5.4) and

$$B = \lim_{r\to\infty}\left[\int_0^r r'v^{(0)}(r')\tilde{\psi}_{11}^{(1)}(r')dr' + \frac{1}{16\pi^2}(\log r + A)r^2 + \frac{d^{(1)}}{2\pi}\log r\right]. \tag{5.19b}$$

If, in particular, $\zeta^{(0)}$ evolves as (4.5), then $P^{(2)} = 0$ and (5.19a) reduces to

$$\dot{Z}^{(2)} = -\frac{3d^{(1)}}{8R_0^3}\left\{\log\left(\frac{8R_0}{\epsilon}\right) + \frac{2}{3}\left[\log\left(\frac{1}{4\sqrt{t}}\right) + \gamma - \frac{7}{4}\right]\right\} - \frac{\pi B}{4R_0^3} - \frac{39}{2^7\pi R_0^3}\hat{\nu}t$$

$$-\frac{2\pi}{R_0}\int_0^\infty \left[\int_0^r r'\zeta_0^{(2)}(r')dr'\right]v^{(0)}(r)dr - \frac{\pi}{2R_0^2}\int_0^\infty \left(2ra + b\tilde{\psi}_{11}^{(1)}\right)rv^{(0)}\tilde{\psi}_{21}^{(2)}dr$$

$$+\frac{\pi}{8R_0^3}\int_0^\infty \left[ra\big(\tilde{\psi}_{11}^{(1)} - 3r\frac{\partial\tilde{\psi}_{11}^{(1)}}{\partial r}\big)\tilde{\psi}_{11}^{(1)} + b\big(\tilde{\psi}_{11}^{(1)} - r\frac{\partial\tilde{\psi}_{11}^{(1)}}{\partial r}\big)\big(\tilde{\psi}_{11}^{(1)}\big)^2\right]dr. \tag{5.20}$$

Our formula is completed with construction of a formal solution for $\zeta_0^{(2)}$, which is the topic of the following section. In a general case, numerical computation of $\tilde{\psi}_{21}^{(2)}$ and $\zeta_0^{(2)}$ and numerical integration in the above formula are required to evaluate $\dot{Z}^{(2)}$. The remaining computation is left for a subsequent paper.

Fortunately the explicit solution is at hand for the Rankine vortex (4.21) at $O(\epsilon^0)$. In this case, $A = 1/4$, $d^{(1)} = -3/(16\pi)$ as read off from (4.23), $R^{(2)} = 0$ and $\zeta_0^{(2)} = 0$. ¿From (5.19b), $B = 11/(3 \cdot 2^7 \pi^2)$. The definition (5.4) gives $P^{(2)} = 3\pi/4$. Introduction of these values, (4.21), (4.23) and (4.34) into the formula (5.19a) gives rise to the third-order correction in Dyson's formula (1.1).

6 The axisymmetric part of second-order vorticity

The vorticity equation at $O(\epsilon^3)$ is solved for $\zeta_{12}^{(3)}$, giving

$$\zeta_{12}^{(3)} = \frac{a}{R_0}\tilde{\psi}_{12}^{(3)} + \frac{r}{v^{(0)}}\left\{-\frac{\partial \zeta_{11}^{(1)}}{\partial t} + \hat{\nu}\left[\left(\frac{\partial^2}{\partial r^2} + \frac{1}{r}\frac{\partial}{\partial r} - \frac{1}{r^2}\right)\zeta_{11}^{(1)} + \frac{1}{R_0}\frac{\partial \zeta^{(0)}}{\partial r}\right]\right\},$$
$$(6.1)$$

where we have defined

$$\psi_{12}^{(3)} = \tilde{\psi}_{12}^{(3)} + R_0 r \dot{R}^{(2)}. \tag{6.2}$$

Substituting $\psi^{(1)} = \psi_{11}^{(1)}\cos\theta$, (4.25), (5.1) and the associated vorticity distribution (4.20) and (4.33) into the vorticity equation at $O(\epsilon^4)$, we get, after some manipulation, a somewhat simple convection–diffusion equation for $\zeta_0^{(2)}$. By virtue of (6.1), a further simplfication is achieved, leaving

$$\frac{\partial \zeta_0^{(2)}}{\partial t} - \hat{\nu}\frac{1}{r}\frac{\partial}{\partial r}\left(r\frac{\partial \zeta_0^{(2)}}{\partial r}\right) = \frac{1}{r}\frac{\partial}{\partial r}\left\{-\frac{r}{2R_0 v^{(0)}}\left[\frac{\partial \zeta_{11}^{(1)}}{\partial t}\right.\right.$$
$$\left.\left. -\hat{\nu}\left(\frac{\partial^2}{\partial r^2} + \frac{1}{r}\frac{\partial}{\partial r} - \frac{1}{r^2}\right)\zeta_{11}^{(1)}\right]\tilde{\psi}_{11}^{(1)} + \frac{\dot{R}^{(2)}r^2}{2R_0}\zeta^{(0)}\right\}. \tag{6.3}$$

The appropriate initial condition is

$$\zeta_0^{(2)}(r, 0) = 0 \quad \text{for } 0 \leq r < \infty. \tag{6.4}$$

A glance at (6.3) shows that, whether viscosity is present or not,

$$\int_0^\infty r\zeta_0^{(2)}(r, t)dr = 0 \quad \text{for all } t \geq 0, \tag{6.5}$$

under the initial condition (6.4).

By using a Green's function, the unique solution of (6.3) for $\hat{\nu} = 1$ may be written out. If, in particular, a δ-function core is assumed at the initial instant, then it admits the following solution:

$$
\zeta_0^{(2)}(r,t) = \frac{1}{2R_0 t} \left(\int_0^t t' \dot{R}^{(2)}(t') dt' \right) \frac{1}{r} \frac{\partial}{\partial r} \left(r^2 \zeta^{(0)} \right)
$$
$$
- \frac{1}{4R_0\hat{\nu}} \int_0^t dt' \frac{\exp\{-r^2/4\hat{\nu}(t-t')\}}{t-t'} \int_0^\infty dr' \exp\left\{ -\frac{r'^2}{4\hat{\nu}(t-t')} \right\} I_0 \left(\frac{rr'}{2\hat{\nu}(t-t')} \right)
$$
$$
\times \frac{\partial}{\partial r'} \left\{ \frac{r' \tilde{\psi}_{11}^{(1)}(r',t')}{v^{(0)}(r',t')} \left[\frac{\partial}{\partial t'} - \hat{\nu} \left(\frac{\partial^2}{\partial r'^2} + \frac{1}{r'} \frac{\partial}{\partial r'} - \frac{1}{r'^2} \right) \right] \zeta_{11}^{(1)}(r',t') \right\}, \qquad (6.6)
$$

where I_0 is the modified Bessel function of zeroth order of the first kind.

References

Caffarelli, L.A., Friedman, A. (1980) 'Asymptotic estimates for the plasma problem', *Duke Math. J.* **47**, 795–742.

Callegari, A. J., Ting, L. (1978) 'Motion of a curved vortex filament with decaying vortical core and axial velocity', *SIAM J. Appl. Maths* **35**, 148–175.

Dyson, F. W. (1893) 'The potential of an anchor ring. – part II', *Phil. Trans. R. Soc. Lond. A* **184**, 1041–1106.

Fraenkel, L. E. (1972) 'Examples of steady vortex rings of small cross-section in an ideal fluid', *J. Fluid Mech.* **51**, 119–135.

Fraenkel, L. E. (1999) *An Introduction to Maximum Principles and Symmetry in Elliptic Problems*, Chapters 3, 4, Cambridge University Press.

Fukumoto, Y., Miyazaki, T. (1991) 'Three-dimensional distortions of a vortex filament with axial velocity', *J. Fluid Mech.* **222**, 369–416.

Fukumoto, Y., Moffatt, H. K. (2000) 'Motion and expansion of a viscous vortex ring. Part 1. A higher-order asymptotic formula for the velocity', to appear in *J. Fluid Mech.*

Jiménez, J., Moffatt, H. K., Vasco, C. (1996) 'The structure of vortices in freely decaying two-dimensional turbulence', *J. Fluid Mech.* **313**, 209–222.

Kelvin, Lord. (1867) 'The translatory velocity of a circular vortex ring', *Phil. Mag. (4)* **35**, 511–512.

Kida, S., Ohkitani, K. (1992) 'Spatiotemporal intermittency and instability of a forced turbulence', *Phys. Fluids A* **4**, 1018–1027.

Klein, R., Knio, O. M. (1995) 'Asymptotic vorticity structure and numerical simulation of slender vortex filaments', *J. Fluid Mech.* **284**, 275–321.

Moffatt, H. K., Kida, S., Ohkitani, K. (1994) 'Stretched vortices – the sinews of turbulence; large-Reynolds-number asymptotics', *J. Fluid Mech.* **259**, 241–264.

Moore, D. W., Saffman, P. G. (1972) 'The motion of a vortex filament with axial flow', *Phil. Trans. R. Soc. Lond. A* **272**, 403–429.

Saffman, P. G. (1970) 'The velocity of viscous vortex rings', *Stud. Appl. Math.* **49**, 371–380.

Shariff, K., Leonard, A. (1992) 'Vortex rings', *Ann. Rev. Fluid Mech.* **24**, 235–279.

Ting, L., Klein, R. (1991) *Viscous Vortical Flows*, Lecture Notes in Physics **374**, Springer.

Ting, L., Tung, C. (1965) 'Motion and decay of a vortex in a nonuniform stream', *Phys. Fluids* **8**, 1039–1051.

Tung, C., Ting, L. (1967) 'Motion and decay of a vortex ring', *Phys. Fluids* **10**, 901–910.

Wang, C.-T., Chu, C.-C., Chang, C. C. (1994) 'Initial motion of a viscous vortex ring', *Proc. R. Soc. Lond. A* **446**, 589-599.

Widnall, S. E., Bliss, D. B., Zalay, A. (1971) 'Theoretical and experimental study of the stability of a vortex pair', In *Aircraft Wake Turbulence and its Detection*, (eds. Olsen, Goldberg & Rogers), 305–338, Plenum.

Stretching and Compression of Vorticity in the 3D Euler Equations

J.D. Gibbon, B. Galanti and R.M. Kerr

1 Introduction

The growth of vorticity in 3D incompressible Euler turbulence is an issue that has been addressed several ways. The theorem of Beale, Kato & Majda (1984) demonstrates that $\int_0^t \|\boldsymbol{\omega}(\tau)\|_\infty \, d\tau$ controls the growth of all quantities and that nothing can blow up without this quantity blowing up. Of equal importance is the direction in which vorticity stretches or compresses. Constantin, Fefferman and Majda (1996) have discussed this question and concluded that the L^∞ norm of $\boldsymbol{\omega}$ can be weakened to L^q for some finite $q \geq 1$, provided certain technical constraints are put on the direction of vorticity and the magnitude of the velocity. The issue of how the various vectors in the problem orient themselves *locally* and what governs their direction is a complicated one that is by no means fully understood. Constantin, Fefferman and Majda (1996) consider the vorticity vector at a point $\mathbf{x}$ and at a neighbouring point $\mathbf{x} + \mathbf{y}$. Angles between vortex lines are determined by the relative orientation of $\boldsymbol{\omega}(\mathbf{x})$ and $\boldsymbol{\omega}(\mathbf{x} + \mathbf{y})$. A singularity can be prevented if angles between vortex lines in a neighbourhood are not allowed to become too large (see also Constantin (1994)). Large curvatures in vortex lines are therefore necessary if a singularity is to form, a point illustrated in Figure 1 involving Kerr's picture of singularity formation showing strong curvature in the region where the vorticity blows up (Kerr (1993)).

This paper looks at how the stretching, compression and direction of vorticity for the 3D incompressible Euler equations might be considered in a different sense than usual. This will involve relations between the strain matrix

$$S_{ij} = \frac{1}{2}\left(\frac{\partial u_i}{\partial x_j} + \frac{\partial u_j}{\partial x_i}\right) \tag{1.1}$$

and the Hessian matrix of the pressure $P = \{P_{ij}\}$ defined by

$$P_{ij} = \frac{\partial^2 P}{\partial x_i \partial x_j} \tag{1.2}$$

and is based upon recent work by Galanti, Gibbon and Heritage (1997). The pressure Hessian is a quantity in which there has been increasing interest. For instance Jeong and Hussain (1995) use it as a diagnostic when asking

the question 'when is a vortex a vortex?' Their answer to this question is the
criterion that P needs to have two positive eigenvalues and one negative one.
This means that the pressure has a local minimum in the plane orthogonal
to the stretching direction. The pressure Hessian has also been used in other
circumstances in Navier–Stokes turbulence (e.g. see Majda (1986, 1991) and
Cantwell (1992)) and references therein.

In §4 this point is illustrated when the example of the Burgers vortex is
used, in tandem with other examples. These examples are relatively simple,
however, when compared to the problem of singularity formation. This will be
our main example (§5) in which we consider the stretching and compression
of vorticity near the formation of the Euler singularity.

2   Stretching equations involving S and P

2.1   The stretching rate α

The idea of the stretching rate has long been known. Consider the incompressible 3D Euler equations

$$\frac{D\mathbf{u}}{Dt} = -\nabla p \tag{2.1}$$

with div $\mathbf{u} = 0$, where p is the pressure and

$$\frac{D}{Dt} = \frac{\partial}{\partial t} + \mathbf{u} \cdot \nabla. \tag{2.2}$$

Then the vorticity $\boldsymbol{\omega} = \operatorname{curl}\mathbf{u}$ obeys

$$\frac{D\boldsymbol{\omega}}{Dt} = \boldsymbol{\omega} \cdot \nabla\mathbf{u} = S\boldsymbol{\omega} \tag{2.3}$$

where the matrix S is the strain matrix defined by $S_{ij} = \frac{1}{2}\left(u_{i,j} + u_{j,i}\right)$. From
(2.3), the scalar magnitude of vorticity ω obeys

$$\frac{D\omega}{Dt} = \alpha\omega \tag{2.4}$$

where the *stretching rate* α is defined by

$$\alpha(\mathbf{x}, t) = \frac{\boldsymbol{\omega} \cdot S\boldsymbol{\omega}}{\boldsymbol{\omega} \cdot \boldsymbol{\omega}}. \tag{2.5}$$

At a point $\mathbf{x}$, when $\alpha > 0$, we have vortex stretching and when $\alpha < 0$ we have
vortex compression. It is also desirable to know how α behaves *directionally*
because α carries within it some relation between $\boldsymbol{\omega}$ and the vortex stretching
vector $\boldsymbol{\sigma} = \boldsymbol{\omega} \cdot \nabla\mathbf{u} = S\boldsymbol{\omega}$ (see Constantin (1994)). In terms of elementary

linear algebra α can be thought of as being an estimate for an eigenvalue of S but to take account of how $\boldsymbol{\omega}$ orientates itself with respect to $\boldsymbol{\sigma}$ it is useful to introduce the vector

$$\boldsymbol{\chi} = \frac{\boldsymbol{\omega} \times S\boldsymbol{\omega}}{\boldsymbol{\omega} \cdot \boldsymbol{\omega}}. \tag{2.6}$$

The object of this article is to see how α and $\boldsymbol{\chi}$ behave in a 3D Euler flow and in particular near the Euler singular region (Kerr (1993)).

2.2   Ohkitani's relation

To pursue the line of thought described above we need firstly to discuss a relation between $S\boldsymbol{\omega}$ and the Hessian matrix of the pressure P. Ohkitani (1993) has shown that the vortex stretching vector $\boldsymbol{\sigma}$ obeys the simple relation

$$\frac{D\boldsymbol{\sigma}}{Dt} = -P\boldsymbol{\omega}. \tag{2.7}$$

The proof of this is simple; consider the total derivative of $\sigma_i = \omega_j u_{i,j}$

$$
\begin{aligned}
\frac{D\sigma_i}{Dt} &= \frac{D\omega_j}{Dt} u_{i,j} + \omega_j \frac{\partial}{\partial x_j} \left(\frac{Du_i}{Dt} \right) - \omega_j u_{k,j} u_{i,k} \\
&= \sigma_j u_{i,j} - \sigma_k u_{i,k} + \omega_j \frac{\partial}{\partial x_j} \left(-\frac{\partial p}{\partial x_i} \right) \\
&= -P_{ij} \omega_j.
\end{aligned}
\tag{2.8}
$$

There are three consequences of this relation; the first is that terms that have exactly cancelled are of order $\boldsymbol{\omega}|\nabla\mathbf{u}|^2$. In fact all the terms that do not include the pressure have vanished identically. Secondly it is clear that since $\boldsymbol{\sigma} = S\boldsymbol{\omega}$

$$\frac{D(\boldsymbol{\omega} \times S\boldsymbol{\omega})}{Dt} = -\boldsymbol{\omega} \times P\boldsymbol{\omega}. \tag{2.9}$$

Hence when $\boldsymbol{\omega}$ aligns with an eigenvector of S, it must also align with an eigenvector of P. Conversely, if $\boldsymbol{\omega}$ aligns with an eigenvector of P, then $\boldsymbol{\omega} \times S\boldsymbol{\omega}$ is a constant of the motion. Thirdly and lastly, using Euler's equation again, we obtain

$$\frac{D^2\boldsymbol{\omega}}{Dt^2} + P\boldsymbol{\omega} = 0. \tag{2.10}$$

This innocuous looking equation is more complicated than it seems as the double derivatives are material derivatives. At or near alignment it is clearly negative eigenvalues of P that cause exponential growth in $\boldsymbol{\omega}$.

2.3   Equations relating α and χ

Ohkitani's relation between $S\boldsymbol{\omega}$ and $P\boldsymbol{\omega}$ can be used to discuss the relationship between α and $\boldsymbol{\chi}$. These two objects have their exact equivalents with

P replacing S

$$\alpha_p = \frac{\omega \cdot P\omega}{\omega \cdot \omega}, \qquad\qquad \chi_p = \frac{\omega \times P\omega}{\omega \cdot \omega}. \qquad (2.11)$$

Differentiation of α and χ and elementary vector algebra then shows that

$$\frac{D\alpha}{Dt} = \chi^2 - \alpha^2 - \alpha_p, \qquad (2.12)$$

$$\frac{D\chi}{Dt} = -2\alpha\chi - \chi_p. \qquad (2.13)$$

Equations (2.12) and (2.13) were first derived by Galanti, Gibbon and Heritage (1997) and are the main equations used in this article; they have the merit of being written in the form where, on particle paths, they appear as a set of four ordinary differential equations driven by the pressure Hessian. Explicitly at least, they are are also independent of ω.

The divergence-free condition div $\mathbf{u} = 0$ has not been used in the derivation of equations (2.12) and (2.13). This is equivalent to $-\Delta p = u_{i,j}\, u_{j,i}$ which converts to a relation between P and S^2

$$\operatorname{Tr} P = -\operatorname{Tr} S^2 + \frac{1}{2}|\omega|^2. \qquad (2.14)$$

3   Angle relations

Dresselhaus and Tabor (1991) used the idea of studying the orientation of the vorticity in terms of angles between ω and eigenvectors of S, which they called a strain basis. Our approach here is somewhat different; we define the angle between ω and $S\omega$ as θ

$$\tan\theta = \frac{\chi}{\alpha} \qquad (3.1)$$

but we do not use a strain basis. Obviously, when ω aligns (anti-aligns) with an eigenvector of S then $\theta = 0$ (π). Then the α–χ equations of (2.12) and (2.13) can be written as

$$\frac{D\tan\theta}{Dt} = -\alpha\tan^3\theta - \left(\alpha - \frac{\alpha_p}{\alpha}\right)\tan\theta - \frac{\tilde{\chi}_p}{\alpha} \qquad (3.2)$$

where $\tilde{\chi}_p = \hat{\chi} \cdot \chi_p$. Because $\tilde{\chi}_p$ is one of the quantities driving (3.2), it is necessary to say something about the relative orientation of χ and χ_p. This can be found by deriving (2.13) and (2.14) an alternative way. Let us consider the unit vector $\hat{\xi} = \omega/\omega$, an arbitrary unit vector $\hat{\Omega}$ orthogonal to $\hat{\xi}$ and a third orthogonal unit vector given by $\hat{\xi} \times \hat{\Omega}$ in a spherical polar co-ordinate[1]

---

[1] To keep to the convention of spherical co-ordinates, the notation here is reversed to that in Galanti *et al.* (1997) where ϕ played the role of θ.

system (r, θ, ϕ). The angles (θ, ϕ) allow us to express any other vector $S\omega$ $(r \equiv |S\omega|)$ lying in this 3-space as

$$
\begin{aligned}
S\omega &= \hat{\Omega}|S\omega|\sin\theta\cos\phi + (\hat{\xi}\times\hat{\Omega})|S\omega|\sin\theta\sin\phi + \hat{\xi}|S\omega|\cos\theta \\
&= \hat{\Omega}\omega\chi\cos\phi + (\hat{\xi}\times\hat{\Omega})\omega\chi\sin\phi + \hat{\xi}\omega\alpha
\end{aligned}
\tag{3.3}
$$

because $\omega\chi = |S\omega|\sin\theta$ and $\omega\alpha = |S\omega|\cos\theta$. Likewise we also have

$$
P\omega = \hat{\Omega}\,\omega\chi_p\cos\phi_p + (\hat{\xi}\times\hat{\Omega})\,\omega\chi_p\sin\phi_p + \hat{\xi}\,\omega\alpha_p.
\tag{3.4}
$$

In (3.3) and (3.4) the angles θ and θ_p are hidden in α, α_p, χ and χ_p. Moreover, χ lies in the horizontal plane at an angle of $\phi + \pi/2$ to the $\hat{\Omega}$-axis so

$$
\chi \cdot \chi_p = \chi\chi_p\cos(\phi - \phi_p).
\tag{3.5}
$$

A simple computation shows that the orthogonal unit vectors $\hat{\Omega}$, $(\hat{\xi}\times\hat{\Omega})$ and $\hat{\xi}$ satisfy[2]

$$
\frac{D}{Dt}\begin{pmatrix}\hat{\Omega}\\ \hat{\xi}\times\hat{\Omega}\\ \hat{\xi}\end{pmatrix} = \begin{pmatrix}0 & \beta & -\chi\cos\phi\\ -\beta & 0 & -\chi\sin\phi\\ \chi\cos\phi & \chi\sin\phi & 0\end{pmatrix}\begin{pmatrix}\hat{\Omega}\\ \hat{\xi}\times\hat{\Omega}\\ \hat{\xi}\end{pmatrix}
\tag{3.6}
$$

where β is an arbitrary function and the variables α, χ and ϕ must obey

$$
\frac{D\alpha}{Dt} + \alpha^2 - \chi^2 + \alpha_p = 0,
\tag{3.7}
$$

$$
\frac{D\chi}{Dt} + 2\alpha\chi = -\chi_p\cos(\phi - \phi_p).
\tag{3.8}
$$

$$
\left(\frac{D\phi}{Dt} - \beta\right)\chi = \chi_p\sin(\phi - \phi_p).
\tag{3.9}
$$

The equations in (3.7)–(3.9) are equivalent to (2.12) and (2.13) but with the additional information on the evolution of ϕ in (3.9). The eigenvalues of the matrix in (3.6) are given by $\lambda_\pm = \pm i\,(\chi^2 + \beta^2)^{1/2}$ so constant values of χ and β represent rotations.

4 Some examples

(1) Burgers solutions for the Euler equations: These solutions have a velocity field of the form (Burgers (1948), Moffatt *et al.* (1994))

$$
\mathbf{u} = \left(-\frac{\gamma}{2}x,\ \frac{\gamma}{2}y,\ \gamma z\right) + (-\Psi_y,\ \Psi_x,\ 0)
\tag{4.1}
$$

---

[2]Equation (3.6) can be reproduced using the Frenet–Serret equations (Constantin, Procaccia & Segel (1995), Constantin (1994) and Galanti, Procaccia & Segal (1996)) although its derivation is not dependent on them.

where γ is a constant or can be a function of time (see Saffman (1993)). Thus $\boldsymbol{\omega} = (0, 0, \omega)^T$ and $\Delta\Psi = \omega$. In this case S and P take on a block diagonal form

$$S = \begin{pmatrix} S_{11} & S_{12} & 0 \\ S_{21} & S_{22} & 0 \\ 0 & 0 & \gamma \end{pmatrix} \tag{4.2}$$

and one eigenvector of S is $(0, 0, 1)^T$. Thus $\boldsymbol{\omega}$ is parallel to this and therefore $\theta = 0$ because $\chi = 0$. The scalar vorticity ω satisfies

$$\omega_t - \gamma(x\omega_x + y\omega_y) + (\Psi_x\omega_y - \Psi_y\omega_x) = \gamma\omega. \tag{4.3}$$

The vorticity is a function of r alone $(r^2 = x^2 + y^2)$, corresponding to the axisymmetric Burgers vortex. Solving $\Delta\Psi = \omega$ we have

$$\frac{d\Psi}{dr} = \frac{1}{r}\int_0^r s\,\omega(s, t)\,ds. \tag{4.4}$$

Now $x\omega_x + y\omega_y = r\omega_r$ and $x\Psi_y - y\Psi_x = 0$ so (4.3) becomes

$$\omega_t = \gamma\left(\omega + \frac{r}{2}\frac{d\omega}{dr}\right) \tag{4.5}$$

which, with initial data $\omega_0 = \omega(r, 0)$, has a solution of the form

$$\omega(r, t) = \exp(\gamma t)\,\omega_0\left(r\exp\left(\frac{\gamma t}{2}\right)\right). \tag{4.6}$$

Hence the support collapses exponentially as the amplitude grows (Majda (1991)). We also find that

$$\begin{aligned} \alpha_p &= -\gamma_0^2 & \chi_p &= 0 \\ \alpha &= \gamma_0^2 & \chi &= 0 \end{aligned} \tag{4.7}$$

In consequence

$$\theta = 0 \tag{4.8}$$

and

$$\frac{D\alpha}{Dt} = 0, \qquad \frac{D\chi}{Dt} = 0. \tag{4.9}$$

Hence the Burgers solutions are like 'equilibrium solutions' of the system in the Lagrangian sense although such solutions are not necessarily stable.

(2) Hill's spherical vortex: This also has $\theta = 0$ for the vorticity field inside the sphere $r^2 + z^2 = a^2$

$$\boldsymbol{\omega} = (0, \, Ar, \, 0) \tag{4.10}$$

$$\begin{aligned} \alpha_p &= \frac{A^2}{50}\left[4r^2 - \frac{10a^2}{3}\right]; & \mu_p &= 0 \\ \alpha &= Az/5 & \chi &= 0. \end{aligned} \tag{4.11}$$

but $D\alpha/Dt \neq 0$, because $\mathbf{u} \cdot \nabla\alpha \neq 0$.

(3) ABC flow: for which $\mathbf{u} = \boldsymbol{\omega}$ with

$$
\begin{aligned}
u_1 &= \omega_1 = \sin z + \cos y, \qquad u_2 = \omega_2 = \sin x + \cos z, \\
u_3 &= \omega_3 = \sin y + \cos x.
\end{aligned}
$$

This flow has $\theta \neq 0$ (nor constant) nor are the material derivatives of α and χ zero.

5   Results on the 3D Euler singularity

Whether or not the 3D Euler equations develops a singularity in the vorticity field in finite time is mathematically still an open problem. Numerically, Kerr (1993) has shown that such a singularity develops from initial data consisting of anti-parallel vortex tubes using a $512^2 \times 256$ grid. The theorem of Beale, Kato and Majda (1984) does not tell us whether such a singularity will occur but says that if it does then it must develop in such a way that that no quantity can become infinite as $t \to t^*$ without

$$
\int_0^t \|\boldsymbol{\omega}\|_\infty \, d\tau \to \infty \tag{5.1}
$$

as $t \to t^*$. A consequence of this result is that if one observes in a simulation that

$$
\|\boldsymbol{\omega}\|_\infty \sim (t_0 - t)^{-\beta} \tag{5.2}
$$

then if the singularity is genuine and not an artefact of the numerical scheme then β must satisfy $\beta \geq 1$. Kerr's singularity (1993) has

$$
\|\boldsymbol{\omega}\|_\infty \sim (t_0 - t)^{-1} \tag{5.3}
$$

which is exactly at the lower end of what is allowed. A more recent version of this idea is that by Constantin, Fefferman & Majda (1996) which takes into account the direction of vorticity as well as its magnitude in terms of angles between vortex lines at points $\mathbf{x}$ and $\mathbf{x}+\mathbf{y}$. Our approach here is to study not only the stretching rate α, which is significant only in terms of stretching and compression, but also the vector χ which takes into account the direction of $\boldsymbol{\omega}$ in terms of its relative orientation with $S\boldsymbol{\omega}$ and $P\boldsymbol{\omega}$.

The purpose of this section is to analyze Kerr's 1993 data in terms of α, χ, α_p and χ_p in order to understand the complicated stretching and compressive processes near the singular region. Kerr's picture of the formation of the singularity (as in Figure 1) is like the elbow of a right arm across a body with the singularity forming near the cap of the elbow (Kerr (1993)).

It is a vorticity isosurface showing one half of a pair of vortices across a symmetry plane ($z = 0$). The initial conditions are two anti-parallel (to the

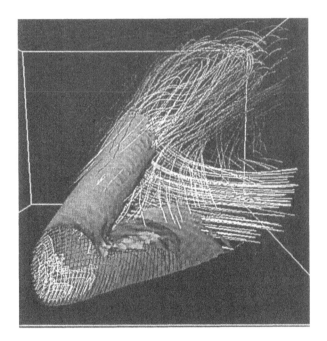

Figure 1: Kerr's picture of singularity formation in the vorticity field $\boldsymbol{\omega}$.

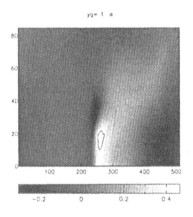

Figure 2: The variable α at the slice $yg = 1$.

y-axis) vortex tubes above and below the $z = 0$ plane (see Kerr (1993)). The co-ordinates are x across the body, y out from the body and z vertically. Subsequent pictures, Figures 2–7, are contrast images of various variables

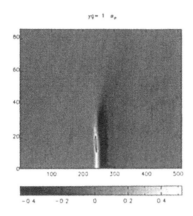

Figure 3: The variable α_p at the slice $yg = 1$.

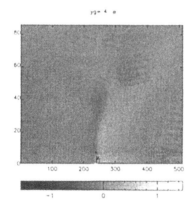

Figure 4: The variable α at the slice $yg = 4$.

in the problem with the contrast scale shown below each one. We have two slices in y; the first is taken at $yg = 1$ (yg refers to the grid point in the y-co-ordinate) at the tip of the elbow where the singularity occurs. This lies in the plane of symmetry; in fact it is on the cap of the elbow – the singular region – where growth develops. $yg = 4$ is at 12 further grid points away from the tip. The $\alpha - \chi$ pictures are made up from data taken at $t = 15$, although the singularity starts to form at $t = 12$.

The structure of vortex stretching and compression around the singular point at $yg = 1$ (in the symmetry plane) is very complicated. For both α and α_p in Figures 2 and 3 there is strong evidence for a violent change in sign in both variables meaning that there is a strong switch from stretching

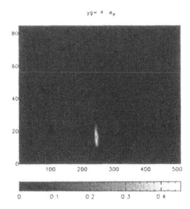

Figure 5: The variable α_p at the slice $yg = 4$.

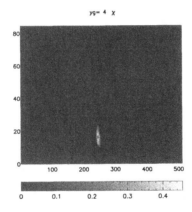

Figure 6: The variable χ at the slice $yg = 4$.

to compression across a line, very much like a vortex jump. Clearly there is no evidence of symmetry in the stretching and compressive processes, supporting the view that the singularity does not have a self-similar structure (Constantin, Fefferman & Majda (1996), Constantin (1994)). For reasons of symmetry, $\chi = \chi_p = 0$ in the symmetry plane so we have exact alignment there. Figure 4 is α taken at $yg = 4$, 12 grid points away from the symmetry plane; the contrast between the stretching and compressive process is weaker with less of a jump than in Figure 2. Figure 5 shows that $\alpha_p \geq 0$ and Figures 6 and 7 show that $\chi = 0$ and $\chi_p = 0$ almost everywhere, except in the singular region. Hence $\theta = 0$ almost everywhere except in the singular region where there is misalignment because χ and χ_p are non equal to zero.

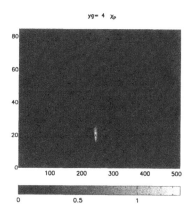

Figure 7: The variable χ_p at the slice $yg = 4$.

We have also computed the RHS of equation (3.2) for $\frac{D\tan\theta}{Dt} = 0$. We have not shown this picture because we find that

$$\frac{D\tan\theta}{Dt} = 0 \qquad (5.4)$$

except in a tiny region around the singular region. We conclude that there are three regions; an outer, an intermediate and an inner, the latter associated with the plane of symmetry.

References

Beale J. T. , Kato T., Majda A. (1984) 'Remarks on the breakdown of smooth solutions for the 3D Euler equations', *Commun. Math. Phys.* **94**, 61.

Burgers J. M. (1948) 'A mathematical model illustrating the theory of turbulence', *Adv. Appl. Math.,* **1**, 1.

Cantwell B. J. (1992) 'Exact solution of a restricted Euler equation for the velocity gradient tensor', *Phys. Fluids A* **4**, 782.

Constantin P. (1994) 'Geometric statistics in turbulence', *SIAM Rev.* **36**, 73.

Constantin P., Fefferman Ch., Majda A. (1996) 'Geometric constraints on potentially singular solutions for the 3D Euler equations', *Commun. Partial Diff. Equations* **21**, 559.

Constantin P., Procaccia I., Segel D. (1995) 'Creation and dynamics of vortex tubes in three-dimensional turbulence', *Phys. Rev. E.* **51**, 3207.

Dresselhaus E., Tabor M. (1991) 'The kinematics of stretching and alignment of material elements in general flow fields', *J. Fluid Mech.* **236**, 415.

Galanti B., Gibbon J. D., Heritage M. (1997) 'Vorticity alignment results for the 3D Euler and Navier-Stokes equations', *Nonlinearity,* **10**, 1675–1694.

Galanti B., Procaccia I., Segel D. (1996) 'Dynamics of vortex lines in turbulent flows', *Phys. Rev. E.* **54** (5), 5122.

Jeong Y., Hussain F. (1995), 'On the identification of a vortex', *J. Fluid Mech.* **285**, 69–94.

Kerr R. (1993) 'Evidence for a singularity of the three-dimensional, incompressible Euler equations', *Phys. Fluids A* **5**, 1725.

Majda A. (1986) 'Vorticity and the mathematical theory of incompressible fluid flow', *Comm. Pure and Appl. Math.,* **39**, 187.

Majda A. (1991) 'Vorticity, turbulence and acoustics in fluid flow', *SIAM Rev.* **33**, 349.

Moffatt H. K., Kida S., Ohkitani K. (1994) 'Stretched vortices – the sinews of turbulence; large-Reynolds-number asymptotics', *J. Fluid Mech.* **259**, 241.

Ohkitani K. (1993) 'Eigenvalue problems in three-dimensional Euler flows', *Phys. Fluids A* **5**, 2570.

P. G. Saffman (1993) *Vortex Dynamics,* Cambridge University Press.

Structure of a New Family of Stretched Non-axisymmetric Vortices

Stéphane Le Dizès

1 Summary

The discovery of concentrated vorticity filaments in turbulent flows and their possible connection to fundamental properties of turbulence has renewed the interest in vortex dynamics. Direct numerical simulations suggest that filaments resemble Burgers vortices. In order to understand the dynamical properties of filaments, it is then natural to construct and analyse asymptotic solutions close to this model. In this framework, Moffatt, Kida and Ohkitani (1994) (referred to hereinafter as MKO94) recently extended the Burgers model to account for a non-axisymmetric correction generated by a stationary external strain field.

In the first part of the article, I generalise the MKO94 analysis by considering a vortex subject to an external multipolar strain field rotating around the vortex axis. This extension is needed in practice as the external strain field of a given vortex is generally created by other vortices rotating around it due to their mutual interactions. The case of a multipolar strain field is also considered in order to describe flows which exhibit fold-symmetry of high order. The analysis will be shown to provide a new family of solutions when the frequency of the rotating strain field is in the range of the angular velocities of the vortex. In that case, the non-axisymmetric correction has indeed a different structure as it exhibits a critical layer at the radial coordinate where the local angular velocity equals the frequency of the strain. This singularity can be resolved by the introduction of nonlinearity or viscosity. However, it is shown that the entire solution depends on whether nonlinearity or viscosity dominates in the critical layer, that is it depends on the value of the so-called Haberman (1972) parameter $h = h_c/(Re\ \varepsilon^{3/2})$, where ε is the amplitude of the non-axisymmetric correction, Re the Reynolds number and h_c a local $O(1)$ parameter. The properties of these new solutions are analysed in detail.

The nonlinear case ($h \ll 1$) is particularly interesting as it gives for a distinguished frequency a non-axisymmetric vortex for which the external strain field can be turned off. In other words, the external strain field is not needed for the vortex to remain non-axisymmetric in that case. It is argued that this solution could correspond to the non-axisymmetric solution recently observed in 2D numerical simulations (Rossi *et al.* 1997) and that it could be a good candidate for the non-axisymmetric vortex structures of turbulence.

2   Introduction

2.1   Vortices in stationary strain field and Lundgren transform

The Lamb–Oseen vortex and the Burgers vortex are two exact solutions of the Navier–Stokes equation which are related to each other by Lundgren's transform (Lundgren 1982). The Lamb–Oseen solution describes the viscous diffusion of a point vortex in a 2D unstretched flow. It is characterised by a circulation Γ, a radius δ which evolves according to $\delta(t) = \sqrt{4\nu t}$, where ν is the kinematic viscosity, and an axial vorticity field given by

$$\omega_0 = \frac{\Gamma}{\pi \delta^2} G\left(\frac{r}{\delta}\right), \qquad (2.1)$$

where $G(x)$ is the Gaussian function:

$$G(x) = \mathrm{e}^{-x^2}. \qquad (2.2)$$

Burgers' vortex is a stationary solution for which viscous diffusion is compensated by a uniform stretching field $(-\frac{\gamma}{2}r, 0, \gamma z)$ along the axis of the vortex. Its vorticity profile is still given by (2.1) but its radius is $\delta = \sqrt{4\nu/\gamma}$.

Lundgren's transform can be used to link both solutions. It is in fact more general as it relates any uniformly stretched solution to an unstretched 2D solution, or more precisely, that the velocity field $\mathbf{u}_s$ of the stretched solution

$$\mathbf{u}_s = (-\frac{\gamma}{2}r, 0, \gamma z) + \left(\frac{1}{r}\frac{\partial \Psi_s}{\partial \theta}, -\frac{\partial \Psi_s}{\partial r}, 0\right), \qquad (2.3)$$

where $\gamma(t)$ is the uniform (in space) stretching rate, is connected to a 2D solution (of a streamfunction Ψ_{2D}) by the relations:

$$\Psi_s(r, \theta, t) = \Psi_{2D}(\xi, \theta, T), \qquad (2.4)$$

$$\xi = \sqrt{S}\, r; \qquad T = \int_0^t S(t')dt'; \qquad S(t) = \exp\left(\int_0^t \gamma(t')dt'\right). \qquad (2.5)$$

These relations let us obtain the evolution of the radius of a Gaussian vortex in any stretching field (see e.g. Eloy & Le Dizès, 1999). They are also useful for obtaining the correct time evolution of the 2D solution when the analogous solution in a constant stretching field is itself time-dependent as in the case treated below.

Moffatt *et al.* (1994) extended Burgers's solution to account for a non-axisymmetric correction generated by an additional external stationary strain field perpendicular to the vortex axis. This extension is equivalent to considering a Gaussian vortex in a non-axisymmetric stretching field. They showed that if the Reynolds number $Re = \Gamma/(\pi\nu)$ associated with the vortex is large and if the strain rate of the external field issmall compared to the maximum vorticity of the vortex that the main features of the vortex are conserved near its core. The additional strain field creates a non-axisymmetric correction which turns out to be linear and non-viscous at leading order [see equation (3.4) below]. This correction modifies the shape of both the streamlines and the vorticity lines which become slightly elliptical. MKO94 computed the eccentricity of the ellipse near the vortex center for a normalized external strain field and also demonstrated that its axis is oriented at 45° with respect to the principal direction of positive strain[1]. Jimenez *et al.* (1996) considered the MKO94 asymptotic solution in two dimensions by applying Lundgren transform and recovered results obtained by Ting & Tung (1965). They compared the model to vortex structures obtained from numerical simulations of 2D turbulence and demonstrated a relatively good agreement concerning the strength and the orientation of the strain field in the vortex core with respect to the external strain field generated by the background flow.

The deformation of a vortex core due to a stationary external strain field is also clearly visible in simple configurations of vortex dynamics such as vortex rings (Fukumoto & Moffatt, this volume), or counter-rotating vortex pairs (Williamson *et al.*, this volume). The streamlines in the vortex core are elongated in the direction of propagation of the vortex ring (or of the vortex pair) which corresponds to a direction 45° to the principal axis of the strain induced by the rest of the vortex ring (or by the other vortex).

It is important to point out that the MKO94 model assumes a stationary external strain field. This assumption limits the use of the model to particular configurations of vortex interactions. In particular, as soon as two vortices no longer have exactly opposite circulations, their mutual interactions make them rotate around each other. This implies that the external strain field generated by one vortex on the other is in general not stationary: it rotates at a fixed angular frequency around the vortex. The article is concerned with the extension of the MKO94 model to this configuration. More generally, we shall consider a Gaussian vortex subjected to a rotating multipolar strain field in order also to describe flows which exhibit a fold symmetry of high order.

2.2 Non-axisymmetric vortices in 2D flows

Wherea MKO94's analysis is concerned with vortices whose non-axisymmetry is generated by an external field, recent numerical simulations have also addressed the existence of non-axisymmetric vortices without the need of external fields (Rossi *et al.* 1997, Dritschel 1998). Rossi *et al.* (1997) demonstrated that a Lamb–Oseen vortex could relax to a tripole structure when perturbed by a large quadrupole perturbation. They claimed that the resulting non-axisymmetric vortex was not subject to the so-called axisymmetrization pro-

---

[1]Hereafter, the strain denotes the additional strain field, i.e. the non-axisymmetric part of the total stretching field. The principal direction of positive strain is then the principal direction (in the plane perpendicular to the vortex axis) of the additional strain field associated with a positive strain rate.

cess (Melander *et al.* 1987). Dritschel (1998) reached the same conclusion with a family of discontinuous vorticity profiles.

However, the axisymmetrization process is active if the non-axisymmetric perturbation is small. Bernoff & Lingevitch (1994) explained that small non-axisymmetric perturbations to a Lamb–Oseen vortex are wound up into spiral structures due to the differential rotation of the vortex and then dissipated on a $O(Re^{1/3})$ time scale by shear-diffusion (Lundgren, 1982). This numerical result was confirmed by Bassom & Gilbert (1998) who proved the linear stability of the Lamb–Oseen vortex with respect to 2D perturbation[2].

This implies that the Rossi *et al.* (1997) non-axisymmetric solution is necessarily due to a nonlinear perturbation of the Lamb–Oseen vortex. In this article, we propose an asymptotic description of such a nonlinear perturbation. This is actually possible because nonlinearity will turn out to be very localized in space. In fact, we will seek a non-axisymmetric perturbation in the form of a 2D linear harmonic perturbation everywhere except around a single critical point where strong nonlinear effects are present. The analysis will be done with a Burgers vortex as base flow, keeping in mind that the stretching field can be suppressed or modified afterwards by Lundgren transform.

3   Framework

Consider a Burgers vortex perturbed by a 2D perturbation of azimuthal wavenumber n and angular frequency w. The velocity field of such a solution is given by (2.3) with a streamfunction of the form

$$\Psi = \Re e\{\Psi_0(r) + \varepsilon\Phi_1(r)e^{in(\theta-wt)} + O(\varepsilon^2)\}. \tag{3.1}$$

where $\Psi_0(r)$ is the axisymmetric streamfunction associated with the Burgers vortex. The small parameter ε in (3.1) measures the amplitude of the correction in the vortex core. It is well defined by the condition (3.7) given below. Hereafter, spatial and time variables are made dimensionless using the radius $\delta = 2\sqrt{\nu/\gamma}$ and the maximum vorticity $\omega_{\max} = \Gamma/(\pi\delta^2)$ of Burgers' vortex. Thus, the incompressible NS equations for Ψ reduce to:

$$\frac{\partial\omega}{\partial t} + J[\omega, \Psi] = \frac{1}{Re}\left(4 + 2r\frac{\partial}{\partial r} + \Delta\right)\omega \tag{3.2}$$

$$\omega = -\Delta\Psi \tag{3.3}$$

where the Jacobian operator $J[f, g]$ is defined in polar coordinates by

$$J[f, g] = \frac{1}{r}\left(\frac{\partial f}{\partial r}\frac{\partial g}{\partial\theta} - \frac{\partial g}{\partial r}\frac{\partial f}{\partial\theta}\right).$$

As above, the Reynolds number Re is defined by $Re = \Gamma/(\pi\nu)$.

---

[2]Bassom & Gilbert (1998) actually proved that a Gaussian vortex is asymptotically stable with respect to 2D inviscid perturbations in a coarse grain sense.

Assuming that both ε and $1/Re$ are small, the leading order equation for Φ_1 in (3.1) is immediately obtained as:

$$(\Omega_0(r) - w)\Delta_n\Phi_1 = \frac{\omega_0'(r)}{r}\Phi_1, \tag{3.4}$$

with

$$\Delta_n = \frac{\partial^2}{\partial r^2} + \frac{1}{r}\frac{\partial}{\partial r} - \frac{n^2}{r^2}, \tag{3.5}$$

where the vorticity ω_0 has been defined in (2.1) (the prime denotes differentiation with respect to the argument) and the angular velocity Ω_0 is

$$\Omega_0(x) = \frac{1 - \exp(-x^2)}{2x^2}. \tag{3.6}$$

Note that in MKO94, the Gaussian vorticity profile of Burgers' vortex was obtained from the calculation at the same order as the equation for Φ_1 by assuming $1/Re = O(\varepsilon)$. Here, we have chosen to postulate the Burgers vortex as base flow right from the beginning in order to leave free the relation between $1/Re$ and ε. It is worth mentioning that if $1/Re \ll \varepsilon$, any axisymmetric vortex (with or without stretching) could have also been considered as a base flow without invalidating equation (3.4). Indeed, viscous and/or stretching effects would have then only introduced a slow evolution on an $O(1/Re)$ time scale of the base flow.

Except for $w \neq \Omega_0(0) = 1/2$, the two possible behaviors near $r = 0$ of solutions to (3.4) are r^n and r^{-n}. Discarding the singular behavior, one can then impose (for any $w \neq \Omega_0(0)$)

$$\Phi_1 \sim r^n \quad \text{as } r \to 0. \tag{3.7}$$

This condition fixes the normalisation of the amplitude of the non-axisymmetric correction and defines in an unambiguous way the parameter ε. Moreover, condition (3.7) also selects a single solution to equation (3.4). The behavior of that solution near infinity is easily shown to be:

$$\Phi_1 \sim s_n r^n + \frac{c_n}{r^n} \quad \text{as } r \to \infty, \tag{3.8}$$

where s_n and c_n are two constants which depends on w and on the way eventual singularities of the equation are resolved. When s_n is non-zero, the total streamfunction Ψ does not vanish for large r and behaves as

$$\Psi \sim \varepsilon|s_n|r^n \cos[n(\theta - wt) + \arg(s_n)] \quad \text{as } r \to \infty. \tag{3.9}$$

This expression represents the external field which is needed to maintain the non-axisymmetric correction Φ_1 in (3.1) with the normalisation (3.7). It is a multipolar strain field with a n-fold symmetry rotating at the angular frequency w. The parameter s_n gives both the strength $|s_n|$ of the external field and its phase shift $\arg(s_n)$ with respect to the orientation of the perturbation in the vortex core. It is also a measure of the 'strain-vortex' interaction. Indeed, without vortex, the external strain field would be valid everywhere and one would have $s_n = 1$. Thus, if $|s_n| < 1$, the strain field is enhanced by the vortex. In the opposite case $|s_n| > 1$ it is weakened.

If $s_n = 0$, the strain field is generated by the vortex itself as no external field is present any more. In that case, Φ_1 represents an eigenmode of the vortex and expression (3.1) is then the first-order approximation of the streamfunction of a non-axisymmetric vortex solution in a free-of-strain environment. Based on the Bassom & Gilbert (1998) results, we already know that s_n cannot vanish if equation (3.4) is valid everywhere. Indeed, it would imply that Ψ_1 is a linear inviscid perturbation which is marginally stable in contradiction with the asymptotic stability proved by Bassom & Gilbert (1998). However, we shall see in the next section that equation (3.4) naturally develops singularities which allow the vanishing of s_n if the singularities are smoothed by nonlinearities. The properties of the perturbations when this occurs are analysed in detail in Section 5.1. The resulting non-axisymmetric vortex is compared to the numerical solution of Rossi *et al.* (1997) in Section 5.2.

4   Vortices in a rotating multipolar strain field

The linear problem (3.4), (3.8), (3.7) is analogous to the one obtained in Moffatt *et al.* (1994) and Jimenez *et al.* (1996) for the stationary elliptic correction of a Gaussian vortex. In particular, equation (2.25) in MKO94 corresponds to (3.4) for the particular values $n = 2$, $w = 0$. Note however that a different normalization has been chosen: the amplitude of Φ_1 is fixed in the vortex core by the condition (3.7) such that the unknown parameter is here $s_n(w)$ and not the strain in the vortex core as in MKO94. As explained above, this choice has been made to allow $s_n(w) = 0$ in order to treat in the same framework non-axisymmetric vortices without external strain field (see Section 5).

The resolution of equations (3.4), (3.7), (3.8) crucially depends on the value of w with respect to the range of Ω_0. If w is in the range of Ω_0, i.e. $\min(\Omega_0) < w < \max(\Omega_0)$, there exists a critical point r_c defined by $\Omega_0(r_c) = w$ where solutions (or their derivatives) to equation (3.4) may exhibit a singularity. In other words, the solution of (3.4) prescribed by (3.7) may become singular at r_c. The resolution of this singularity would require the introduction of higher order effects such as viscosity or nonlinearity. Before treating this case which will turn out to be the most interesting, let us first consider the case without a critical layer.

4.1   Cases without critical layer

The configurations without a critical layer correspond to frequencies satisfying $w < \min(\Omega_0)$ or $w > \max(\Omega_0)$. For these frequencies, the integration of (3.4) with condition (3.7) at zero is possible from 0 to ∞ and provides an

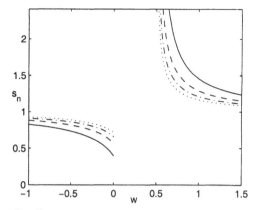

Figure 1: Coefficient s_n versus the angular frequency w. Solid line: $n = 2$; dashed line: $n = 3$; dash-dotted line: $n = 4$; dotted line: $n = 5$.

everywhere differential function Φ_1. The coefficient $s_n(w) = \lim_{r \to \infty}(\Phi_1/r^n)$ is then perfectly defined in a linear inviscid framework.

In Figure 1 the variation of $s_n(w)$ with respect to w for $n = 2, 3, 4, 5$ is represented. The frequency interval $(0, 0.5)$ has been excluded as it corresponds to configurations with a critical layer. As in MKO94, for $n = 2$, the streamlines are elliptical with major axis oriented at $45°$ with respect to the principal direction of the strain. Moreover, the eccentricity of the streamline in the vortex center is given by $1/s_2$ when the external strain field is normalized at infinity. Figure 1 shows that, when no critical layers are present, the vortex eccentricity is maximum for $w = 0$: the effect of rotation of the strain field is then to diminish the strength of the strain in the vortex core. Note also that the non-axisymmetric distortion is stronger in the vortex core than outside for anticyclonic vortices ($s_n < 1$ if $w < \min(\Omega_0) = 0$) while it is the opposite for strongly cyclonic vortices ($s_n > 1$ if $w > \max(\Omega_0) = 1/2$).

For the critical value $w = 1/2 = \Omega_0(0)$, condition (3.7) does not apply. The adequate behavior of Φ_1 near zero is $r^{\sqrt{n^2+8}}$ and not r^n in that case (Bassom & Gilbert 1998). Enforcing condition (3.7) close to $w = 1/2$ then leads to very large values of s_n which explains the divergence of s_n observed in Figure 1 near $w = 1/2$.

It is also worth mentioning that the results presented in figure 1 are qualitatively the same for all n. We shall see below that it is no longer the case when there are critical layer singularities.

4.2 Critical layer analysis

Critical layers are well-known in the parallel shear flow framework and refer to regions close to the (critical) points where the basic flow velocity equals the phase velocity of the perturbation. They have been the subject of numerous works since the fifties and are now recognized to play an essential role in shear flows and boundary layer transitions (see Maslowe 1986, Cowley & Wu 1994, and references therein). Here the critical points are the radial positions r_c where the angular velocity of the vortex equals the angular frequency of the perturbation, that is where $\Omega_0(r_c) = w$. The critical layers around these points are the same as in the parallel framework. We shall assume that the critical layers are in equilibrium such that classical results from Benney & Bergeron (1969), Haberman (1972), Brown & Stewartson (1978) and Smith & Bodonyi (1982) can be used. A brief discussion of time-dependent effects is given at the end of the article.

Starting to integrate equation (3.4) from $r = 0$ with condition (3.7) ultimately leads to a solution which exhibits at the critical point r_c, for $r < r_c$, a weak singularity of the form

$$\Phi_1(r) \sim \alpha + \beta(r - r_c) + \alpha\kappa_c(r - r_c)\ln(r_c - r) \quad \text{as } r \to r_c^-, \qquad (4.1)$$

where α and β are real constants and $\kappa_c = \omega_0'(r_c)/(r_c\Omega_0'(r_c))$. This weak singularity is usually solved in a 'critical layer' by considering higher-order terms such as viscous or nonlinear terms (Lin 1955, Benney & Bergeron 1969, Haberman 1972). The main result, which is what concern us, is that an azimuthal velocity jump is created across the critical layer such that the amplitude Φ_1 in (3.1) has an expansion for $r > r_c$ of the form

$$\Phi_1(r) \sim \alpha + \beta(r - r_c) + \alpha\kappa_c(r - r_c)[\ln|r - r_c| + i\chi] \quad \text{as } r \to r_c^+. \qquad (4.2)$$

The value of the real parameter χ depends on the nature of the critical layer, that is, whether nonlinearity or viscosity dominates near $r = r_c$. Lin (1955) showed that $\chi = -\pi$ if $\mathrm{sgn}(n\Omega_0'(r_c)) < 0$ ($\chi = +\pi$ if $\mathrm{sgn}(n\Omega_0'(r_c)) > 0$) when the critical layer is viscous. Benney & Bergeron (1969) obtained $\chi = 0$ for a purely nonlinear critical layer. Haberman (1972) considered both effects simultaneously and proved that χ varies continuously from $-\pi$ to 0 when nonlinearity is progressively increased. Moreover, he showed that χ depends on a single parameter[3] h given by

$$h \equiv \frac{h_c}{\varepsilon^{3/2}Re}, \qquad (4.3)$$

where the coefficient h_c is here

$$h_c = \frac{|\Omega_0'(r_c)|^{1/2}}{n}\left|\frac{r_c}{\Phi_1(r_c)}\right|^{3/2}. \qquad (4.4)$$

---

[3]This parameter is here designated by the letter h instead of λ as is usually done in critical layers works (see Maslowe 1986) because λ often refers to the non-axisymmetric part of the strain in the vortex literature (see e.g. Moffatt *et al.* 1994). Moreover, this parameter was sometimes called Haberman parameter (Goldstein & Hultgren 1988), so 'h' seems adequate.

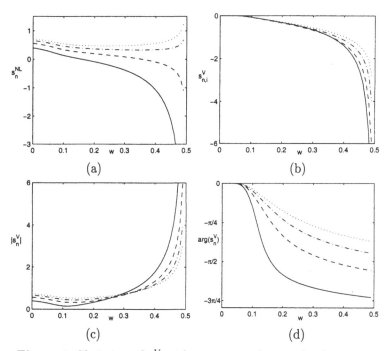

Figure 2: Variation of s_n^V with respect to the angular frequency w. Solid line: $n = 2$; dashed line: $n = 3$; dash-dotted line: $n = 4$; dotted line: $n = 5$. (a) Real part (s_n^{NL}), (b) Imaginary part $(s_{n,i}^V)$, (c) Modulus $(|s_n^V| = [(s_n^{NL})^2 + (s_{n,i}^V)^2]^{1/2})$, (d) Phase $(arg(s_n^V) = \arctan(s_{n,i}^V/s_n^{NL}))$.

The algebraic manipulations leading to (4.3) and (4.4) are given in the Appendix, as well as the curve $\chi(h)$ obtained by Haberman (1972) and Smith & Bodonyi (1982). With this definition of h, the critical layer is then viscous if $h \gg 1$, and purely nonlinear if $h \ll 1$.

As soon as h is computed, $\chi(h)$ is known from Figure 8, so that equation (3.4) can be integrated from r_c to $+\infty$ with condition (4.2). Near infinity, the solution behaves according to (3.8). The form of expansion (4.2) guarantees that s_n can be written as

$$s_n(w, h) = s_n^{NL}(w) - i\frac{\chi(h)}{\pi}s_{n,i}^V(w) \qquad (4.5)$$

where s_n^{NL} and $s_{n,i}^V$ are real functions defined as limit values for large r of $f(r)/r^n$ and $g(r)/r^n$ respectively. Here, the functions f and g are solutions of (3.4) prescribed by the following expansions near r_c^+

$$f(r) \sim \alpha + \beta(r - r_c) + \alpha\kappa_c(r - r_c)\ln|r - r_c| \quad \text{as } r \to r_c^+, \qquad (4.6)$$

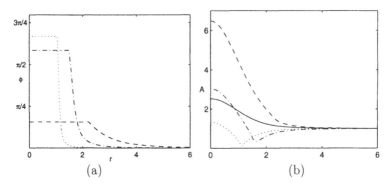

Figure 3: Phase ϕ (a) and amplitude A (b) of the elliptic distortion (as defined in (4.8)) for a viscous critical layer ($h \gg 1$). Solid line: $w = 0$; dashed line: $w = 0.1$; dash-dotted line: $w = 0.2$; dotted line: $w = 0.3$.

and

$$g(r) \sim \alpha \kappa_c \pi (r - r_c) \quad \text{as } r \to r_c^+. \tag{4.7}$$

It immediately follows that $s_n = s_n^{NL}$ for a purely nonlinear critical layer ($\chi = 0$), and $s_n = s_n^V = s_n^{NL} + i s_{n,i}^V$ for a purely viscous critical layer ($\chi = -\pi$).

The variations of s_n^{NL} and $s_{n,i}^V$ with respect to w are displayed in figures 2(a),(b) for $n = 2, 3, 4, 5$. These figures let us construct the missing part of the curves in Figure 1 for any value of h using expression (4.5) and Figure 8. The first important point to note is that s_n is in general complex, which means that both the orientation and the strength of the non-axisymmetric distortion varies with respect to the radial coordinate. The change of orientation in the vortex core between the critical layer and no critical layer cases is measured by the phase of s_n which reaches its maximum for the viscous case ($\chi = -\pi$). The dependence of this maximum with respect to w is shown in Figure 2(d). The corresponding modulus of s_n is displayed in Figure 2(c).

For $n = 2$, Figure 2(d) implies that the elliptical streamlines near the center have no longer a major axis oriented at $45°$ with respect to the external strain but that there is an additional angular shift $\phi_2 = -\arg(s_2)/2$. This additional shift is maximum for viscous critical layers and grows with w up to $|\phi_2|_{\max} \approx 1.15$. This effect can also be seen in Figures 3 and 4. In Figure 3 are shown variations of the phase and amplitude of the non-axisymmetric distortions as a function of the radial coordinate r for different frequencies, where the phase and amplitude are defined from Φ_1 by

$$\Phi_1 = s_2 r^2 A(r) e^{i\phi(r)}. \tag{4.8}$$

In Figure 4 are displayed the streamlines of the elliptically perturbed vortex in a frame corotating with the external strain field for 3 different frequencies

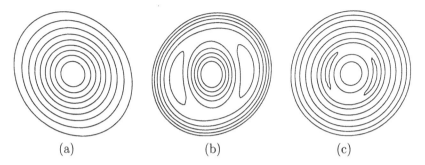

Figure 4: Streamlines of the perturbed Gaussian vortex in the frame corotating with the external strain field with a viscous critical layer. ($\varepsilon = 0.005$) (a) $w = 0$, (b) $w = 0.15$, (c) $w = 0.3$.

and for a fixed ε. One clearly sees that from left to right the orientation of the elliptic streamlines in the vortex core rotates clockwise. Two regions of recirculation located outside the vortex axis are also present in Figures 4(b) and 4(c). They are associated with the critical layer. Their position moves towards the vortex center as the frequency increases as expected from the displacement of the critical point r_c towards the center. In Figure 5 is plotted the amplitude A of the correction Φ_1 as defined in (4.8) when the critical layer is purely nonlinear ($\chi = 0$) for the same values of w as in Figure 3(b). Comparing Figures 3(b) and 5 demonstrates that the nature of the critical layer has a strong influence on the structure of the non-axisymmetric distortion everywhere.

5 Non-axisymmetric vortices without external field

5.1 Characteristic properties

An important consequence of viscous effects in the critical layer is that they preclude the vanishing of s_n. This is not the case when the critical layer is dominated by nonlinearity. Indeed, it can be seen in Figure 2(a) that both s_2 and s_3 vanish for a particular value of the frequency while s_4 and s_5 never vanish. This means that there does exist a nonlinear eigenmode for both $n = 2$ and $n = 3$. It follows that a perturbed Gaussian vortex can remain non-axisymmetric without any external field only if the azimuthal wavenumber of the perturbation is either 2 or 3.

Characteristic quantities of the nonlinear eigenmodes are given in Table 1. The nonlinear eigenfrequency w_n^{NL} is given in the first column. The second column gives the radial position of the critical layer. At this position, the streamfunction of the eigenmode exhibits a vertical tangent (see Figure 6). If the critical layer is in equilibrium as assumed, a vorticity jump is also created across the critical layer which has the form $\varepsilon^{1/2}\delta w$ at leading order.

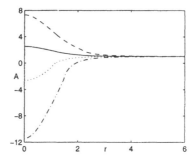

Figure 5: Amplitude A of the elliptic distortion (as defined in (4.8)) with a nonlinear critical layer ($h \ll 1$). Solid line: $w = 0$; dashed line: $w = 0.1$; dash-dotted line: $w = 0.2$; dotted line: $w = 0.3$.

	w_n^{NL}	r_c	h_c	$\delta\omega$	$s_{n,i}^V$	w_n^V
$n = 2$	0.1596	1.7241	0.9371	-0.2138	0.6584	$0.1113 - 0.0397i$
$n = 3$	0.3581	0.8422	2.2064	-0.9749	3.6956	$0.1128 - 0.0896i$

Table 1: Characteristic quantities of the nonlinear eigenmode for $n = 2$ and $n = 3$.

The expression of this jump is (see Appendix)

$$\delta\omega = -2C\omega'_{0_c}\left|\frac{\Phi_{0_c}}{r_c\Omega'_{0_c}}\right|^{1/2}, \qquad (5.1)$$

where the subscript c indicates values taken at r_c and $C \approx 1.3$. Its value for the two nonlinear eigenmodes is given in the fourth column of Table 1.

The parameter h_c in the third column is the local Haberman parameter which characterises the nature of the critical layer. To remain nonlinear, the critical layer must satisfy:

$$\varepsilon^{3/2}Re \gg h_c. \qquad (5.2)$$

Thus, h_c indicates the amplitude threshold of the nonlinear eigenmode (the parameter ε) for a given Reynolds number, or the minimum Reynolds number required for the non-axisymmetric vortex solution formed with the nonlinear eigenmode at a given amplitude ε to survive.

If condition (5.2) is not satisfied, the nonlinear eigenmode does not exist. This means that either an external field is needed to maintain the non-axisymmetric disturbance, or the latter does not survive. The parameter $s_{n,i}^V$ (fifth column of Table 1) is the strength of the external strain that is required

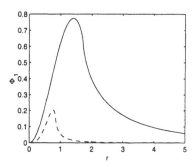

Figure 6: Radial profile of the nonlinear eigenmodes streamfunction. Solid line: $n = 2$, dashed line: $n = 3$.

to maintain a non-axisymmetric perturbation with the same frequency but with a viscous critical layer. This estimate for the external strain applies if $h > 5$, i.e. $\varepsilon^{3/2} Re < 0.2 h_c$. When $\varepsilon^{3/2} Re > 0.2 h_c$, the external strain is smaller. An estimate for its strength can be obtained from $|s_n| = |\frac{\chi(h)}{\pi}| s_{n,i}^V$ and Figure 8 for $\chi(h)$.

If condition (5.2) is not met and no external strain is present, one expects the non-axisymmetric distortion to be damped. In such a case, the critical point moves in the complex plane and it becomes much more difficult to take into account nonlinear effects. However, if one assumes that nonlinear effects are negligible, the damped rate can be obtained from equation (3.4) alone by integrating along a path in the complex plane that contours the critical point from above. This procedure is also possible when the critical point is on the real axis and is equivalent to assuming that the critical layer is purely viscous (see Lin 1955, for instance). For the Gaussian vortex, the complex frequency of the linear viscous eigenmode is given in the last column of Table 1. It is worth pointing out that these frequencies strongly differ from the frequencies of the nonlinear eigenmodes: the damping rate is non-negligible in both cases and the gap between frequency real parts is 45% for $n = 2$ and 227% for $n = 3$.

5.2 Comparison with 2D numerical evidence

The non-axisymmetric solutions of the previous section can be recast in a 2D framework by eliminating the stretching field. As explained in Section 2.1, this is possible using Lundgren's transform. Expression (3.1) is the streamfunction of a solution in a uniform stretching field with constant stretching rate $\gamma^* = 4/Re$. The corresponding solution without stretching is given by relations (2.4) and (2.5). Thus, the streamfunction $\Psi_{2D}(\xi, \theta, T)$ of the 2D unstretched solution is given by (3.1) via the change of variables

$$t = Re \ln(4T/Re + 1)/4; \qquad r = \xi/\sqrt{4T/Re + 1}. \qquad (5.3)$$

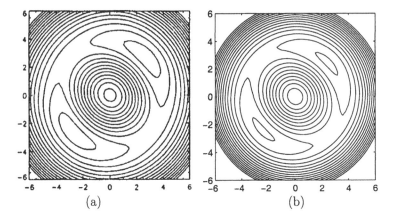

Figure 7: Streamlines in the corotating frame of the non-axisymmetric solution obtained from a Lamb vortex. (a) Numerical solution (from Rossi *et al.*, 1997), (b) Asymptotic solution with $\varepsilon = 0.02$.

Note that the time dependence of the 2D solution is much more complex and in particular that the non-axisymmetric disturbance is no longer harmonic. However, for small time $T \ll Re$, $t \sim T$ and $r \sim \xi$ so stretched and unstretched streamfunctions are identical. This implies that the nonlinear eigenmode calculated in the previous section for the Burgers vortex is also a nonlinear eigenmode of the Lamb–Oseen vortex as long as viscous diffusion remains negligible, i.e. $T \ll Re$.

Rossi *et al.* (1997) numerically demonstrated the existence of a non-axisymmetric vortex solution by perturbating with a localized elliptic distortion a Lamb-Oseen vortex. We claim that their solution could be the non-axisymmetric state obtained here by neglecting viscous diffusion. Figure 7(a) is taken from Rossi *et al.* (1997) and shows the streamlines of their vortex in a corotating frame for $Re = 10^4$. Figure 7(b) are the streamlines of the Gaussian vortex perturbed by the distortion $\varepsilon \Phi_1(r) \cos(2(\Theta - wt))$ computed in Section 5.1 in the frame rotating at the angular frequency w. The value $\varepsilon = 0.02$ has been chosen in order to obtain the best qualitative correspondence between the two plots since Rossi *et al.* (1997) did not provide the amplitude of the perturbation for the final state solution. The shapes of the streamlines for both solutions are surprisingly similar. Note that for their Reynolds number ($Re = 10^4$) and our ε, one has $\varepsilon^{3/2} Re \approx 28$ which is consistent with the nonlinear critical layer condition (5.2). A good agreement is also found for the angular frequency which is $w \approx 0.01$ in Rossi *et al.* (1997) and $w \approx 0.012$ in our calculation with their normalization ($\omega_{\max} = 1/(4\pi + t/Re) \approx 0.075$ at $t = 500$). Rossi *et al.* (1997) also showed that their non-axisymmetric solution is rapidly eroded and modified when Re is divided by a factor 10. This could be associated with the apparition of viscous effects in the critical layer which in particular would explain the weak damping and the modification of the orientation of the distortion observed by Rossi *et al.* (1997). Unfortunately, Rossi *et al.* (1997) did not report results with higher Reynolds numbers. From the present study, one can conjecture that the threshold in perturbation amplitude for the existence of the non-axisymmetric state decreases with increasing Reynolds numbers.

6 Discussion

In this article, the non-axisymmetric correction to a Burgers vortex generated by an additional rotating multipolar strain field has been calculated. We have shown that this extension to MKO94 who only considered a stationary strain field is not trivial when the angular frequency of the external field is in the range of the angular velocity of the Burgers vortex due to the presence of a critical point singularity. We have resolved that singularity by the introduction of viscous and nonlinear effects and shown that the entire structure of the correction depends on the value of the Haberman parameter $h = h_c/(\varepsilon^{3/2} Re)$ which characterizes the nature of the critical layer. For a purely nonlinear critical layer ($h \ll 1$), we have found that the non-axisymmetric correction can survive without an external field for specific angular frequencies when the perturbation azimuthal wavenumber is $n = 2$ and 3. The resulting non-axisymmetric vortex has been compared to the numerical solutions obtained by Rossi *et al.* (1997) for $n = 2$ and a good agreement has been found.

No transient effects have been considered in the present analysis. They were considered in Bernoff & Lingevitch (1994) and Bassom & Gilbert (1998) for the simple 2D linear unforced case. They showed that during the evolution from an arbitrary initial condition, strong transients characterized by the generation of inertial waves are present (Bernoff & Lingevitch, 1994). These inertial waves eventually die on a $O(Re^{1/3})$ time scale everywhere except in the vortex center where they become singular and generate an algebraically decreasing eigenmode (Bassom & Gilbert, 1998).

In the linear regime, the presence of an external forcing does not modify this scenario. This means that if there is no critical layer or if the critical layer is purely viscous, the non-axisymmetric correction created by the external field is not influenced by transients. It eventually appears after the transient has died out even if it may take a very long time.

However, as soon as nonlinear effects are present, transients may affect the non-axisymmetric correction. Indeed, transients are expected first to be important in the critical layer. It modifies the nature of the critical layer which becomes time-dependent and consequently it also modifies some essential properties such as the phase jump χ across the critical layer which defines the non-axisymmetric correction. In certain conditions, the time-dependent critical layer may slowly evolve into an equilibrium critical layer as explained by Goldstein & Hultgren (1988) but this is not always the case. In particular, several other scenarios including finite time singularity have already been identified (see Cowley & Wu, 1994, and references therein).

Nevertheless, the good agreement obtained in Section 5.2 between the stationary solution and the numerical solution obtained from an initial value problem tend to show that a quasi-equilibrium critical layer regime could indeed be achieved.

A   Jumps and scalings in an equilibrium critical layer

In this appendix, the algebraic manipulations leading to the definitions of h and h_c are given. Classical results from Haberman (1972) and Smith & Bodonyi (1992) concerning the jumps across a equilibrium critical layer are presented.

The distinguished scaling in the critical layer is obtained when both viscous and nonlinear effects are of the same order, that is when $\varepsilon^{1/2}$ and $Re^{-1/3}$ are of same order. In that case, the critical layer singularity is resolved on a characteristic viscous-nonlinear scale $\delta = \sqrt{\varepsilon} = O(1/Re^{1/3})$.

In the frame rotating with the angular frequency w, the streamfunction outside the critical layer is

$$\Psi = \Psi_c - \int_{r_c}^r (\Omega_0(s) - w)s\,ds + \varepsilon \mathfrak{Re}(\Phi_1(r)e^{in\theta}) + \ldots \tag{A.1}$$

By examining this expression near $r = r_c$, one reaches the conclusion that the expansion for the streamfunction in the critical layer has the form (assuming without restriction $\Psi_c = 0$)

$$\Psi = \varepsilon \Psi_1(\tilde{r}, \theta) + \varepsilon^{3/2} \ln(\varepsilon)\Psi_2(\tilde{r}, \theta) + \varepsilon^{3/2}\Psi_3(\tilde{r}, \theta) + \ldots \tag{A.2}$$

The first two terms Ψ_1 and Ψ_2 are immediately obtained by rewriting the outer solution with the local variable

$$\begin{aligned}
\Psi_1 &= -\frac{r_c\Omega'_{0_c}}{2}\tilde{r}^2 + \Phi_{1_c}\cos(n\theta) \\
\Psi_2 &= -\frac{\omega'_{0_c}\Phi_{1_c}}{2r_c\Omega'_{0_c}}\tilde{r}\cos(n\theta),
\end{aligned} \tag{A.3}$$

where the suffix c indicates that values are taken at r_c.

The qquation for Ψ_3 reads

$$\frac{\partial \Psi_1}{\partial \tilde{r}}\frac{\partial^3 \Psi_3}{\partial\theta\partial\tilde{r}^2} - \frac{\partial \Psi_1}{\partial\theta}\frac{\partial^3 \Psi_3}{\partial\tilde{r}^3} = \frac{r_c}{\delta_c\varepsilon^{3/2}Re}\frac{\partial^4 \Psi_3}{\partial\tilde{r}^4} + \frac{1}{r_c}\frac{\partial \Psi_1}{\partial\theta}\frac{\partial^2 \Psi_1}{\partial\tilde{r}^2} \tag{A.4}$$

i.e.

$$-\Omega'_{0_c} r_c \tilde{r} \frac{\partial^3 \Psi_3}{\partial \theta \partial \tilde{r}^2} + n\Phi_{1_c} \cos(n\theta) \frac{\partial^3 \Psi_3}{\partial \tilde{r}^3} = \frac{r_c}{\varepsilon^{3/2} Re} \frac{\partial^4 \Psi_3}{\partial \tilde{r}^4} + n \sin(n\theta) \Phi_{1_c} \Omega'_{0_c}, \quad (A.5)$$

with the matching condition

$$\Psi_3 \sim (\Omega'_{0_c} - w'_{0_c}) \frac{\tilde{r}^3}{6} + \frac{w'_{0_c} \Phi_{1_c}}{r_c \Omega'_{0_c}} \tilde{r} \ln |\tilde{r}| \cos(n\theta) \quad \text{as } r \to -\infty \quad (A.6)$$

If one uses the following change of functions:

$$\Psi_3 = -w'_{0_c} \left| \frac{\Phi_{1_c}}{r_c \Omega'_{0_c}} \right|^{3/2} \overline{\Psi}_3 + \Omega'_{0_c} \frac{\tilde{r}^3}{6},$$

$$R = \left| \frac{r_c \Omega'_{0_c}}{\Phi_{1_c}} \right|^{1/2} \tilde{r}, \quad (A.7)$$

$$X = n\theta,$$

we get a single parameter problem for $\overline{\Psi}_3$

$$R \frac{\partial^3 \overline{\Psi}_3}{\partial X \partial R^2} + \sin(X) \frac{\partial^3 \overline{\Psi}_3}{\partial R^3} = h \frac{\partial^4 \overline{\Psi}_3}{\partial R^4} \quad (A.8)$$

$$\overline{\Psi}_3 \sim \frac{R^3}{6} + R \ln |R| \cos(X) \quad \text{as } R \to -\infty, \quad (A.9)$$

with

$$h = \frac{|\Omega'_{0_c}|^{1/2}}{n Re} \left| \frac{r_c}{\varepsilon \Phi_{1_c}} \right|^{3/2}. \quad (A.10)$$

This problem was first studied by Haberman (1972), then re-examined and corrected by Brown & Stewartson (1978) and Smith & Bodonyi (1982). The conclusion they reached is that the behavior of $\overline{\Psi}_3$ for large R is of the form

$$\overline{\Psi}_3 \sim \frac{R^3}{6} + R \ln |R| \cos(X) + \eta(h) \frac{R^2}{2} + U(h, X) R \quad \text{as } R \to +\infty, \quad (A.11)$$

where $\eta(h)$ and $U(h, X)$ are vorticity and velocity jumps respectively. The velocity jump $U(h, X)$ is 2π-periodic with respect to X and can be written as

$$U(h, X) = \sum_{p=0}^{\infty} (U_p(h) \sin(nX) + V_p(h) \cos(nX)). \quad (A.12)$$

It follows that the phase jump appearing in (4.2) is given by

$$\chi(h) = -U_1(h) = -\frac{1}{\pi} \int_0^{2\pi} U(h, X) \sin(X) dX. \quad (A.13)$$

Smith & Bodonyi (1982) computed the function $\chi(h)$ and obtained a result comparable to Haberman's calculation although the latter did not consider the harmonics in (A.12). Their result is reproduced in Figure 8.

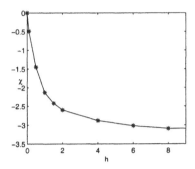

Figure 8: Sketch of the phase jump χ versus h (after Smith & Bodonyi 1982).

Haberman (1972) showed that the vorticity jump $\eta(h)$ can be computed from $\chi(h)$ by the formula:

$$\eta(h) = \frac{\chi(h)}{4h}. \tag{A.14}$$

For a given value of h ($\neq 0$), several harmonics are in general present in (A.12) and $\eta(h)$ is in general nonzero which means that in addition to the phase jump, the critical layer generates in the outer region an $O(\varepsilon^{1/2})$ axisymmetric correction as well as $O(\varepsilon)$ harmonic corrections.

For a purely viscous critical layer, Smith & Bodonyi (1982) proved that

$$U(h, X) \sim \pi \sin(nX) \quad \text{as } h \to \infty. \tag{A.15}$$

which guarantees that $\chi(h) \sim -\pi$ as $h \to \infty$ but also that the $O(\varepsilon)$ harmonic corrections and the $O(\varepsilon^{1/2})$ axisymmetric correction are absent in that limit. For a purely nonlinear critical layer ($h \ll 1$), they showed that $U(k, X) = o(h)$ which means the $O(\varepsilon)$ harmonic corrections are also negligible in that case. By contrast, the $O(\varepsilon^{1/2})$ axisymmetric corrections does not disappear since the vorticity jump $\eta(h)$ goes to a constant. Smith & Bodonyi (1982) obtained the following estimates for $\chi(h)$ and $\eta(h)$

$$\chi(h) \sim -4Ch \quad \text{as } h \to 0, \tag{A.16}$$

$$\eta(h) \sim -C \quad \text{as } h \to 0, \tag{A.17}$$

with $C \approx 1.3$.

The implications of these results can be summarized as follows. Although there may exist corrections of same order or larger, the fundamental distortion is perfectly defined in the outer region by equations (3.4) and (4.2). The only relevant quantity from the critical layer is the phase jump $\chi(h)$ which is sketched in Figure 8. However, expansion (3.1) is in principle not an approximation of the total field as $O(\varepsilon^{1/2})$ terms generated by the critical layer should be taken into account. The only exception is for viscous critical layers.

References

BASSOM, A.P. & GILBERT, A.D. 1998 The spiral wind-up of vorticity in an inviscid planar vortex. *J. Fluid Mech.*, **371**, 109–140.

BENNEY, D.J. & BERGERON, R.F. 1969 A new class of nonlinear waves in parallel flows. *Stud. Appl. Math.*, **48**, 181–204.

BERNOFF, A.J. & LINGEVITCH, J.F. 1994 Rapid relaxation of an axisymmetric vortex. *Phys. Fluids*, **6**(11), 3717–3723.

BROWN, S.N. & STEWARTSON, K. 1978 The evolution of the critical layer of a Rossby wave. Part II. *Geophys. Astrophys. Fluid Dyn.*, **10**, 1–24.

COWLEY, S.J. & WU, X. 1994 Asymptotic Approaches to Transition Modelling. *Agard Report*, **793**, 3.1–3.38.

DRITSCHEL, D.G. 1998 On the persistence of non-axisymmetric vortices in inviscid two-dimensional flows. *J. Fluid Mech.*, **371**, 141–155.

ELOY, C. & LE DIZÈS, S. 1999 Three-dimensional instability of Burgers and Lamb–Oseen vortices in a strain field. *J. Fluid Mech.*, **378**, 145–166.

GOLDSTEIN, M.A. & HULTGREN, L.S. 1988 Nonlinear spatial evolution of an externally excited instability wave in a free shear layer. *J. Fluid Mech.*, **197**, 295–330.

HABERMAN, R. 1972 Critical layers in parallel flows. *Stud. Appl. Math.*, **51**, 139–161.

JIMÉNEZ, J., MOFFATT, H.K. & VASCO, C. 1996 The structure of the vortices in freely decaying two dimensional turbulence. *J. Fluid Mech.*, **313**, 209–222.

LIN, C.C. 1955 *The Theory of Hydrodynamics Stability*. Cambridge University Press.

LUNDGREN, T.S. 1982 Strained spiral vortex model for turbulent fine structure. *Phys. Fluids*, **25**(12), 2193–2203.

MASLOWE, S.A. 1986 Critical layers in shear flows. *Annu. Rev. Fluid Mech.*, **18**, 405–432.

MELANDER, M.V., MCWILLIAMS, J.C. & ZABUSKY, N.J. 1987 Axisymmetrization and vorticity-gradient intensification of an isolated two-dimensional vortex through filamentation. *J. Fluid Mech.*, **178**, 137–159.

MOFFATT, H.K., KIDA, S. & OHKITANI, K. 1994 Stretched vortices – the sinews of turbulence; large-Reynolds-number asymptotics. *J. Fluid Mech.*, **259**, 241–264.

ROSSI, L.F., LINGEVITCH, J.F. & BERNOFF, A.J. 1997 Quasi-steady monopole and tripole attractors for relaxing vortices. *Phys. Fluids*, **9**(8), 2329–2338.

SMITH, F.T. & BODONYI, R.J. 1982 Nonlinear critical layers and their development in streaming-flow stability. *J. Fluid Mech.*, **118**, 165–185.

TING, L. & TUNG, C. 1965 Motion and decay of a vortex in a nonuniform stream. *Phys. Fluids*, **8**(6), 1039–1051.

Core Dynamics of a Coherent Structure: a Prototypical Physical-space Cascade Mechanism?

Dhoorjaty S. Pradeep & Fazle Hussain

1 Introduction

Large scale *coherent structures* (CS) have been the primary focus of turbulence research for the past three decades. This has been motivated by the expectation that CS dominate turbulence effects of technological interest (drag, heat transfer, entrainment, mixing, aerodynamic noise, etc.), and that understanding CS dynamics is essential for developing turbulence models and flow control strategies. The key feature of CS is that they are vortical; *i.e.* underlying the random vorticity field that constitutes turbulence, there is an instantaneously phase-correlated vorticity field over the extent of a CS. This naturally encourages the application of vortex dynamics concepts as these are commonly regarded as amenable to *physical-space*, not merely statistical, interpretation.

We hope that some turbulence physics – involving complexities such as three-dimensionality, couplings among a wide range of scales, and reconnection – can be understood in terms of vortex dynamics. It is possible that the evolution of complexity in turbulence can be captured in simple, idealized problems. In this paper, we demonstrate one such scenario: the growth of axial variations of core vorticity in a cylindrical vortex (*core dynamics*, CD) in the presence of uniform shear, leading to transition and hence to fully-developed fine-scale turbulence. CD is particularly of interest for the following reasons:

(a) Strained two-dimensional vortices are known to be unstable to three-dimensional perturbations, which can be classified into bending waves and CD. While bending waves have been extensively studied (*e.g.* [7, 13, 14]), little is known about the stability of CD or its possible role in the transition to turbulence.

(b) Vortices in turbulent flows, in general, have *internal* vorticity structure characterized by varying core size and vorticity magnitude (generated by nonuniform axial stretching) along their axes (*i.e.* CD). Reconnection also produces vortices which have nonuniform core size, hence CD.

To understand the role of nonuniform core vorticity in vortex evolution, we first review CD in the *absence* of an external strain field (§2). We show that CD is associated with a standing wave oscillation of core area involving vorticity wavepacket motion along the vortex axis. CD-induced oscillation, which decays in the absence of shear, is *amplified* via resonant forcing in the presence of an external shear flow. This instability causes transition to a turbulent flow (§3) with fine-scale vortices, which themselves appear to have, in turn, their own CD. Finally, we demonstrate the importance of CD instability in a practical flow, *viz.* a plane mixing layer, where this mechanism, triggered via helical pairing of rolls, dominates transition to turbulence (§4).

Therefore, we discuss here a prototypical physical-space cascade mechanism of relevance to turbulence. Examples of vortices with CD include bridges formed during vortex reconnection, vortices formed by localized pairing (such as helical pairing in a mixing layer), and 'worms' of intense vorticity that have been observed in homogeneous turbulent flows [5, 6].

2 Vortex Core Dynamics

We first analyze CD in an isolated rectilinear axisymmetric vortex [1] (*i.e.* in the absence of any external strain). Rectilinearity eliminates the complicating effects of self-induced motion, and the axisymmetry of the flow allows a clear representation of vortex evolution in terms of vortex line geometry (see §2.2, 2.5). While vortex lines are material (and, therefore, have dynamic sense) only in an *inviscid* flow, the physical-space mechanisms deduced in an inviscid setting can still be useful in obtaining insight into *viscous* dynamics (if interpreted carefully). In this paper, we often utilize the concepts of inviscid vortex dynamics to help us understand vortical events in a viscous flow.

2.1 Swirl-meridional flow formulation

For axisymmetric flows, the Navier–Stokes equations in cylindrical coordinates (r, θ, z) can be written in terms of $\xi \equiv ru_\theta$, which is proportional to the circulation associated with the swirl (u_θ is a measure of swirl), and $\eta \equiv \omega_\theta/r$, which is a measure of the torsion of vortex lines, as

$$\frac{D\xi}{Dt} = \frac{1}{Re}\left(\frac{\partial^2 \xi}{\partial r^2} - \frac{1}{r}\frac{\partial \xi}{\partial r} + \frac{\partial^2 \xi}{\partial z^2}\right), \tag{2.1}$$

$$\frac{D\eta}{Dt} = \frac{1}{r^4}\frac{\partial \xi^2}{\partial z} + \frac{1}{Re}\left(\frac{\partial^2 \eta}{\partial r^2} + \frac{3}{r}\frac{\partial \eta}{\partial r} + \frac{\partial^2 \eta}{\partial z^2}\right), \tag{2.2}$$

where $\frac{D}{Dt} \equiv \frac{\partial}{\partial t} + u_r\frac{\partial}{\partial r} + u_z\frac{\partial}{\partial z}$ is the material derivative. Since ξ defines axisymmetric vortex surfaces on which vortex lines lie, this representation is very useful in interpreting vortex line geometry, meridional flow, and core oscillation.

2.2 Inviscid coupling between swirl and meridional flows

Let us briefly ignore the purely diffusive viscous terms in equations (2.1) and (2.2). In equation (2.2), the geometry of vortex surfaces (described by ξ) dynamically influences the coiling of vortex lines (described by η) through the coupling term $(1/r^4)\partial\xi^2/\partial z$. The physics represented by this coupling term are easily understood by its Taylor series expansion for two points A and B on a vortex surface $\xi = \xi_0$, given by

$$\frac{1}{r^4}\frac{\partial\xi^2}{\partial z}|_A = \omega_z(z_A, r_A) \times lim_{B \to A}\frac{\dot{\theta}_B - \dot{\theta}_A}{z_B - z_A}. \tag{2.3}$$

Here $\dot{\theta}_B = \xi_0/r_B{}^2, \dot{\theta}_A = \xi_0/r_A{}^2$. Equation (2.3) indicates that the rate of η generation at a point is given by the differential rotation along the axisymmetric vortex surface passing through that point weighted by the axial vorticity value.

Consequently, where an axisymmetric surface converges in the z-direction (*i.e.* $\partial\xi^2/\partial z > 0$), local vortex line twisting generates positive η. Likewise, negative η generation occurs where the vortex tubes diverge (*i.e.* $\partial\xi^2/\partial z < 0$) in the z-direction.

While the distribution $\xi(r, z)$ determines the evolution of η, the shape of a vortex surface is itself modified by the instantaneous geometry of vortex lines through the meridional advection terms in the material derivative in equation (2.1); note that the meridional flow, *i.e.* u_r and u_z, is completely determined by the instantaneous η distribution. Equations (2.1) and (2.2) therefore describe the mutually coupled dynamics between the shape of the vortex surface and the geometry of vortex lines that lie on the surface.

2.3 Initial Conditions

To study CD, we initialize a vortex with a sinusoidal variation of its core radius, so that the axial vorticity profile is self-similar at any z. The initial radial vorticity component is determined such that the vorticity field is divergence free and the azimuthal component is specified to be zero, so that no meridional flow exists initially (*i.e.* all vortex lines are initially uncoiled). Specifically, the initial vorticity distribution is given by

$$\omega_r(r, \theta, z) = P\left(\frac{r}{r_0 s(z)}\right)\frac{r}{s^3(z)}\frac{ds}{dz}, \tag{2.4}$$

$$\omega_\theta(r, \theta, z) = 0, \tag{2.5}$$

and

$$\omega_z(r, \theta, z) = P\left(\frac{r}{r_0 s(z)}\right)\frac{1}{s^2(z)}, \tag{2.6}$$

where P is a compact Gaussian function, r_0 is the unperturbed core radius, and $s(z) \equiv 1 - \mu \cos(2\pi z/\lambda)$ is the shape of the initial axisymmetric vortex surfaces (λ is the wavelength of the perturbation and μ the amplitude). The evolution of this vorticity field is obtained via Direct Numerical Simulations (DNS) using periodic boundary conditions.

The flow geometry is depicted in Figure 1, which also provides an overview of the fairly complex internal dynamics resulting from this simple core size perturbation. In particular, note the coiling of initially straight vortex lines, and the associated generation of meridional flow (absent in the initial condition).

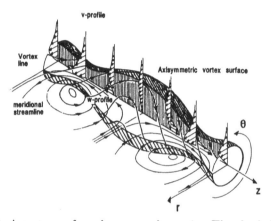

Figure 1. A cartoon of no-shear core dynamics. The shaded surface is an axisymmetric vortex surface (constant ξ surface) with a 90° wedge cut away. On this surface six vortex lines are shown, all of which are identical. Also shown are radial profiles of u_θ and u_z at various z stations. Streamlines of the meridional flow (constant ψ lines) are shown in one meridional plane.

2.4 Vorticity wavepacket motion

The essence of CD is best reflected by the evolution of vorticity magnitude (Figure 2) for $Re \equiv \Gamma/\nu = 665$, where Γ is the vortex circulation and ν is the kinematic viscosity. Clearly, the initial ω peak splits into two equal peaks of considerably lower(by a factor of 3) amplitude (compare Figures 2a–b and recall the periodic boundary condition in z). The vorticity transport at these times occurs as traveling wavepackets, so that mass and vorticity are transported in opposite directions (compare contours of ψ, the meridional streamfunction, in Figure 3 with vorticity contours in Figure 2). Enstrophy production P_ω, which allows variations in vorticity magnitude, changes sign near each wavepacket – enstrophy is produced at the front of each packet and annihilated at the back of the packet. The low-enstrophy bubble observed in Figure 2h occurs at the location where P_ω is negative for a significant time. The disappearance of this bubble at later times is due to a change in the sign of P_ω and also 'fill-in' by viscous diffusion.

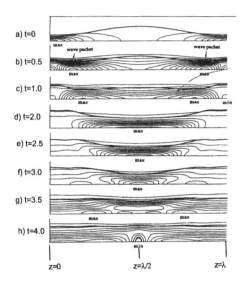

Figure 2. Instantaneous vorticity magnitude for no-shear CD in a meridional plane at various times. The thick overlaid line corresponds to $N_k = 1$, and defines the boundary of the vortex core. Along the axis some local vorticity maxima and minima have been indicated to enable ease of dynamic interpretation.

Significant insight into the evolution of P_ω can be obtained by studying the ψ distributions shown in Figure 3. From equation (2.1), constant ξ surfaces are convected with the meridional velocity in an inviscid flow. Hence the local cross section of a vortex tube shrinks to increase the axial vorticity where the meridional flow converges on the axis, and vice versa. In Figure 3b, P_ω is thus negative near $z = 0$ at early times ($t < 2$), because the meridional flow is directed away from the axis here. The meridional flow converges on the axis near $z = \lambda/2$, making P_ω positive there due to vortex stretching.

In Figure 3b we see two counter circulating cells – *primary cells* – per wavelength of the vortex. If these cells were to persist indefinitely, the vortex would continue to expand near $z = 0$ and contract near $z = \lambda/2$. However, the emergence of secondary cells (see Figure 3d), with a reversed meridional flow, arrests and eventually reverses the core size variation.

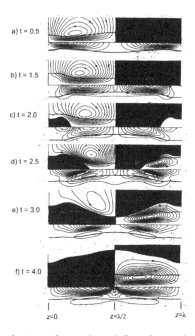

a) t = 0.5

b) t = 1.5

c) t = 2.0

d) t = 2.5

e) t = 3.0

f) t = 4.0

z=0   z=λ/2   z=λ

Figure 3. The evolution of meridional flow (u_r, u_z) in a meridional plane. Each frame includes both streamlines (above the axis) and contours of meridional velocity magnitude (below the axis). Also shown in each frame is the $N_k = 1$ contour (thick line).

2.5   Evolution of vortex line geometry

The generation and evolution of the meridional flow inherent to CD is qualitatively understood in terms of the evolution of patterns of vortex lines starting from a radial rake (Figure 4). At $t = 0$, all vortex lines lie in meridional planes and therefore have no torsion. However, the axial variation of the swirling motion causes vortex lines to become coiled, as implied by the coupling term in equation (2.2). Coiled vortex lines induce the meridional motion.

The net effect of the coupling between swirl and meridional flows is best summarized in terms of vortex line evolution on an *inviscid* vortex. Consider an axisymmetric inviscid vortex tube with core area variation as shown in Figure 5a. For brevity, we hereinafter refer to planes $z = 0$, $\lambda/4$ and $\lambda/2$ as z_0, $z_{\pi/2}$ and z_π respectively (where the subscripts denote the value of $2\pi z/\lambda$). Note that initially all vortex lines are uncoiled. As the circulation enclosed by the vortex surface at cross sections z_0 and z_π is the same, the swirl velocity is greater at z_π than at z_0. This differential swirl immediately generates coiling of vortex lines on the vortex surface. Meridional flow (shown by dashed lines in the figure), induced by the coiled vortex lines, distorts the vortex tube

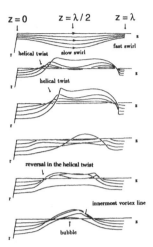

Figure 4. The evolution of vortex line geometry illustrated by four typical vortex lines starting on a radial rake at selected instants. Note that vorticity magnitude varies significantly along vortex lines.

shape via vortex stretching/compression; core size nonuniformity is thereby reduced and the vortex tube becomes cylindrical (Figure 5c). However, as the sign of $D\eta/Dt$ remains the same throughout the evolution from Figure 5a to Figure 5c, the vortex lines are now highly coiled. Therefore the direction of the circulatory meridional motion is preserved and $\partial\xi^2/\partial z$ changes sign (from positive to negative) while continuing to decrease; *i.e.* the core area variation now begins to reverse from that in Figure 5a. Since the sign of $\partial\xi^2/\partial z$ has now changed, the vortex lines begin to uncoil (equation 2.2), but, clearly, there will be a time delay before the differential rotation fully uncoils the vortex lines. The persisting meridional flow distorts the vortex tube *away* from being cylindrical until the vortex lines are fully uncoiled (Figure 5e). Note that the core area variation in Figure 5e is opposite to that in Figure 5a. The evolution described here corresponds to half-a-cycle of CD oscillation; and as there is no mechanism for either damping or amplifying this oscillation, CD will persist indefinitely in an *inviscid* flow.

2.6   Core size oscillation in viscous CD

To provide a quantitative estimate of the core size evolution in viscous CD, we first define the vortex core as the spatial region where the *kinematic vorticity number* [10] $N_k \equiv (\omega^2/2S_{ij}S_{ij})^{1/2} > 1$ (for a discussion of methods of vortex identification, see [18]). Note that in regions where $N_k > 1$, rotation dominates strain. The vortex core cross-sectional area $A(z,t)$ can then be used to define

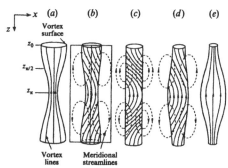

Figure 5. Schematic of half-a-cycle of CD oscillation at a finite amplitude on an inviscid axisymmetric vortex tube. The vortex lines lie on one axisymmetric vortex surface, and the dashed lines represent streamlines of the meridional flow.

the instantaneous local core size as

$$\sigma(z,t) = [A(z,t)/\pi]^{1/2}. \tag{2.7}$$

The evolution of core size nonuniformity is shown in Figure 6 by the standard deviation of $A(z,t_o)$ normalized by the instantaneous average area $\overline{A}(t_o) \equiv \pi\overline{\sigma}^2(t_o)$ for $\mu = 0.5$ and for different Re. It is clear that the axial core size variations progressively disappear with increasing viscosity, confirming that the damping is purely viscous. The decay of axial variations is not monotonic, but behaves rather like a damped oscillation, where both the amplitude and frequency depend on Re. The oscillation frequency increases with Re, but appears to have a finite limit as $Re \to \infty$; the 'super-viscosity' simulation, in which an artificial viscosity of form $\nu\nabla^4$ is used, is indicative of this limiting behavior. Therefore, CD in this isolated, axisymmetric vortex case behaves like a neutrally-stable, oscillatory, inviscid mode, damped at finite Re by viscous effects.

3 Core dynamics instability (CDI) of a vortex in shear

We now study how CD is modified when a vortex with varying core area is embedded in an externally imposed shear flow [3].

The initial condition for DNS comprises an axisymmetric vortex column (as described in §2.3) placed in a uniform shear flow, given by $u(y) = -\gamma y$, oriented perpendicularly to the vortex axis z (hereinafter, we use also an xyz coordinate system wherever convenient). The shear and the unperturbed vortex have vorticities of the same sign, and $s \equiv \gamma/\omega_0$ (ω_0 being the peak vorticity in the unperturbed vortex) is taken as a parameter of the problem

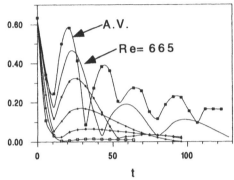

Figure 6. Evolution of the standard deviation of core area normalized by the instantaneous average core area. The different curves are for $Re = 43, 83, 166, 332,$ and 665. Core area variation decay rate increases monotonically with decreasing Re. The curve marked 'A.V.' corresponds to a simulation using artificial viscosity.

(in the simulations reported here $0 < s < 0.25$). Note that we only specify the initial condition and allow it to evolve, *i.e.* no modes are forced. This particular orientation of the shear with respect to the vortex is chosen for its relevance to CS in shear flows, which are oriented perpendicularly to the mean shear (e.g. rolls in mixing layers). Additionally, such an alignment is also relevant to fully-developed turbulent flows where intense vorticity preferentially aligns with the eigenvector corresponding to the intermediate eigenvalue of the net strain field; in the simple case of a two-dimensional flow, this alignment corresponds to that in our base flow.

It is now well-known that an elliptic vortex is susceptible to a three-dimensional instability (called 'elliptic instability' [13, 14]). Thus we expect that CD perturbations of a vortex in shear will also grow, since shear deforms the circular vortex core into an elliptic shape [12]. The elliptic instability is related to the instability of Kelvin bending waves on a vortex in a straining flow [7]. Kelvin waves comprise modes that deflect the vortex axis (bending waves), and modes that either change the core size or deform the core shape without deflecting the vortex axis (CD). We address the latter case in the following.

3.1 Core dynamics instability: small-amplitude perturbations

Consider the evolution of small core area perturbations ($\mu \sim 10^{-3}$, in $s = 1 - \mu \cos(2\pi z/\lambda)$) on a vortex column embedded in a shear flow. By choosing such small perturbations, we ensure that nonlinear interactions between three-dimensional modes are negligible compared with the interactions between the three-dimensional perturbations and the base flow. Thereby, we can determine the linear stability characteristics via DNS.

Even in the presence of shear, CD evolves as a standing wave oscillation of core area variation, showing coupling of swirl and the meridional flow. While the flow formulation described in §2.1 is not applicable here, it can be shown [2] that even for a nonaxisymmetric flow the rate of change of the meridional flow strength (quantified by an integral measure, $H \equiv \int_x \int_y \int_{z=0}^{z=\lambda/2} \eta dx dy dz$; $\eta = \omega_\theta/r$) is related to the differential u_θ along a vortex tube, and vice versa. The crucial difference between no-shear CD and CD in the presence of shear is that in the latter case the energy associated with three-dimensional perturbations amplifies exponentially (Figure 7a); *i.e.* CD now constitutes an *instability*. The meridional flow strength oscillates as in no-shear CD, but the oscillation amplitude also increases exponentially (Figure 7b).

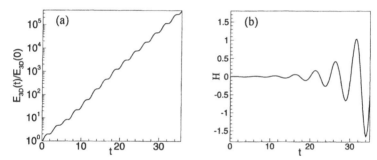

Figure 7. CD in the presence of shear ($s = 0.1$, $Re = 3000$, $k = 2.0$) showing instability: *(a)* Evolution of volume-integrated energy in all 3D modes, and *(b)* Evolution of the strength of meridional flow (quantified by $H = \int_x \int_y \int_{z=0}^{z=\lambda/2} \eta \, dx dy dz$), showing amplifying oscillations.

Defining the growth rate of the instability as

$$\sigma = \frac{1}{2} \frac{d}{dt} \ln \frac{E_{3D}(t)}{E_{3D}(0)},$$

E_{3D} here being the energy of all three-dimensional modes, it can be deduced from Figure 7a that the growth rate varies with time (the growth rate oscillates at a fixed frequency which is twice that of CD oscillations; twice because E_{3D} is velocity squared). We estimate the average (stationary) growth rate by fitting an exponential curve through $E_{3D}(t)$.

We now consider the dependence of σ on s $(= \gamma/\omega_0)$, $Re \equiv \Gamma/\nu$ (where Γ is the *vortex* circulation) and the axial wavenumber, $k \equiv 2\pi/\lambda$; note that k is nondimensionalized by the unperturbed core radius (r_0).

Figure 8a shows that the growth rate σ increases monotonically with s; *i.e.* for a given ω_0, the growth rate σ increases with increasing shear strength, γ. However, below a certain low-shear cutoff value, the instability cannot be sustained, and CD oscillations are damped via viscous mechanisms, as in the no-shear case. This cut-off value depends on Re and k. When s and k are held constant, and Re is varied (Figure 8b), the growth rate first increases sharply with increasing Re, and then levels off to an asymptotic value; there also exists a low-Re cutoff for the instability. When s and Re are held constant, and k is varied (Figure 8c), we find that the growth rate peaks at a finite k, with both longer and shorter waves being less unstable.

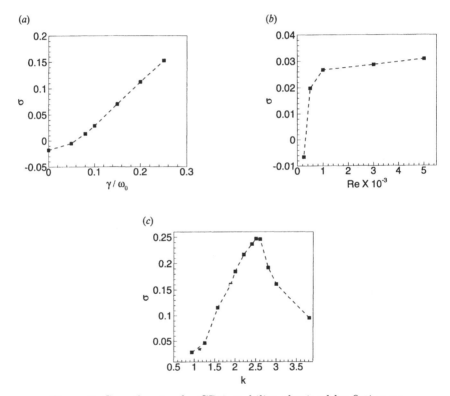

Figure 8. Growth rates for CD instability obtained by fitting exponential curves through the 3D energy evolution: *(a)* σ vs. s $(=\gamma/\omega_0)$ for $Re = 3000$ and $k = 0.95$; *(b)* σ vs. Re for $s = 0.1$ and $k = 0.95$; and *(c)* σ vs. k for $s = 0.1$ and $Re = 3000$. In all three, $\omega_0 = 1.0$.

Thus, CD is essentially a short-wave inviscid instability that requires the presence of a sufficiently strong shear in a viscous flow. The strain component of the shear causes vortex stretching, which is necessary for 3D instability. In a viscous flow, the stabilizing effect of viscous diffusion competes with the destabilizing effect of vortex stretching (which increases with increasing shear strength). This explains the existence of a low-shear cutoff, a low-Re cutoff, and high-k cutoff (high-k produces large vorticity gradients, which are damped by viscosity).

3.2   A model for the instability mechanism

To understand how external shear can amplify the CD oscillation, we simplify the problem by considering an axisymmetric perturbation to an (*inviscid*) Rankine vortex (note that the oscillatory nature of CD discussed in §2.5 does not involve core vorticity profile; hence we transit to this simpler profile). We focus on the dynamics in the cross-sectional plane $z_{\pi/2}$, which has the following symmetries (that simplify the governing equations):

$$u_r' = u_\theta' = \partial u_z'/\partial z = 0, \quad \text{and} \quad \partial w_r'/\partial z = \partial w_\theta'/\partial z = \omega_z' = 0.$$

(In this subsection, primes denote perturbation quantities and mean quantities are denoted by uppercase letters.) Note that the components of perturbation vorticity that are stretched by the strain field, *viz.* ω_r' and ω_θ', take their maximum values in this plane (see Figures 5a,c).

First consider CD in the absence of any external strain. In the $z_{\pi/2}$ plane, the linearized Euler equations take the form

$$\frac{D\omega_r'}{Dt} = -\omega_\theta' \frac{U_\theta}{r} + \Omega_z \frac{\partial u_r'}{\partial z},$$

$$\frac{D\omega_\theta'}{Dt} = \omega_r' \frac{\partial U_\theta}{\partial r} + \Omega_z \frac{\partial u_\theta'}{\partial z}, \quad \text{and}$$

$$\frac{D\omega_z'}{Dt} = 0.$$

Referring to Figure 5, note that when the vortex tube is cylindrical and the vortex lines are maximally coiled (Figure 5c), there is no ω_r' in the $z_{\pi/2}$ plane, and that when the vortex lines are fully uncoiled and the core size variation is maximum, there is no ω_θ' in $z_{\pi/2}$ (Figure 5e). Based on this observation of out-of-phase oscillations of ω_r' and ω_θ' in CD (§2.5, and Figure 5), we assume that r and θ components of the above equations can be written as follows:

$$D\omega_r'/Dt = f\omega_\theta' \quad \text{and} \quad D\omega_\theta'/Dt = -f\omega_r'.$$

Here, f is the oscillation frequency of ω_θ' and ω_r' associated with a fluid particle.

We now subject this 'CD oscillation' to a strain field so that the net mean velocity field is given by

$$U_r = -\frac{\gamma}{2}r \sin 2\theta, \quad \text{and} \quad U_\theta = \dot{\Theta}r - \frac{\gamma}{2}r \cos 2\theta.$$

The irrotational part of this flow is the straining component of uniform shear in our DNS. To understand the instability mechanism in the presence of strain, it is sufficient to consider γ to be a small parameter (*i.e.* $\gamma/\Omega_z \ll 1$), and expand perturbation quantities as

$$\{\omega_r', \omega_\theta'\} = \{\omega_r'^{(0)}, \omega_\theta'^{(0)}\} + \frac{\gamma}{2}\{\omega_r'^{(1)}, \omega_\theta'^{(1)}\} + \cdots .$$

Note that due to the resonant character of the instability, a regular expansion can only capture the initial behavior of perturbation vorticity; however, such an expansion suffices to determine the condition for instability.

Using a Lagrangian formulation (which makes interpretation easy) in which the position of a fluid particle at time t is given by

$$\theta = \dot{\Theta}t + \phi + O(\gamma), \quad \text{and} \quad r = O(\gamma),$$

where ϕ is the Lagrangian tag, it can be shown that $\omega_\theta'^{(1)}$ satisfies

$$\frac{d^2}{dt^2}\omega_\theta'^{(1)} = -f^2\omega_\theta'^{(1)} + 2\dot{\Theta}\cos(2\dot{\Theta}t + 2\phi - ft),$$

which is the simple forced harmonic oscillator equation. The natural frequency in the oscillator equation coincides with the forcing frequency when

$$f = \dot{\Theta},$$

upon which ω_θ', and, consequently, ω_r' and ω_z', will amplify.

Note that f is the frequency with which the perturbation vorticity vector of a particle rotates in the clockwise direction, while $\dot{\Theta}$ is the angular velocity of the particle as it is swept in the counter-clockwise direction in the vortex core (Figure 9). Therefore, when frequency-matching occurs, the vorticity vector of a particle maintains a fixed orientation with respect to the principal axes of the strain field, and if the vorticity vector is aligned with the stretching direction, sustained stretching of perturbation vorticity can occur (Figure 9), causing growth of this vorticity perturbation, leading to instability and growth of three-dimensional energy.

Clearly, due to the varicose nature of perturbation, simultaneous vortex stretching is not possible at all azimuthal locations in the core. Therefore, the growing perturbation is azimuthally localized with the same azimuthal (π-periodic) symmetry as the strain flow, and only material patches of fluid can undergo sustained stretching, even when the frequency-matching condition is satisfied. That this mechanism is responsible for the instability is supported by a comparison of the perturbation vorticity field from the model with that from DNS (Figure 10): in both cases large perturbation vorticity is localized azimuthally (near $\theta = 3\pi/4$ and $-\pi/4$ when ω_θ' is nearly zero – core area variation is maximum, and near $\theta = \pi/4$ and $-3\pi/4$ when ω_r' is nearly zero – vortex lines are maximally coiled).

The arguments in this section also indicate that CD instability is to be expected in the inviscid limit even for a strain field of infinitesimal strength.

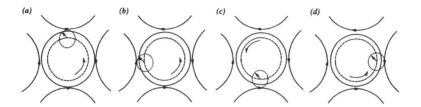

Figure 9. Physical interpretation of the instability mechanism: locking of the perturbation vorticity vector (vector shown for one material particle) direction with the stretching direction of the imposed strain flow (shown by solid hyperbolic curves with arrows). The same particle is shown at four different times as it is swept in the counter-clockwise direction in the vortex core (solid circle). The dashed circle represents the particle path.

3.3 CDI: nonlinear evolution

We now consider large-amplitude core area perturbations ($\mu = 0.2$ in $s = 1 - \mu\cos(2\pi z/\lambda)$). The simulations presented in this section are for $Re = 3000$, $s = 0.1$, and $k = 2$.

First we observe that even when the perturbation amplitude is large, CD is characterized by an amplifying oscillation of axial vorticty (Figure 11), as can be seen from the ω contours in the symmetry planes z_0 and z_π. This core size oscillation is associated with the dynamics of meridional flow, as in the no-shear case (Figure 12). The initial variation of core area leads to the generation of azimuthal vorticity ω_θ (eqn. 2.2), and the magnitude of this vorticity component itself oscillates in response to the oscillation of the core area. High-amplitude CD oscillation also causes variation of *core shape*: where the core vorticity magnitude is large, the core becomes compact and nearly axisymmetric; and where the core vorticity magnitude is small, the vortex core becomes highly elliptic. This is because the ellipticity of a vortex core decreases with increasing magnitude of its vorticity relative to the strain field [12]. Furthermore, it can be seen that in regions where vortex compression occurs, a thin sheath of intense vorticity that surrounds a low-enstrophy 'bubble' is formed (Figure 11d). It is interesting to note that a similar low-enstrophy bubble is seen even in no-shear CD (Figure 2h), where CD was triggered at a high amplitude. (It may be worthwhile to note that the low-enstrophy bubble formation is the underlying mechanism of vortex breakdown.) This shows that the dominant effect of the shear is to amplify CD oscillations.

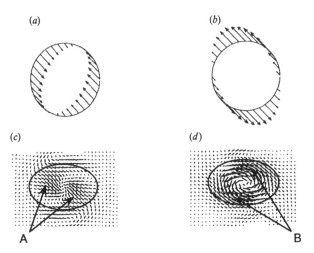

Figure 10. Perturbation vorticity vectors (for small-amplitude CD instability) in the vortex core for the model ((a) and (b)), and from DNS ((c) and (d)) in the $z_{\pi/2}$ plane. (a) and (c) correspond to the phase of oscillation when core area variation is maximum, and (b) and (d) when meridional flow is strongest. In (a) and (c) $\omega_r' >> \omega_\theta'$ (see region marked 'A' in DNS figure), and in (b) and (d) $\omega_\theta' >> \omega_r'$ (see region marked 'B' in DNS figure). Note that large perturbation vorticity at both phases of the oscillation is thus aligned with the stretching direction ($\theta = 3\pi/4, -\pi/4$). In (c), (d), the $\omega = 0.2\omega_0$ contour gives the core shape.

3.3.1   Sheath formation and dynamics

As the formation and dynamics of the sheath surrounding the low-enstrophy bubble turn out to be crucial for transition, we focus on these processes in the following. The formation of the sheath is a nonlinear-instability effect resulting from the advection of azimuthal vorticity, via the $u_z \partial \omega_\theta / \partial z$ term in the vorticity equation. Note that both u_z and ω_θ are perturbation quantities. The axial advection of the azimuthal vorticity ω_θ in Figure 12c–d is like the motion – along the vortex column and towards each other – of two co-axial opposite-signed 'vortex rings'. The induced velocity field of the rings stretches them, increasing their diameter and their ω_θ vorticity magnitude, and causes pile-up of ω_θ (see azimuthal vorticity distribution near z_π in Figure 12d). Simultaneously, the azimuthal vorticity pile-up also leads to strong axial vorticity compression near the vortex axis, and to *stretching* near the core periphery. The net effect is that core vorticity is compressed on the axis, advected radially outward, and stretched near the core periphery, leading to the formation of thin sheaths of intense vorticity (as in Figure 11d). These sheaths

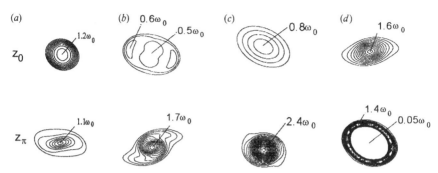

Figure 11. Evolution of ω_z contours in planes z_0 (top) and z_π (bottom) at four different times spanning nearly a cycle of nonlinear CDI oscillation. *(a)* $t = 0.9$, *(b)* $t = 2.7$, *(c)* $t = 4.5$, and *(d)* $t = 6.3$. Contour scale is same for all figures with $\delta\omega = 0.1\omega_0$.

subsequently collapse into compact cores (not shown) when the meridional flow, responsible for sheath formation, reverses.

Since CDI comprises amplifying oscillations, sheaths collapse and form with every meridional flow reversal inherent to global CD. As sheath-Re ($\equiv \omega\delta^2/\nu$, where δ is the sheath thickness) is larger for each successive sheath, due to the successively increasing strength of the meridional flow, it follows that eventually a Kelvin–Helmholtz-like instability develops on a sheath, and leads to sheath roll-up. Such a roll-up into several discrete smaller-scale vortex filaments is apparent in the mid-plane in Figure 13.

Sheath formation and roll-up is clearly a process associated with the transfer of energy from large to smaller scales. This transfer occurs in parallel with transfer of energy in the opposite direction via the pairing of such rolled-up vortices. Pairing occurs as a result of meridional flow reversal which causes radially converging flow which brings the fine-scale vortices towards one another. Simultaneously, vorticity stretching begins on the axis, where it generates strong axial vorticity which 'wraps' the rolled-up vortices into one compact core. This example captures well the simultaneous occurence of cascading and anticascading events – a process typical of turbulent flows.

3.3.2 Vortex line folding and reconnection

It is observed that during the time of sheath formation, opposite-to-mean (negative) axial vorticity is generated within the low-enstrophy bubble. This process is an important indicator of the onset of transition because while initially ω_z is positive everywhere, a fully-developed turbulent flow that has evolved from the initial state must contain vorticity of both signs in ω_z. The presence of negative ω_z implies a qualitative change in the geometry of vortex lines: from lines that only run in the positive z direction to those

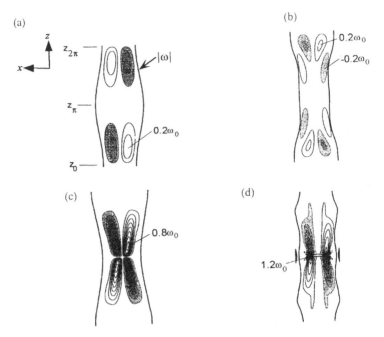

Figure 12. Contours of ω_y in a y plane containing the vortex axis showing oscillation of meridional flow coupled to core area variation (see Figure 11). Four different times shown are: *(a)* $t = 0.9$, *(b)* $t = 3.6$, *(c)* $t = 5.4$, and *(d)* $t = 7.2$. In all figures $\delta\omega = 0.1\omega_0$. Regions of $-\omega_y$ are shaded; also shown is the $|\omega| = 0.2\omega_0$ contour.

containing kinks with segments of the lines in the negative z direction. In addition, negative ω_z also involves topology changes such as reconnection (shown below).

As negative vorticity first appears within the low-enstrophy bubble, its generation must be related to the pile-up of azimuthal vorticity that also forms the sheath. In regions containing large azimuthal vorticity magnitudes (see Figure 12d), the geometry of meridional flow streamlines is shown in Figure 14a. In an inviscid flow, such a meridional flow will advect an inner-core vortex line towards the core periphery. This vortex line is stretched in the outer-core but pressed together in the inner-core, forming S-shaped kinks. This process is shown schematically in Figure 14a-c. This inviscid mechanism explains the geometry of vortex lines, observed in DNS, passing through the states of opposite-to-mean vorticity (Figure 14d).

As the folded vortex filament is pressed together by the meridional flow, it will reconnect (in a viscous flow) and form a closed filament loop, thereby generating opposite-to-mean vorticity in the z_π plane (Figure 14a). Note that

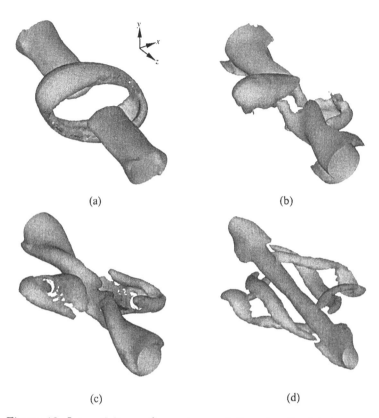

(a) (b)

(c) (d)

Figure 13. Isovorticity surfaces at $\omega = 1.2\omega_0$ visualized between planes $z_{\pi/2}$ and $z_{3\pi/2}$ showing roll-up of the vortex sheath (seen in mid-plane in *(a)*). Figures correspond to: *(a)* $t = 11.6$, *(b)* $t = 12.4$, *(c)* $t = 13.3$, and *(d)* $t = 14.2$.

in z_0 and z_π, $u_z = \omega_x = \omega_y = 0$ and therefore opposite-to-mean vorticity can only be produced by a viscous mechanism (*i.e.* reconnection).

3.3.3 Onset of transition

Transition to turbulence from CD results via these two nonlinear-instability effects, *viz.* sheath formation and its roll-up, and vortex line folding and re-connection. Both these mechanisms are due to the interaction between core vorticity and the amplifying meridional flow. In the nonlinear stage, the CD-induced axial velocity in the vortex core even exceeds peak u_θ! The onset of transition is characterized by the simultaneous roll-up of vortex sheaths near *both* the symmetry planes, z_0 and z_π (whereas prior to this, the formation of a vortex sheath near z_π is concurrent with the collapse of a sheath near z_0,

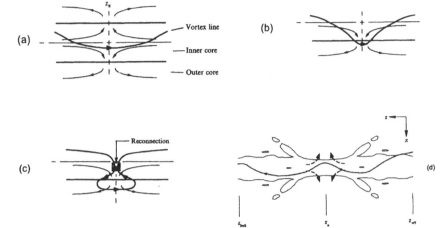

Figure 14. Schematic mechanism of vortex line folding and reconnection $((a)-(c))$ due to strong meridional flow, and vortex
lines started from a fixed point in DNS showing the progressive
evolution of vortex line (thick line) geometry, becoming kinked.
(a) $t = 21.3$, (b) $t = 22.2$, and (c) $t = 23.1$. Also shown is the
$\omega = 0.2\omega_0$ contour.

and vice versa).

Flow evolution during transition is illustrated in terms of isovorticity surfaces in Figure 15. The oscillation of core vorticity magnitude can be deduced
from the isovorticity surfaces, e.g. compare mid-planes in Figures 15a–b. A
number of azimuthally-oriented tube-like surfaces are seen (Figure 15c), indicating that azimuthal vorticity magnitude (a perturbation quantity) becomes
comparable to the axial vorticity magnitude. These vortices are generated by
the instability of the sheath, where vortex lines have hair-pin like geometry.
The roll-up of sheet-like strutures into tube-like structures can be discerned
(Figure 15d–e) both near z_0 (axial planes at the ends of the visualization)
and z_π (axial plane at the middle of the visualization). A turbulent flow
with smaller-scale tube-like structures is seen at the end of the time-sequence
shown here.

Vortex line bundles started in regions of intense vorticity at the time corresponding to that in Figure 15f reveal (see Figure 16, where the vortex
lines are gray-scaled on vorticity magnitude) that these smaller-scale vortices
themselves contain core area variations and hence their own CD, suggesting a
self-similar CDI-cascade if Re were high enough for the smaller-scale filaments
to be unstable.

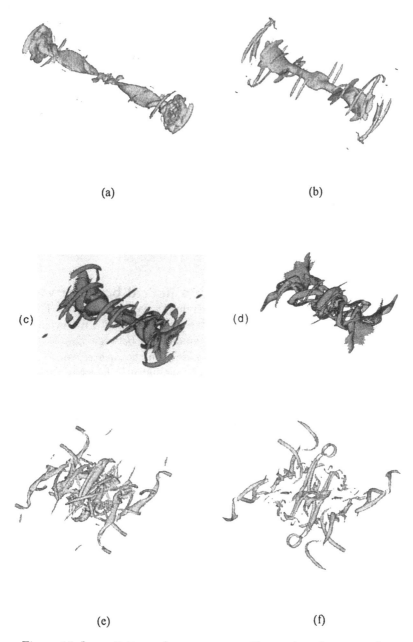

(a)

(b)

(c)

(d)

(e)

(f)

Figure 15. Isovorticity surfaces at $\omega = \omega_0$ illustrating the onset of transition by the appearance of fine-scale filaments. *(a)* $t = 23.1$, *(b)* $t = 24.9$, *(c)* $t = 26.7$, *(d)* $t = 28.4$, *(e)* $t = 30.2$, and *(f)* $t = 32$.

Figure 16. Vortex lines plotted through regions of intense vorticity at $t = 32$. The vortex lines are gray-scaled on vorticity magnitude; vortex line bundles (i.e. tubes) show large vorticity variation, hence CD.

3.4   Comparison of CD with a helical bending wave

An elliptic vortex is also unstable to bending waves [13, 14, 17]. Therefore, the transition to turbulence of a two-dimensional vortex could occur due to both bending-wave and CD modes. A comparison between E_{3D} for CD and bending wave instabilities growing separately on the sheared vortex (Figure 17a) shows that the bending wave grows significantly faster, a reason why bending waves received far more attention so far.

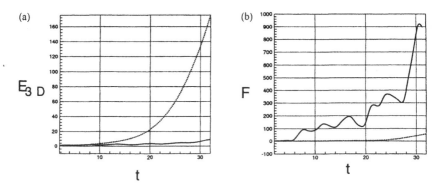

Figure 17. Comparison of CDI (solid line) with bending wave instability (dotted line) growing on the same base flow: *(a)* 3D energy vs. time, *(b)* Φ (vorticity intermittency measure) vs. time.

However, the nonlinear evolution of the bending waves does not lead to the explosive appearance of fine-scale vortices, in contrast to CDI. To quantify the appearance of fine-scale structure in the two cases, we use $\Phi \equiv ||\nabla \vec{\omega}||$. The evolution of Φ for bending wave and CD instabilities is compared in Figure 17b. It appears that CDI transfers energy to fine-scales more rapidly than the bending wave does, which indicates that even in the presence of the faster growing bending wave, CD may play a crucial role in elliptic-vortex transition.

4 Core dynamics instability (CDI) and mixing layer transition

To demonstrate the significance of CDI as a transition mechanism in shear flows, we review here mixing layer transition [2].

In a mixing layer, CDI can occur within spanwise rolls. This transition mechanism reconciles an important anomaly (observed but not explained) in numerical and experimental results regarding mixing layer transition. The conventional, widely-accepted transition mechanism – the entrapment of ribs within pairing rolls [15, 16] – has been shown by DNS [16] to require artificially large initial rib perturbations (5 times more energetic than the 2D modes responsible for roll-up and pairing). If the 3D perturbation amplitude is moderate (comparable to that of the 2D modes initially), transition does not occur even after two pairings, and, in fact, small-scale vorticity becomes less prominent after the second pairing. In contrast, experimental studies clearly indicate that transition is delayed, but not forever suppressed, by weaker initial three-dimensionality. Hence, a different transition mechanism (such as our CDI) must be operating when the 2D and 3D initial perturbation amplitudes are comparable, or when 2D modes are amplified by forcing. In the following, we will demonstrate that CDI does in fact initiate transition when the 2D and 3D modes are initially comparable, and is the likely generic transition mechanism of vortices, mixing layers and structures in turbulence.

4.1 Excitation of CDI

CDI is an unstable mode of the steady Stuart vortex, which models mixing layer vortices (rolls) formed by Kelvin–Helmholtz instability [8]. (Remarkably, the perturbation vorticity field associated with this Stuart vortex eigenmode is virtually identical to that resulting from CD in an initially axisymmetric vortex in shear, indicating the insensitivity of the underlying instability mechanism to the finer details of the core vorticity and strain field distributions.)

Two distinct routes of CDI excitation exist in a mixing layer which can occur during unsteady vortex roll-up and roll pairing: (i) fundamental – CDI within fundamental-wavelength rolls after roll-up, and (ii) subharmonic – CDI of subharmonic-wavelength coalesced rolls which commences after pairing. Fundamental CDI results from a core size perturbation, much like that considered here in §3. The process of subharmonic CDI excitation is shown schematically in Figure 18, with the 3D perturbation amplitude exaggerated for clarity. At roll-up, the so-called 'helical pairing' class of secondary instability causes opposite spanwise undulations of neighboring rolls. Between roll-up and pairing, these undulations amplify while, at the same time, pairs of rolls approach due to (nonlinear) 2D pairing growth. Once the rolls begin to coalesce, their opposite undulations create the meridional flow cells characteristic of CD, thus initiating CDI of the composite vortex pair. This subharmonic route of CDI excitation eventually produces an intense meridional flow within rolls that triggers transition. In flows which undergo pairing, subharmonic CDI is more energetic than its fundamental mode counterpart and thus will be addressed in the following.

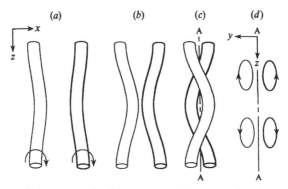

Figure 18. Schematic of subharmonic-CDI excitation by the combination of roll-pairing and subharmonic oblique modes (helical pairing).

4.2   Nonlinear CDI evolution

We now consider the nonlinear, post-pairing evolution of a subharmonic CDI mode, initialized with half the energy of the 2D modes which excite roll-up and pairing. The flow geometry is defined in Figure 19, which contains two views of the evolution. Constant $|\omega|$ surfaces are shown for one-half of each original vortex and two bundles of vortex lines passing through the two $|\omega|$ peaks in z_0 are shown in the other half. The locations of these distinct $|\omega|$ peaks are tracked in time and the corresponding vortex line bundles are distinguished by their line thickness (heavy and light for the downstream and upstream rolls respectively).

The $|\omega|$ field at $t = 18$ illustrates the geometry of the pairing rolls, which have been bent by a helical pairing-type instability. The meridional flow induced by these bent portions of the rolls excites finite-amplitude CDI after pairing. In the early stages of evolution (Figures 19a–b), the rolls simply rotate around each other as distinct vortices, with a planar 3D deformation of

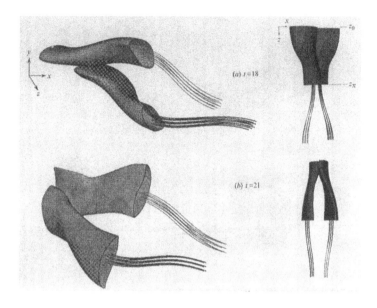

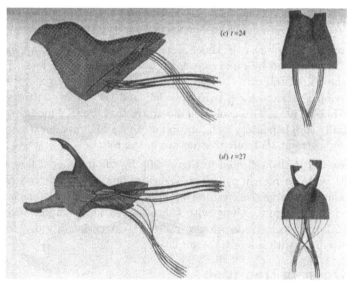

Figure 19. Two views of the evolution of isovorticity surfaces and vortex lines for finite-amplitude subharmonic-CDI at times: *(a)* $t = 18$, *(b)* $t = 21$, *(c)* $t = 24$, *(d)* $t = 27$, *(e)* $t = 30$ and *(f)* $t = 33$. Each bundle of vortex lines starts as a rake centered around the vorticity peak in z_0; the two peaks are tracked in time and associated bundles are distinguished by thin and heavy lines.

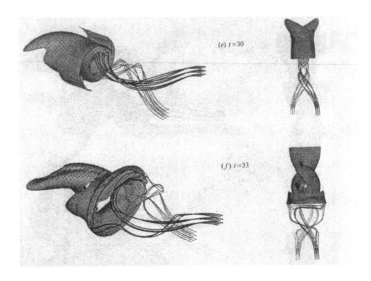

Figure 19 (continued).

roll vortex lines. However, in time, a compact paired roll first appears near z_π ($t = 30$), followed by rapid expansion and formation of a low-$|\omega|$ bubble within the paired roll core ($t = 33$). At the same time, the roll vortex lines first develop a strong helical twist ($t = 30$) and then sharp outward flaring near the bubble in z_π. This evolution illustrates that the CDI mode excited by this initially simple pairing configuration attains a high nonlinear amplitude, producing a strong 3D internal structure in the roll.

Intense CD is also reflected in Figure 20b by the peak axial flow velocity u_z at the vortex center in $z_{\pi/2}$. This spanwise flow within the roll oscillates and attains very large peak values, which even *exceed* the free-stream U, as indicated in the figure. Along with the observed local amplification of ω_z (Figure 20a), this illustrates intense roll three-dimensionality due to a CDI mode initialized with a moderate amplitude.

4.3   Onset of transition

Although the local measures of velocity and vorticity in Figure 20 reflect the generation of small-scale vorticity, their large values alone do not necessarily imply roll vorticity's internal intermittency – reflected by a granular vorticity distribution characteristic of fully turbulent flows. The appearance of significant opposite-to-mean vorticity magnitude, while a local measure, is however a good indicator of the onset of transition. Figure 20 demonstrates that the development of significant $+\omega_z$ (note that in this simulation, mean axial vor-

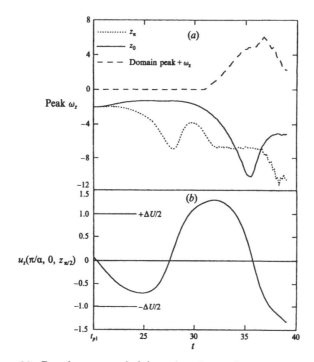

Figure 20. Development of *(a)* peak values of ω_z in z_0 and z_π, and the domain peak of opposite-to-mean $+\omega_z$, and *(b)* spanwise velocity u_z at the paired roll center in $z_{\pi/2}$. Observe that peak u_z eventually exceeds the free-stream velocity $\Delta U/2$.

ticity is negative) begins around $t = 31$ and eventually reaches a maximum of $-3\omega_0$.

To ensure that this is in fact a turbulent transition, the initial 3D energy of this CDI mode was doubled. In this case, similar CD occurs, and the locally entangled vortex lines and significant opposite-to-mean vorticity in Figure 21 indicate an explosive transition to turbulence. The presence of large regions of opposite-to-mean vorticity (shown as cross-hatched surfaces in Figure 21d) indicates that internal intermittency develops, and ensures that the fine scales in Figure 21c are not numerical noise.

To summarize, accelerated small-scale and internal intermittency generation indicate the onset of localized transition due to nonlinear evolution of a single CDI mode. As mentioned earlier, this scenario is significant in that it occurs with moderate initial 3D perturbation energy, 5 times smaller than that required for rib-induced transition. It is also important to note that this CDI evolution and transition are not reliant on the presence of special

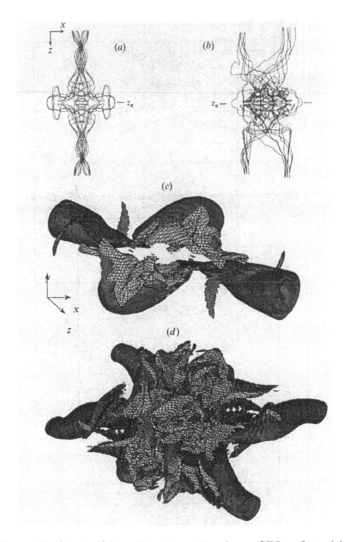

Figure 21. Onset of transition in mixing layer CDI, reflected by core vortex lines *(a)*, *(b)* and isovorticity surfaces with cross-hatched surfaces of opposite-to-mean vorticity *(c)*, *(d)*. The initial 3D perturbation is $E_{3D}(0) = 0.5E_{2D}(0)$ in *(a)*, *(c)* and $E_{3D}(0) = E_{2D}(0)$ in *(b)*, *(d)*. The times are $t = 37.5$ in *(a)*, *(c)* and t=39 in *(b)*, *(d)*.

symmetries and have been shown to occur similarly even in the presence of moderate-strength ribs. We therefore conclude that CDI is a more generic transition mechanism in mixing layers as well as in all turbulent flows.

5   Concluding remarks

We have shown that a core area perturbation triggers a meridional flow, which in turn modifies the vortex core area. This mutual coupling between swirl and meridional flow leads to a standing wave oscillation of core area that constitutes CD. Such a perturbation is unstable if the vortex is embedded in a sufficiently strong shear flow. The instability mechanism is the sustained stretching of perturbation vorticity that can occur when the frequency of the CD oscillation matches the angular velocity of the fluid in the vortex.

The nonlinear development of this instability causes transition to turbulence via two distinct mechanisms: i) the formation and subsequent roll-up of thin sheaths of intense vorticity, and ii) the generation of folded vortex filaments, which can undergo vortex reconnection, by the meridional flow. The post-transitional flow contains finer-scale vortices, whose core sizes are also nonuniform – hence a self-similar cascade is possible if Re is sufficiently high.

We believe that this instability and the ensuing transition process are relevant to cascade in turbulent flows wherein vortices with nonuniform internal structure and exposed to strain are expected to be the norm. The dominant role of CDI in mixing layer transition has already been demonstrated, and it is possible that CDI plays a similar crucial role in the transition to turbulence of an elliptic vortex. Our preliminary comparison of nonlinear bending-wave and CD instability evolution supports this view.

Acknowledgements

Some of the research described here was performed in collaboration with M.V. Melander and Wade Schoppa. We are grateful to Praveen Ramaprabhu for carefully reviewing this manuscript. A part of this research was supported by NSF grant CTS-9622302. Computer time for the simulations reported here was provided by NASA Ames Research Center.

References

[1] Melander, M.V. & Hussain, F. 1994. Core dynamics on a vortex column. *Fluid Dyn. Res.* **13**, 1.

[2] Schoppa, W., Hussain, F., & Metcalfe, R.W. 1995. A new mechanism of small-scale transition in a plane mixing layer: core dynamics of spanwise vortices. *J. Fluid Mech.* **298**, 23–80.

[3] Pradeep, D.S. & Hussain, F. 1999. Core dynamics of a strained vortex column: linear instability, nonlinear evolution and transition. *AIAA paper no. 99-3768.*

[4] Melander, M.V. & Hussain, F. 1989. Cross-linking of two antiparallel vortex tubes. *Phys. Fluids A* **1**, 633.

[5] Jimenez, J., Wray, A.A., Saffman, P.G. & Rogallo, R.S. 1993. The structure of intense vorticity in isotropic turbulence. *J. Fluid Mech.* **255**, 65.

[6] She, Z.S., Jackson, E. & Orszag, S.A. 1990. Intermittent vortex structures in homogeneous isotropic turbulence. *Nature* **344**, 226.

[7] Tsai, C. -Y. & Widnall, S.E. 1976. The stability of short waves on a straight vortex filament in a weak externally imposed strain field. *J. Fluid Mech.* **255**, 65.

[8] Pierrehumbert, R.T. & Widnall, S.E. 1982. The two- and three-dimensional instabilities of a spatially periodic shear layer. *J. Fluid Mech.* **114**, 59.

[9] Melander, M.V. & Hussain, F. 1993. Polarized vorticity dynamics on a vortex column. *Phys. Fluids A* **5**, 1992.

[10] Truesdell, C. 1954. *The Kinematics of Vorticity* (Indiana University Publ. Science Series No. 19).

[11] Melander, M.V. & Hussain, F. 1993 Coupling between a coherent structure and fine-scale turbulence. *Phys. Rev. E* **48**, 2669.

[12] Kida, S. 1981. Motion of an elliptic vortex in a uniform shear flow. *J. Phys. Soc. Jpn.* **50** (10), 3517.

[13] Pierrehumbert, R.T. 1986. Universal short-wave instability of two dimensional eddies in an inviscid fluid. *Phys. Rev. Lett.* **57**, 2517.

[14] Bayly, B.J. 1986. Three-dimensional instability of an elliptical flow. *Phys. Rev. Lett.* **57** 2160.

[15] Huang, L.-S. & Ho, C.M. 1990. Small-scale transition in a plane mixing layer. *J. Fluid Mech.* **210**, 475.

[16] Moser, R.D. & Rogers, M.M. 1993. The three-dimensional evolution of a plane mixing layer: pairing and transition to turbulence. *J. Fluid Mech.* **247**, 275.

[17] Lundgren, T.S. & Mansour, N.N. 1996. Transition to turbulence in an elliptic vortex. *J. Fluid Mech.* **307**, 34.

[18] Jeong, J. & Hussain, F. 1995. On the identification of a vortex. *J. Fluid Mech.* **285**, 69.

Fundamental Instabilities in Spatially-Developing Wing Wakes and Temporally-Developing Vortex Pairs

C.H.K. Williamson, T. Leweke and G.D. Miller

Abstract

In this article we include selected results which have originated from vortex dynamics studies conducted at Cornell, in collaboration with IRPHE, Marseille. These studies concern, in particular, the spatial development of delta wing trailing vortices (Miller & Williamson, 1998), and the temporal development of counter-rotating vortex pairs (Leweke & Williamson, 1998a, 1998b). There are clear similarities in the instabilities of both of these basic flows, as shown in our laboratory-scale studies. At full-scale, the trailing vortices behind wings can reach considerable strengths, and represent a danger to following aircraft, especially smaller ones, due to the rolling moment they induce.

In the case of the spatial development of vortex pairs in the wake of a delta wing, either in free flight experiments, or towed from an XY carriage system in a towing tank, we have found three distinct instability length scales as the trailing vortex pair travels downstream. The first (smallest-scale) instability is found immediately behind the delta wing, involving 'braid wake' instability vortices, which scale on the thickness of the two shear layers separating from the upper and lower surfaces of the wing trailing edge. The second (short-wave) instability, at an intermediate distance downstream, scales on the primary vortex core dimensions. The third (long-wave) instability far downstream represents the classical 'Crow' instability (Crow, 1970), scaling on the distance between the two primary vortices. By imposing disturbances on the delta wing incident velocity, we find that the large-scale instability is receptive to a range of wavelengths. Our DPIV measurements of instability growth rate, taking into account directly the axial velocity profile as well as the circumferential velocity profile, represents the first time that experimental growth rates have been compared with the theory of Widnall *et al.* (1971), without the imposition of ad hoc assumptions regarding the vorticity distribution. The agreement with theory appears to be very good.

In the case of the temporal growth of vortex pairs, formed by the closing of a pair of long flaps underwater, we find two principal instabilities; namely, a long-wavelength Crow instability, and a short-wavelength 'elliptic' instability. Comparisons between experiment and theory for the growth rates of the long-wave instability, over a range of perturbed wavelengths, appears to be very good. The vortex pair 'pinches off', or reconnects, to form vortex rings in the manner assumed to occur in contrails behind jet aircraft. Our work has also involved clear observations of short-wave vortex instability. We discover a symmetry-breaking phase relationship for the disturbances growing in the two vortices, which we show to be consistent with a kinematic matching condition between the two disturbances. Further results demonstrate that this instability is a manifestation of an elliptic instability, which is here identified for the first time in a real open flow. We therefore refer to this flow as a 'cooperative elliptic' instability. The long-term evolution of the flow involves the inception of secondary miniscule vortex pairs, which are perpendicular to the primary vortex pair and develop near the leading stagnation line of the pair.

1   Introduction

The dynamics of a pair of parallel counter-rotating vortices has been the object of a large number of studies in the last three decades. The continued interest in this flow is, to a great extent, due to its relevance to the problem of aircraft trailing vortex wakes, whose far-field is composed of such a pair. The vortices from one aircraft also represent a danger to other manoeuvring aircraft, due to the rolling moments imposed on a wing by such a vortex. In addition to this practical aspect, the counter-rotating vortex pair also represents one of the simplest flow configurations for the study of elementary vortex interactions, which can yield useful information for our understanding of the dynamics of more complex transitional or turbulent flows.

A prominent feature of this flow is a long-wavelength instability, which can frequently be observed in the sky, and which is shown in laboratory scale here in Figure 1c. The first theoretical analysis of this phenomenon was made by Crow (1970). He showed that the mutual interaction of the two vortices can lead to an amplification of sinusoidal displacement perturbations, whose axial wavelength is typically several times the vortex separation distance. The vortex displacements are symmetric with respect to the mid-plane between the two vortices, and they lie in planes roughly 45° with respect to the line joining the vortices. The origin of this instability is linked to the balance between the stabilizing effect of self-induced rotation of the perturbations, and the destabilizing influence of the strain field that each vortex induces at the location of its neighbour (see for example, Widnall *et al.*, 1974). When the instability grows large enough, portions of the displaced vortices can approach each other, 'pinch off', or reconnect, into an array of vortex rings.

(a)

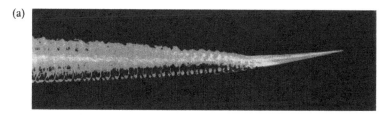

(b)

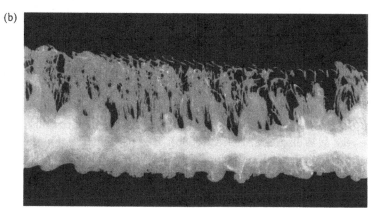

(c)

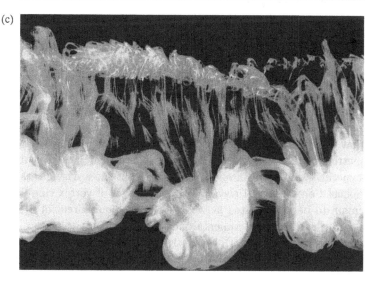

Figure 1. The wake of a delta wing in free-flight, showing three distinct instabilities. (a) The near wake involves a smallest-scale instability which scales on the wake velocity profile of the separating shear layers from the trailing edge of the wing. (b) Intermediate 'short-wave' instability, at 30 chordlengths downstream (left edge of photograph). (c) Far downstream 'long-wave' instability, at 70 chordlengths from the wing.

Figure 1 is available for download in colour from
www.cambridge.org/9780521175128

Widnall *et al.* (1974) and Tsai & Widnall (1976) proposed a second mechanism for instability in flows with strained concentrated vortices, of which the counter-rotating vortex pair is one example. It involves more complex perturbations leading to internal deformations of the vortex cores; their axial wavelength scales on the vortex core diameter and is typically less than the initial separation of the pair. As for the Crow instability, no comprehensive observations or flowfield measurements concerning this short-wave instability in controlled laboratory experiments can be found in the literature, although there are several studies in which a waviness in the vortices may be observed (for example, Thomas & Auerbach, 1994).

In the present article, we shall describe some principal results from two fundamental flow investigations carried out at Cornell, namely the spatial development of instabilities of trailing vortices behind delta wings (Miller & Williamson, 1998), and the temporal development of instabilities for a vortex pair (Leweke & Williamson, 1998a, 1998b). It is particularly of interest that one finds, not unexpectedly, the long-wavelength instability in both of these flows, but one also finds evidence of the short-wave instability in the spatially-growing wing wake, which matches well the lengthscale predicted on the basis of the temporally-growing vortex pair study.

2 Delta Wing Trailing Vortex Instabilities

Previous studies of the wake of a delta wing have centered on the flow right over the wing, with very few studies looking at the wake development further downstream. Among the few existing studies of delta wing wakes, Hackett & Thiesen (1971) and Sarpkaya (1983) found in careful towing tank experiments that the development in the trailing vortex pair often involves growth of symmetric waviness. However, although it is evident that a symmetric 'Crow'-type instability will develop in this flow, previous visualizations are mainly focused on the initial waves. The final stages of vortex ring development remain to be investigated as well as a precise measurement of growth rates related directly to experimental-measured core size, velocity distributions (axial and circumferential), circulation and vortex spacing. These form the principal measurements of the present study. The most clear observations of the late development of the long-wave instability to-date come from full-scale visualizations of the 'Crow' instability acting on the contrails of a jet aircraft observed in the sky (see for example, Crow, 1970).

Regarding downstream measurements of delta wing vortices, it is remarkable that there has been no comprehensive comparison of the wavelengths of instability and the growth rate, between theory and experiment, without recourse to ad hoc assumptions regarding the distribution of vorticity within the vortices, and assumptions regarding axial flow within the cores. Almost all the studies have, for example, made the assumption that the vortex core has a uniform-vorticity core, and that there is no axial flow. It has not always been obvious in previous studies whether one should expect the core axial flow to be upstream or downstream. Many of the previous trailing vortex results, and the extensive and insightful works relating to the near wake of the delta wing, may be found in the excellent review papers of Widnall (1975), Lee & Ho (1989), Rom (1992), Spalart (1998), and the reader is referred there for these results. The wakes of conventional wings of higher aspect ratio (than delta wings) have been studied extensively and much has been learnt about their development through analytical work, flight test and laboratory experiments, as outlined in the review by Donaldson & Bilanin (1975).

The brief descriptions of our results presented here are based on the more comprehensive study in Miller & Williamson (1998), where extensive details may be found. In this study, we measure all of the pertinent parameters describing the vorticity field of the primary vortex core, which are essential as input into the stability analysis of Widnall *et al.* (1971). The flow in question is that found downstream of a delta wing, with a 75° leading edge sweep angle, thus having an apex angle of 30°, and flying at angles of incidence of 10°–20°. The delta wing is towed in the computer-controlled XY Towing Tank at Cornell, with a Reynolds number of order 10^3 to 10^4, based on chord.

Typical visualizations of the developing trailing vortices in side view are shown in Figure 1, where it may be noted that all the photographs are to the same scale. This particular flow is due to free-flight experiments of a delta wing, rather than towing experiments. The smallest-scale instability in (a) is essentially a vortex street instability of the wake profile formed by the separating shear layers from the upper and lower surfaces of the wing at the trailing edge. In (b), at 30 chordlengths downstream of the wing, one can observe a short-wave instability riding on the primary vortices, although it is clearly larger than the initial vortex street wavelength in (a). Finally, in (c), at 70 chordlengths downstream, the long-wave instability of the vortex pair has ultimately caused a reconnection of the pair to form vortex rings which are inclined to the flight path. Streamwise wavelengths associated with these instabilities (λ) are shown in Figure 2, where they are all normalized by the spacing between the vortices (b). The smallest-scale is around $0.3b$, which is reasonable based on a predicted wavelength one would expect from measured separating shear layer thickness at the wing trailing edge. The short-wave instability further downstream is around $1.0b$, and it is not clear in advance what governs this length scale. We shall expand on this point below. Finally, the long-wave instability is around $4.5b$, which is less than the length $8.0b$ that is normally associated and often assumed for the well-known Crow instability.

This study has been made possible by a combination of experimental approaches. To determine the core velocity field, we have used a Digital Particle

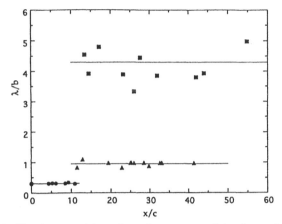

Figure 2. Fundamental length scales measured in the wake; normalized wavelength (λ/b) versus normalized downstream distance (x/c): Circles: Braid wake (vortex street) instability; Triangles: Short-wave instability; Squares: Long-wave 'Crow'-type instability.

Image Velocimetry (DPIV) technique. The data representing the axial and circumferential velocity fields are seen in Figure 3. To study the flow response to a range of imposed wavelengths, we have perturbed the incident velocity on the wing, by computer-control of the carriage in our XY Towing Tank. To interpret the wavelength and growth rate of the primary vortex instability, we have developed a novel image processing technique, which tracks the core displacements (illuminated by selectively dyeing specific parts of the wing wake using Laser-Induced Fluorescence), and determines the spectral content of the core waviness by spatial FFT. Figure 4 shows the development of vortex core displacements, without image processing.

It is of interest to know under what conditions of perturbation amplitude (measured as the fluctuation in wing velocity (u') normalized by the mean free stream (U)) the response of the vortex pair matches the input disturbance wavelength. As might be expected, the envelope in Figure 5 for the range of λ over which the vortex pair locks onto the imposed disturbance wavelength, becomes larger as the perturbation amplitude increases. At low amplitudes, the envelope is naturally centered around the non-perturbed wavelength $\lambda = 4.5b$. What is, however, particularly interesting is that the high-wavelength edge of the envelope is quite a strong function of amplitude, whereas the low-wavelength edge of the envelope is highly insensitive to perturbation amplitude. This will be compared with the stability analysis below.

We have found that the circumferential velocity distribution is very well

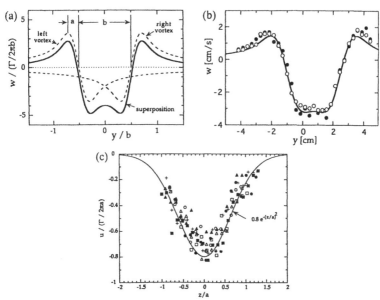

Figure 3. Circumferential and axial velocity profiles from DPIV measurements. (a) A pair of Oseen vortices superposed, to yield a good representation of the flowfield due to the vortex pair. $w =$ vertical velocity. (b) Results of a typical 'best-fit' of experimental data to the Oseen vortices representation. Symbols are from DPIV measurements; Curve is a vertical velocity profile from the best-fit Oseen pair. (c) Axial velocity within the vortex core centre, well represented by a best-fit Gaussian expression.

represented by a pair of Oseen vortices, whose velocity fields are superposed on one another. By an iterative minimum-error method, we can deduce the best-fit values, in a given vortex pair, for the circulation (Γ), vortex spacing (b), and vortex core radius (a), in the case where the velocity distribution is given by :

$$V_\theta = \frac{\Gamma}{2\pi r}\left(1 - e^{-k\left(\frac{r}{a}\right)^2}\right).$$

The axial velocity profile is also reasonably well fitted by a Gaussian, as shown in Figure 3(c), and we find that there is a wake-like velocity defect, with flow upstream within the cores. We find that the core radius (a), the maximum vorticity (ω_{max}) in the vortices, and the circulation (Γ) vary functionally with the downstream distance (x) in the following form:

$$\alpha \sim \sqrt{x} \qquad \omega_{max} \sim \frac{1}{\alpha^2} \sim \frac{1}{x} \qquad \Gamma \sim \text{ constant}.$$

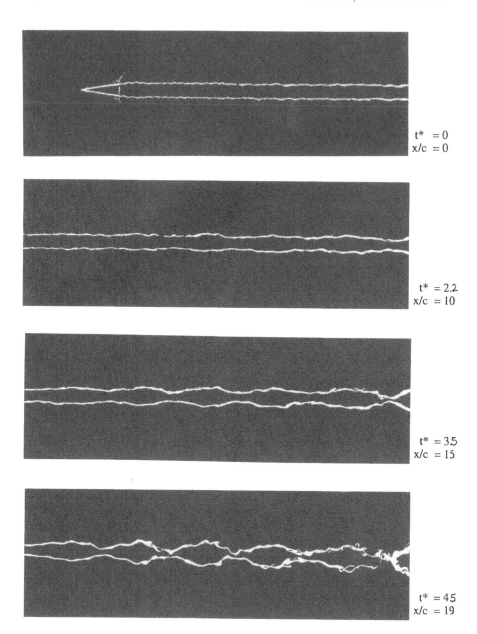

t* = 0
x/c = 0

t* = 2.2
x/c = 10

t* = 3.5
x/c = 15

t* = 4.5
x/c = 19

Figure 4. Development of long-wavelength instability. In this
sequence of top views of a towed delta wing, only the primary
vortex cores are marked. In this case, a small-amplitude perturbation was applied ($u'/U = 0.01$), encouraging the periodicity in
the waves. Such sequences were used to analyze the time evolution
of amplitude. $\alpha = 15°$; $Re = 10,000$.

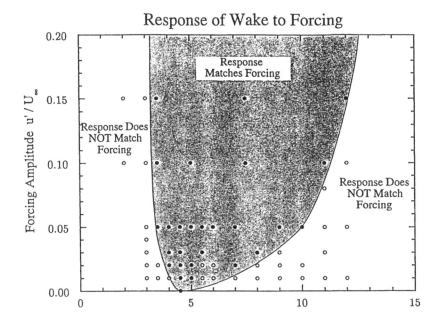

Figure 5. Response envelope plot showing regions where the response matches the perturbations. The extent of the synchronization is measured by the shaded range of normalized wavelengths (λ/b), as a function of the perturbation amplitude (u'/U). Outside this region, the principal wavelengths of trailing vortices do not match the forcing wavelength, and one finds principally the natural wavelength $(\lambda/b = 4.5); \alpha = 15°; Re = 10,000$.

The result that cores grow as $\tilde{x}$ agrees with in-flight measurements of Mc-Cormick *et al.* (1968) and others.

These measurements of core size and axial flow of Figure 3 are critical to our use of the theoretical predictions for the long-wavelength instability. Widnall *et al.* (1971) incorporate the effect of these parameters in an ingenious manner through an 'effective' core size a_{eff}, which redefines the cut-off distance used in the Biot–Savart integral for vortex motion in this problem. The effective core size for the case here, where we have a distributed vortex plus axial flow, can then be used in conjunction with existing solutions corresponding with a uniform core with no axial velocity, as computed in the earlier paper of Crow (1970). This was a major contribution from the work of Widnall *et al.*, yet it has apparently not been comprehensively compared to experiment for the three decades since that time, where the distributed nature of the velocity field with axial flow is taken into account. The definition of 'effective' core size is

$$a_{\text{eff}} = ae^{\frac{1}{4}-A+C}$$

$$A = \lim_{r/a \to} \left[\left(\frac{2\pi a}{\Gamma}\right)^2 \int_0^{r/a} r'(V_\theta))^2 dr' - \ln r/a \right] \qquad C = \left(\frac{2\pi a}{\Gamma}\right)^2 \int_0^\infty r'u^2 dr'.$$

Parameters A and C are determined by the form of the circumferential and axial velocity distributions, which reduce to $A \sim 0.056$ for an Oseen vortex, and $C \sim 0.255$, for our Gaussian axial velocity profile of Figure 3(c). This permits us to find the effective core size, and so to use this in the theories of Widnall et al. (1971) and Crow (1970). In Figure 6, we show the growth rate σ versus wavelength, measured using our image-processing and spatial FFT method (symbols), compared with the stability analysis (the curve), which is computed using experimental input for the effective core size. The agreement appears to be very good. We may conclude from this work that it is critical that we properly characterize the effective core radius. Had we chosen the physical core size of $0.25b$, rather than the computed effective core size of $0.45b$, we would have predicted a most unstable wavelength that is 50% larger than what we both measured and predicted in Figure 6. Our study provides the first clear experimental confirmation for the stability theory in Widnall et al. (1971), without ad-hoc assumptions being made about core vorticity distribution, and even in the case of a spatially-developing wing wake flow.

3   Temporally-Developing Instabilities in Vortex Pairs

3.1   Long-wave instability

In the experiments concerning the temporal development of vortex pairs, these vortices were generated in a water tank at the sharpened edges of two flat plates, hinged on one side to a common base, and moved in a prescribed symmetric way by a computer-controlled motor. The vortex pair is characterized by circulation (G), core radius (a) and vortex spacing (b) as for the delta wing study, but with the absence in this case of the axial flow component. Also consistent with the delta wing vortices, the pair is well represented by Oseen vortices, with $a/b \sim 0.20\text{-}0.25$, and 'vortex' Reynolds numbers $(Re = \Gamma/n)$ of 1000–3000. Further details are found in Leweke & Williamson (1998a, 1998b).

The dye visualization of Figure 7 and 8 show the general features of both the long and short wavelength instabilities. In the case of the long-wave instability, one can see that the initially-straight and parallel vortices develop

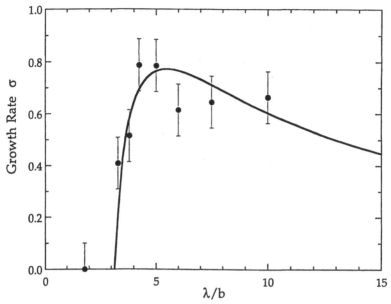

Figure 6. Growth rate versus wavelength. Measured growth rates σ (symbols) versus the analytical result (curve). The stability analysis receives, as input, the experimental initial conditions. The theoretical curve depends on *effective* core size, which was computed from axial and circumferential velocities to be $0.45b$, near $t^* = 1$; $\alpha = 15°$; $Re = 10,000$; $t^* = \text{time} \times \Gamma/2\pi b^2$.

a growing symmetric waviness, of a wavelength that is several times the vortex spacing. Simultaneous views (see Leweke and Williamson, 1998b) were processed digitally to yield 3D views of the vortex pair waviness, showing clearly that the plane of the wavy perturbation is inclined by approximately 45° with respect to the plane of the initial vortex pair, confirming Crow's (1970) prediction. The amplitude of the waves continues to increase until the cores touch and cross-link to form an array of 3D vortex rings. Figure 8(a) shows the flow shortly before the end of the reconnection process. At later times, the rings elongate into oval vortices, seen in (b) and (c) from two orientations. The oval rings exhibit a well-known oscillatory behaviour (see review by Lim & Nickels, 1995), due to their varying curvature. Despite their dramatic change of topology with respect to the initial configuration, it is interesting that the translation speed of the vortices remains practically unchanged during and after the reconnection.

The reconnection process has also been studied in detail. We find a characteristic reconnection time which is found to be approximately the time it takes the vortex pair to travel one vortex spacing, b. This may be usefully

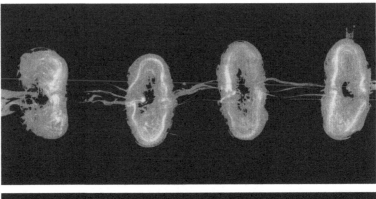

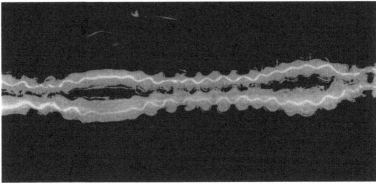

Figure 7. Overview of the long and short wave instabilities in a vortex pair. In these photographs, the vortex pair is convecting towards us, showing the long-wavelength instability (upper picture), and the short-wave instability coexisting with the long-waves (lower picture).

Figure 7 is available for download in colour from
www.cambridge.org/9780521175128

compared with the characteristic time found from simulations (e.g., by Melander & Hussain, 1989). Interestingly, the threads that remain connecting the vortex rings after reconnection, actually contain a noticeable fraction of the initial circulation. Further details of the reconnection process are described in Leweke & Williamson (1998b).

The linear growth rate of the long-wave instability was measured for different axial wavelengths l, which could be imposed by very slight modulation of the vortex-generating plate edges. The measured growth rates in Figure 9 are compared with the theoretical predictions, which may be derived from the theories of Crow (1970) and Widnall $et\ al.$ (1971) for the present conditions. This prediction involves precise information about the initial ve-

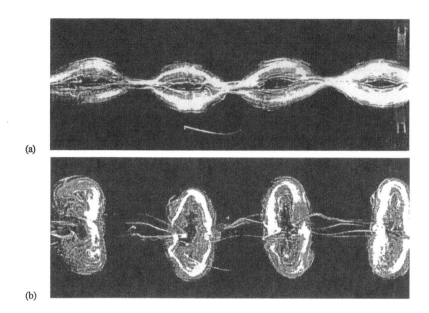

(a)

(b)

Figure 8. Visualization of the large-scale vortex pair instability for $Re = 1450$, $a/b = 0.23$, $\lambda/b = 5.4$. (a) At $t^* = 5.3$, the pair is moving towards the observer. (b) View : $t^* = 9.3$, where $t^* =$ time$\times \Gamma/2\pi b^2$.

locity profiles of the vortices, obtained by DPIV, and we find good agreement with the experimental data. A complete comparison between experimental measurements and theoretical predictions of the initial growth rate for this flow was achieved for the first time, yielding a comparable agreement to that found in the wing trailing vortex study shown above. We may conclude that good agreement is found between the theory and both a temporally- and a spatially-developing flow.

3.2   Short-Wave Instability: 'Cooperative elliptic' instability

Above a critical Reynolds number, a short-wave instability develops in addition to the long-wave instability. From Figure 7, we can see that the short-wave and long-wave instabilities usually develop simultaneously. A closer view of the short-wave instability under more uniform conditions in Figure 10 illustrates immediately that the short-wave instability involves a modification of the internal structure of the vortex cores. Although the vortex centres, marked by the bright dye filaments, are again perturbed into a wavy

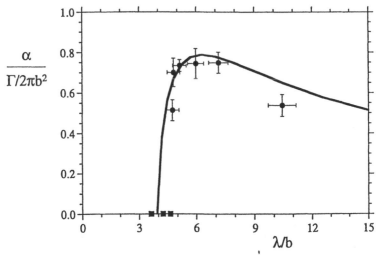

Figure 9. Growth rate of the Crow instability as a function of axial wavelength. Round symbols represent experimental measurements made in the range $1500 < Re < 2500$. The instability could not be forced at wavelengths marked by a square symbol. The line shows the theoretical prediction for $a/b = 0.25$, which is representative of all the experiments.

shape, one may also observe a dye layer around the core which remains unchanged. Inside and outside of this 'invariant stream tube', fluid is displaced in opposite radial directions. Simultaneous visualizations from perpendicular orientations reveal that the sinusoidal displacements of the vortex centres again lie in inclined planes at 45° to the plane of the pair—with orientations in the direction of the mutually induced strain rate in the vortices.

These observations strongly suggest this short-wave instability is linked to an instability occurring in flows with elliptic streamlines, which is a flow resulting from the interaction of a rotational flow with plane strain. In the present flow, it is found in the cores of the vortices. From theoretical studies treating this elliptic instability phenomenon (Landman & Saffman, 1987; Waleffe, 1990) one can show that the spanwise shape of the disturbed streamtubes representing one vortex takes the form depicted in Figure 10(b). The experimental and theoretical spanwise instability shape, including the 'invariant surface' mentioned earlier, bear a strong resemblance to each other. From the theoretical work, one can also deduce the following expression for the growth rate σ of the short-wave instability for the vortex pair flow (see

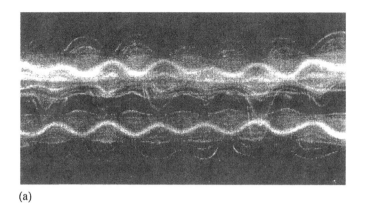

(a)

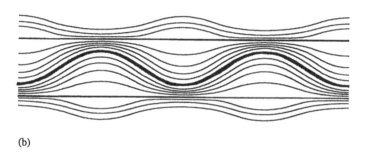

(b)

Figure 10. Close-up views of the short-wavelength perturbation, from experiment and from elliptic instability theory. We can see the characteristic internal deformations, as well as the distinct phase relationship between the two vortices. $Re = 2750$; $t^* = 6.2$. $t^* = \text{time} \times \Gamma/2\pi b^2$.

Leweke & Williamson, 1998a):

$$\frac{\sigma}{\Gamma/2\pi b^2} = \frac{9}{8} - \frac{32\pi^2}{(\lambda/b)^2 Re},$$

This relation shows that, for a given wavelength λ, the instability only occurs above a critical Reynolds number:

$$Re_c = \frac{2^8 \pi^3}{9(\lambda/b)^2} \approx \frac{16\pi^3}{9(a/b)^2}.$$

For the vortex pair, our measurements from visualizations and DPIV have shown that

$$\lambda/a \approx 4.0$$

which is close to theoretical prediction. Quantitative comparisons of growth rate are also satisfactory, taking into account the idealized nature ot the theoretical flow.

An interesting and unexpected phenomenon relates to the phase relationship between the perturbations on the two vortices. In Figure 10(a), the vortex centres at each axial location are displaced in the same transverse direction. This symmetry-breaking is further illustrated by the cross-sectional view in Figure 11(a): the vortex centre is displaced to the lower right in the left vortex, and to the upper right in the right vortex. DPIV measurements in the same plane shown in Figure 10(b) confirm that this is not simply an effect caused by the dye technique. The maxima of vorticity are displaced in the same way as the dye images, and in close agreement with theoretical predictions for the elliptic instability of a strained vortex (Waleffe, 1990). An example of the unstable flowfield from theory is shown in Figure 10(c). The observed phase relationship can be explained by a kinematic matching condition for the perturbations on each vortex (Leweke & Williamson, 1998a). In this manner, this instability might suitably be termed a 'cooperative elliptic' instability of a vortex pair.

3.3   Relation between the short-wave instability in the wing trailing vortices and the vortex pair

Having discovered the nature of the short-wave instability in the temporally-growing vortex pair problem, how then does it relate to the short-wave instability found in the delta wing trailing vortices? To provide some evidence to link these instabilities, we can replot the intermediate lengthscale of Figure 2, in the form of wavelength normalized with respect to the local value of vortex core radius (a), shown in Figure 12(a). In the early part of the development of the short-wave instability the wavelength in the trailing vortices is: $\lambda \sim 3\text{-}4a$, which is of the order of the measured and predicted wavelength for the elliptic instability of the temporally-growing vortex pair, $\lambda \approx 4a$. One might suggest that the two instabilities are essentially similar, with both of them scaling on the vortex cores.

Further evidence of a similar cooperative instability in the wing trailing vortices is found from visual images of the flow in Figure 12(b) compared with a vortex pair flow in Figure 12(c). (The photograph in (c) is taken from a vertical vortex pair flow in a stratified fluid, which thereby inhibited the Crow instability; Williamson & Chomaz, 1998.) In the case of the vortex pair flow, near the leading stagnation point, one finds the growth of secondary miniscule vortex pairs (arranged perpendicular to the primary vortex pair), as a result of the cooperative nature of the instability, and the swapping and stretching of fluid between each primary vortex in this region of the flowfield. One

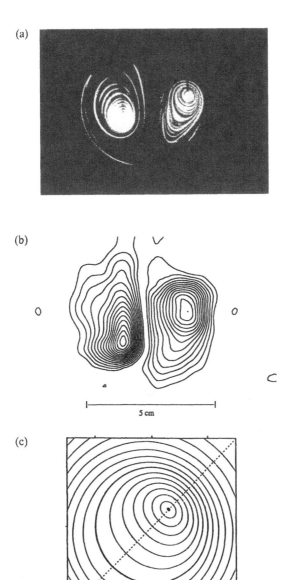

Figure 11. Visualization, vorticity measurements and theoretical flowfield of elliptic instability of the short-wave perturbation in a cross-cut plane. The visualization is for $Re = 2400$, $t^* = 7.5$. The DPIV measurements are for $Re = 2660$, $t^* = 7.1$. Contours are separated by $\Delta\omega_z = 0.86(\Gamma_o/2\pi b^2)$.

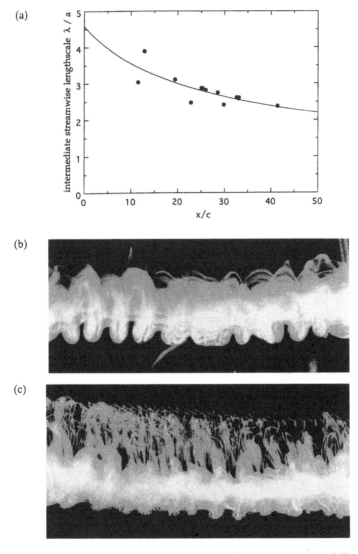

Figure 12. (a) Downstream evolution of short-wave instability wavelength, normalized by local core radius, a. (b) A vortex pair travelling downwards, showing the leading-edge miniscule vortex pairs that are arranged perpendicular to the primary pair (2 per wavelength of the elliptic instability). (c) Similar perpendicular miniscule vortex pairs are found for the delta wing trailing vortices at intermediate distances downstream, suggesting that it is also subject to a distinct phase relationship for the short-wave disturbances in each vortex.

finds that, for each wavelength of the short-wave instability, two secondary vortex pairs are generated at the lower (leading) edge of the primary vortex pair. The same miniscule secondary vortex pairs are found in the case of the trailing vortices in (b), providing further evidence that not only the short wave instabilities of the two flows scale on the vortex cores, as suggested earlier, but also that a similar mutual phase relationship exists between the primary vortices, despite the fact that an axial upstream flow exists inside the cores, in the case of the wing trailing vortices. Further work is needed to confirm these results.

4 Conclusions

In this article we have investigated the short-wave and long-wave instabilities in both the trailing vortices downstream of a delta wing, and also in a vortex pair. For the first time, and in both flows, we have made a complete comparison of the instability growth rates for the long-wave instability measured experimentally with those predicted from theory (Widnall *et al.*, 1971), without recourse to previously-employed assumptions regarding the axial or circumferential velocity distributions. The use of known disturbances in both flows, coupled with extremely sensitive flow visualizations and DPIV measurements, have enabled us to determine these growth rates of instability as a function of perturbation wavelength. The agreement between the temporal stability theory and experiment is very good, not only for the temporally-developing vortex pair flow, but also in the case of the spatially-developing wing trailing vortex flow.

The short-wave instability has been identified as an elliptic instability in an open flow, and we have discovered it to be a 'cooperative' instability in that there is a specific symmetry-breaking phase relationship between the instabilities on each vortex. This is explained by a kinematic matching condition for the disturbances. Evidence is shown that the short-wave instability of the wing trailing vortices is indeed the same instability, scaling on the vortex cores, as we find in the temporally-developing vortex pair flow, although further confirmation is needed on this point.

Acknowledgments

Many thanks are due to Chantal Champagne, Raghuraman Govardhan, Anil Prasad and Sylvie Petit for superb help and enthusiasm with this work in its many stages. T.L. and C.H.K.W. acknowledge the support of NATO Collaborative Grant CRG-970259, which has enabled much of the interactions between us. The support from the Ocean Engineering Division of the O.N.R., monitored by Tom Swean, is gratefully acknowledged (O.N.R.Contract No.

N00014-94-1-1197 and N00014-95-1-0332). T.L. acknowledges the support from the Deutsche Forchungsgemeinschaft under Grant Number: Le 972/1-1.

References

Crow, S.C. (1970) 'Stability theory for a pair of trailing vortices', *AIAA Journal* **8**, 2172.

Donaldson, C.P. & Bilanin, A.J. (1975) 'Vortex wakes of conventional aircraft', *AGAR-Dograph*, Number 204.

Hackett, J.E. & Thiesen, J.G. (1971) 'Vortex wake development and aircraft dynamics'. In *Aircraft Wake Turbulence and its Detection*, J.H. Olsen *et al.*, eds., Plenum Press, 243–263.

Lee, M. & Ho, C.-M. (1989) 'Vortex dynamics of delta wings'. In *Frontiers in Experimental Fluid Mechanics*, M. Gad-el-Hak, ed., 365–427.

Landman, M.J. & Saffman, P.G. (1987) 'The three-dimensional instability of strained vortices in a viscous fluid', *Phys. Fluids* **30**, 2339.

Leweke, T. & Williamson, C.H.K. (1998a) 'Cooperative elliptic instability in a vortex pair', *J. Fluid Mechanics* **360**, 85.

Leweke, T. & Williamson, C.H.K. (1998b) 'Long-wavelength instability and reconnection of a vortex pair', *J. Fluid Mechanics* submitted.

Lim, T.T. & Nickels, T.B.. (1995) 'Vortex rings'. In *Fluid Vortices*, S.I. Green, ed., Kluwer Academic Publishing, 95–153.

Melander, M.V. & Hussain, F.. (1989) 'Cross-linking of two anti-parallel vortex tubes', *Phys. Fluids* **A1**, 633.

Miller, G.D. & Williamson, C.H.K. (1998) 'Instabilities in the wake of a delta wing', *J. Fluid Mechanics*, submitted.

Rom, J. (1992) *High Angle of Attack Aerodynamics*. Springer-Verlag.

Sarpkaya, T. (1983) 'Trailing vortices in homogenous and density-stratified media', *J. Fluid Mechanics* **136**, 85.

Spalart, P.R. (1998) 'Airplane trailing vortices', *Ann. Rev. Fluid Mech.* **30**, 107.

Thomas, P.J. & Auerbach, D. (1994) 'The observation of the simultaneous development of a long-and short-wave instability mode on a vortex pair', *J. Fluid Mechanics* **265**, 289.

Tsai, C.-Y. & Widnall, S.E. (1976) 'The stability of short waves on a straight vortex filament in a weak externally imposed strain field', *J. Fluid Mechanics* **73**, 721.

Waleffe, F. (1990) 'On the three-dimensional instability of strained vortices', *Phys. Fluids* **A2**, 76.

Widnall, S.E. (1975) 'The structure and dynamics of vortex filaments', *Ann. Rev. Fluid Mech.* **7**, 141.

Fundamental instabilities in spatially-developing wing wakes 103

Widnall, S.E., Bliss, D.B. & Tsai, C.-Y. (1974) 'The instability of short waves on a vortex ring', *J. Fluid Mechanics* **66**, 35.

Widnall, S.E., Bliss, D.B. & Zalay, A. (1971) 'Theoretical and experimental study of the stability of a vortex pair'. In *Aircraft Wake Turbulence and its Detection*, J.H.Olsen *et al.*, eds., Plenum Press, 339–354.

Williamson, C.H.K. & Chomaz, J-M. (1998) 'A new zig-zag instability for a vertical cortex pair in stratified fluid', *Phys. Fluids*, submitted.

Vortex Lines and Vortex Tangles in Superfluid Helium

Carlo F. Barenghi

1 Introduction

Superfluid turbulence takes place in a fluid called *Helium II*. Helium II is traditionally studied within the low temperature physics and condensed matter physics communities. In this contex the general motivation is that, as the temperature is reduced, there is less thermal disorder and the fundamental properties of matter become more apparent. From this point of view superfluid turbulence is important because it represents a form of disorder in the limit of absolute zero. However, since it is the disorder of a *fluid* system rather than a solid, it is useful to approach the subject using the methods and the ideas of classical fluid mechanics. The key ingredients which are required are the concepts of *viscous flow, inviscid flow* and *vortex lines*, which are more familiar to fluid dynamicists and applied mathematicians than to low temperature physicists.

The mutual interaction of viscous flow, inviscid flow and vortex lines makes the hydrodynamics of Helium II a very rich subject, of which relatively little is known. The few *laminar* flows which have been studied are often very different from the flows of a classical fluid. This is why the recent experimental evidence that the *turbulent* flow of Helium II is similar in many respects to classical turbulence is surprising.

The aim of this article is to introduce the subject of vortex tangles in superfluid turbulence from the point of view of classical fluid mechanics. The organization is the following: Sections 2 and 3 introduce the basic physics of Helium II and Landau's two-fluid model. Section 4 illustrates some non classical flows which result from the existence in Helium II of two separate fluid components. Section 5 and 6 introduce vortex lines and laminar vortex flows. Section 7 is devoted to the main issue of this article: the observed classical aspects of superfluid turbulence. Section 8 describes the vortex dynamics method used to model the vortex tangle. Sections 9 and 10 show some results obtained using this method. Section 11 is devoted to the problem of superfluid turbulence at absolute zero. Finally, Section 12 attempts to draw some conclusions and points at directions of future research.

2 Liquid Helium I and Liquid Helium II

Helium has two isotopes: He^4, which is common, and He^3, which is very rare. He^3 is a fermion, not a boson like He^4, so its low temperature properties require a very different discussion. Hereafter we shall deal only with He^4. Helium is a gas at ordinary room temperatures and pressures. To obtain a liquid it is necessary to cool it to few degrees above absolute zero, typically 4K. At this temperature every other known substance is solid. For example nitrogen and oxygen become solid at 63K and 54K respectively, and even hydrogen solidifies at 14K. A remarkable property of liquid Helium is that, unless a very high pressure is applied, it does not become solid but remains liquid down to absolute zero. Solid state physicists therefore routinely use liquid Helium to cool their samples. Liquid Helium is also used as a coolant for devices ranging from superconducting magnets in particle physics accelerators to infrared detectors in astrophysics.

Liquid Helium exists in two phases: a high temperature phase, called Helium I, which is an ordinary, classical fluid governed by the Navier–Stokes equations, and a low temperature phase, called Helium II, which is a quantum fluid. Helium I and Helium II are separated by a phase transition called the *lambda transition*, which takes place at the critical temperature $T_\lambda = 2.172K$ at saturated vapour pressure. The lambda transition marks the onset of *Bose Einstein condensation* (BEC) and quantum order.

Before dealing exclusively with Helium II it is worth noticing the importance of Helium I in the study of classical turbulence (Barenghi, Swanson and Donnelly 1995). Helium I has a very small kinematic viscosity ν, about sixty times smaller than water's, which makes it possible to reach very large Reynolds numbers $Re = VD/\nu$, where D and V are the length and speed scales of the flow. Moreover, the ratio $\alpha/\nu\kappa$, where α is the expansion coefficient and κ is the thermal diffusivity, is three thousand times bigger than in water, so it is possible to study Benard convection at very high Rayleigh number $Ra = g\alpha\Delta T D^3/\nu\kappa$ where ΔT is the temperature gradient and g is the acceleration due to gravity.

3 The Two-Fluid Model

The aim of this section is to introduce the basic framework of the hydrodynamics of Helium II, the *two-fluid model* of Landau (Donnelly 1991a). In this model Helium II is considered as the intimate mixture of two fluid components, the *normal fluid* and the *superfluid*. Each fluid component has its own density and velocity field, ρ_n and $\mathbf{v}_n$ for the normal fluid and ρ_s and $\mathbf{v}_s$ for the superfluid. The total density of Helium II is $\rho = \rho_n + \rho_s$. The superfluid is related to the quantum mechanical ground state, carries neither entropy nor viscosity and is similar to a classical, inviscid Euler fluid. The normal fluid corresponds to thermally excitated states, called *phonons* and *rotons* depending on the wavenumber. The normal fluid carries the entire entropy and viscosity of Helium II and is similar to a classical, viscous Navier–Stokes fluid.

The relative proportion of normal fluid and superfluid is determined by temperature. At $T = 0K$ Helium II is entirely superfluid: $\rho_s/\rho = 1$ and $\rho_n/\rho = 0$. If the temperature is increased the superfluid fraction decreases and the normal fluid fraction increases, until, at $T = T_\lambda$, Helium II becomes entirely normal: $\rho_s/\rho = 0$ and $\rho_n/\rho = 1$. Note that the temperature dependence of the superfluid and normal fluid fractions is very nonlinear: ρ_n/ρ drops from 100% at $T = T_\lambda$ to 55% at 2.0K and to 7.5% at 1.4K, and is effectively negligible at temperatures below 1K.

The mathematical formulation of the two-fluid model is the following. First of all we have the law of conservation of mass, $\partial\rho/\partial t + \nabla \cdot \mathbf{j} = 0$ where $\mathbf{j} = \rho_n\mathbf{v}_n + \rho_s\mathbf{v}_s$ is the total mass current. Then there is the energy equation, which expresses the fact that it is the normal fluid which carries the entropy: $\partial(\rho S)/\partial t + \nabla \cdot (\rho S\mathbf{v}_n) = 0$ where S is the entropy per unit mass. If we assume incompressibility we have $\nabla \cdot \mathbf{v}_n = 0$ and $\nabla \cdot \mathbf{v}_s = 0$ and the equations of motion of the two fluids are

$$\frac{\partial\mathbf{v}_n}{\partial t} + (\mathbf{v}_n \cdot \nabla)\mathbf{v}_n = -\frac{1}{\rho}\nabla P - \frac{\rho_s}{\rho_n}S\nabla T + \nu\nabla^2\mathbf{v}_n, \tag{1}$$

$$\frac{\partial\mathbf{v}_s}{\partial t} + (\mathbf{v}_s \cdot \nabla)\mathbf{v}_s = -\frac{1}{\rho}\nabla P + S\nabla T \tag{2}$$

where P is the pressure, η is the viscosity and $\nu = \eta/\rho_n$. Finally we have Landau's condition that the superfluid is irrotational

$$\nabla \times \mathbf{v}_s = 0. \tag{3}$$

4   Helium II without Vortices

The two-fluid model explains many apparently paradoxical observations which puzzled the early researchers (Donnelly 1991a). It is useful to review some of these observations because they illustrate the fact that the flow of Helium II is usually very different from a classical flow *provided that the Reynolds number is not high*. The first example is the viscosity paradox. Consider a long, thin capillary tube of given length through which Helium II is pushed at constant volume flow rate by means of bellows. Pressure values are measured at the ends of the capillary and the pressure gradient ΔP is determined. The flow speed V is obtained from the imposed volume flow rate and the diameter of the tube. If $T > T_\lambda$ (Helium I) a nonzero pressure gradient is observed, from which one deduces a finite value of the viscosity η. If $T < T_\lambda$ (Helium II), the induced pressure gradient ΔP is found to be zero for all flow velocities V smaller than a certain critical velocity V_c. The critical velocity V_c is larger the smaller the diameter of the tube is. This experiment clearly indicates that, for $V < V_c$, Helium II is *inviscid* and flows without any friction. The critical velocity V_c denotes the onset of some dissipative mechanism (caused by superfluid vortex lines, see Section 7). The experiment also suggests that η has a discontinuity at $T = T_\lambda$, at which η becomes zero.

The evidence for frictionless motion is contradicted by other experiments which show that Helium II does have viscosity. An example is the flow between two concentric cylinders (Couette flow). Suppose that the inner cylinder is held fixed and that the outer cylinder rotates at constant angular velocity Ω_2. Then the fluid contained between the cylinders transmits a torque G to the inner cylinder and the measured ratio G/Ω_2 yields the value of the viscosity. If the working fluid is Helium one finds that the viscosity η varies continuosly from Helium I to Helium II across the lambda transition, and is nonzero for Helium II, in contradiction with viscosity measurements in capillary tubes.

The two-fluid model explains the viscosity paradox. In flows through capillaries or narrow channels the normal fluid is clamped by its own viscosity and does not move: the mass flow is carried entirely by the inviscid superfluid. In the Couette apparatus the normal fluid, which satifies classical no-slip boundary conditions, is carried along by the outer cylinder and transmits a torque to the inner cylinder because it is viscous.

A great success of the two-fluid theory was the prediction of the existence of a non-classical wave motion which is different from ordinary sound, called *second sound*. In Helium II ordinary sound is called simply *first sound*. A first sound wave is an oscillation of the density ρ and the pressure P in which the temperature T and the entropy S remain almost constant. Conversely, a second sound wave is an oscillation of T and S in which ρ and P remain almost constant. In first sound $\mathbf{v}_n$ and $\mathbf{v}_s$ move in phase; in second sound they move in antiphase, so second sound is an oscillation of the relative velocity $\mathbf{v}_n - \mathbf{v}_s$. The speed of propagation of first sound ranges from about 240m/s at low temperature to about 220m/s near the lambda point; the second sound speed is about ten times smaller than the first sound speed and tends to zero as $T \to T_\lambda$.

A simple way to generate second sound is to consider a membrane which is covered by many microscopic holes. If the membrane oscillates, the viscous normal fluid cannot flow through the small holes and is pushed back and forth by the membrane. On the contrary the inviscid superfluid moves freely through the holes. In this way a relative motion $\mathbf{v}_n - \mathbf{v}_s$ is set up.

Finally, the two-fluid model explains the extraordinary ability of Helium II to transport heat, which is important in engineering applications. Consider a channel which is closed at one end and open to the Helium bath at the other. At the closed end a resistor dissipates a known heat flux W. In an ordinary fluid, such as Helium I, heat is transferred away from the resistor by conduction, provided that one is careful to prevent convective motion, so the heat flux W is proportional to the temperature gradient ∇T and there is a well defined thermal conductivity at small W. Experiments performed in Helium II however indicate that the thermal conductivity is virtually infinite in the limit of small W. The two-fluid model explains why: the heat is carried by the normal fluid away from the resistor, $W = \rho ST|\mathbf{v}_n|$. Because of the presence of the closed end, however, the mass flux is zero, $\mathbf{j} = \rho_s \mathbf{v}_s + \rho_n \mathbf{v}_n = \mathbf{0}$, so some superfluid must flow toward the resistor to conserve mass, $\mathbf{v}_s = -(\rho_n/\rho_s)\mathbf{v}_n$. In this way a *counterflow* $\mathbf{v}_{ns} = \mathbf{v}_n - \mathbf{v}_s$ is generated which is proportional to the applied heat flux, $|\mathbf{v}_{ns}| = W/(\rho_s ST)$. The counterflow is essentially zero frequency second sound. It follows that if an AC voltage is applied to the resistor a second sound wave is generated. Note that if W is larger than a critical value then this (laminar) counterflow breaks down, and a (turbulent) tangle of superfluid vortex lines appears which limit the heat transfer property of Helium II (see Section 7).

5   Vortex Lines

What makes the hydrodnamics of Helium II particularly interesting is the existence of superfluid vortex lines. Hereafter the term *vortex line* will refer exclusively to the superfluid, but we should remember that regions of concentrated vorticity can exist in the normal fluid too.

Vortex lines appear when Helium II rotates or moves faster than a critical velocity. The critical velocity at which the first vortex line appears is a major problem of superfluidity, but it is not addressed here since we are concerned with turbulent flows containing many vortices. Vortex lines can be spatially organized (laminar vortex flows, see Section 6) or disorganized (turbulent *vortex tangles*, see Section 7). Before discussing vortex tangles it is useful to review the basic physics of vortex lines.

Superfluid vortex lines were discovered to solve the rotation paradox (Donnelly 1991a). Consider a classical fluid contained inside a cylindrical container which rotates at constant angular velocity Ω around its axis of symmetry z. In a steady state situation the fluid rotates like a solid body. Using cylindrical cooordinates (r, ϕ, z) the velocity is $\mathbf{v} = \Omega r \hat{\phi}$ and the vorticity is $\boldsymbol{\omega} = \nabla \times \mathbf{v} = 2\Omega \hat{\mathbf{z}}$ where $\hat{\mathbf{z}}$ and $\hat{\phi}$ are the unit vectors in the z and ϕ direction, and the free surface of the fluid has a parabolic shape of height z above

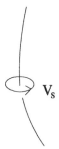

Figure 1: Feynman's vortex line.

the level $z = 0$ at rest

$$z = \frac{\Omega^2 r^2}{2g}.$$ (4)

Let us now replace the classical fluid with Helium II. We expect that the normal fluid component is driven to rotation by viscous forces at the boundaries and that the superfluid component does not rotate because of (3), hence the shape of the free surface of Helium II should be

$$z = \left(\frac{\rho_n}{\rho}\right)\frac{\Omega^2 r^2}{2g},$$ (5)

which reduces to (4) in the classical limit $T \to T_\lambda$. When this experiment was carried out, however, it was found that the surface always obeys (4), not (5). More experiments confirmed that when Helium II rotates the distinction between normal fluid and superfluid is still valid. Clearly the two-fluid model requires modification.

The puzzle was solved by Onsager and Feynman who showed that the superfluid forms isolated *vortex lines* which are aligned along the axis of rotation. The key property is that the superfluid velocity around a single vortex line has fixed circulation

$$\int_C \mathbf{v}_s \cdot \mathbf{dl} = \frac{h}{m} = \Gamma,$$ (6)

where C is a circular path around the axis of the vortex, m is the mass of the helium atom, h is Planck's constant and $\Gamma = 9.97 \times 10^{-4} \mathrm{cm^2/s}$ is the *quantum of circulation* – see Figure 1. Feynman argued that in a rotating countainer the vortices form a uniform distribution with areal density $n = 2\Omega/\Gamma$. Typically n is high: if the period of rotation of the container is one second, then there are $n \approx 12,000$ vortex lines per square centimeter. Experiments in which individual vortex lines were visualized at increasing Ω confirmed Feynman's idea.

The discovery of vortex lines explained the observation that rotating Helium II has a free surface which obeys equation (4) rather than (5). Landau's condition (3) that $\boldsymbol{\omega}_s = \nabla \times \mathbf{v}_s = 0$ is microscopically correct: the velocity field around each vortex is potential. However the net effect of a uniform distribution of superfluid vortex lines with density $n = 2\Omega/\Gamma$ is that, in a coarse-grained sense, the superfluid has the same vorticity $2\Omega\hat{\mathbf{z}}$ of the normal fluid.

Let us consider an individual vortex line again. It follows from (6) that the superfluid velocity around the core is

$$\mathbf{v}_s = \frac{\Gamma}{2\pi r}\hat{\phi}. \tag{7}$$

The question then arises of what happens as $r \to 0$, because here, unlike what happens in classical fluid dynamics, there is no viscosity to prevent a singularity. To understand what happens at the vortex core one must leave the two-fluid model and recall the underlying N-body quantum mechanics. From the Hartee approximation one can derive the following Gross–Pitaevskii nonlinear Schroedinger equation (NLSE) which describes a gas of N bosons which interact with each others via a delta function repulsive potential of strength V_0:

$$i\hbar\frac{\partial\psi}{\partial t} = -\frac{\hbar^2}{2m}\nabla^2\psi - mE\psi + V_0\psi|\psi|^2. \tag{8}$$

Here E is the energy per unit mass, $\hbar = h/2\pi$ and the condensate's wavefunction ψ is normalized by the volume integral $\int_\mathcal{V} d\mathcal{V}|\psi|^2 = N$. The wave function can be written as $\psi = Ae^{iF}$ in terms of an amplitude A and a phase F so that the density of the condensate is $\rho_{BEC} = mA^2$ and its velocity is $\mathbf{v}_{BEC} = (\hbar/m)\nabla F$, which confirms Landau's irrotational condition (3). Using ρ_{BEC} and $\mathbf{v}_{BEC}$ the relation between the NLSE and the Euler equation becomes apparent: the NLSE can be turned into the usual continuity equation and a Bernoulli's equation modified by the presence of an extra quantum pressure term. This extra term is important only in region where A changes significantly.

Equation (8) has a vortex solution: if we let $F = \phi$ we obtain $\mathbf{v}_{BEC} = \Gamma/(2\pi r)\hat{\phi}$ which is Feynman's vortex (7). Substitution into the NLSE yields a differential equation for ρ_{BEC} around the vortex core. One finds that ρ_{BEC} tends to the bulk value $m^2 E/V_0$ for $r \to \infty$, and that $\rho_{BEC} \to 0$ for $r \to 0$, see Figure 2. The characteristic distance over which ρ_{BEC} changes from its bulk value to zero is $a_0 \approx 10^{-8}$cm. This distance is called the *vortex core parameter*. We conclude that the superfluid vortex is hollow at the

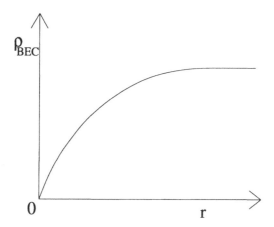

Figure 2: Density of the condensate near the vortex core.

core. There is no singularity at $r = 0$: the velocity increases but the density decreases so that the momentum is finite. Geometrically, a vortex line makes the volume occupied by the superfluid multiply connected. This is a very important difference between the superfluid and an Euler fluid.

Hereafter we identify ρ_{BEC} with ρ_s at absolute zero and $\mathbf{v}_{\mathrm{BEC}}$ with $\mathbf{v}_s$. It must be noted that this identification is convenient from the point of view of the hydrodynamics but is not entirely correct. The reason is that Helium II is a dense fluid, not the weakly interacting Bose gas described by the NLSE, so the condensate is not exactly the same as the superfluid component. Moreover, the NLSE fails to describe the observed dispersion relation of Helium II at high momenta. Despite these shortcomings, the NLSE is much used as a model of Helium II at $T = 0K$.

Let us now consider what happens at finite temperature. The superfluid vortex lines interact with the normal fluid and couple together the two fluid components of Helium II. This force of interaction, called *mutual friction*, has been the subject of many theoretical and experimental investigations (Barenghi, Donnelly and Vinen 1983). Normal fluid flowing with velocity $\mathbf{v}_n$ past a vortex core exerts a frictional force $\mathbf{f}_D$ per unit length on the superfluid in the neighborhood of the core given by

$$\mathbf{f}_D = -\alpha \rho_s \Gamma \mathbf{s}' \times [\mathbf{s}' \times (\mathbf{v}_n - \mathbf{v}_{sl})] - \alpha' \rho_s \Gamma \mathbf{s}' \times (\mathbf{v}_n - \mathbf{v}_{sl}). \qquad (9)$$

Here $\mathbf{s}'$ is the unit vector along the direction of the vortex line (see Figure 3); $\mathbf{v}_{sl}$, the superfluid velocity in the vicinity of the core, is the sum of any superfluid velocity $\mathbf{v}_s$ imposed externally plus the superfluid velocity $\mathbf{v}_i$ which the vortex line induces onto itself because of its own curvature: $\mathbf{v}_{sl} = \mathbf{v}_s + \mathbf{v}_i$.

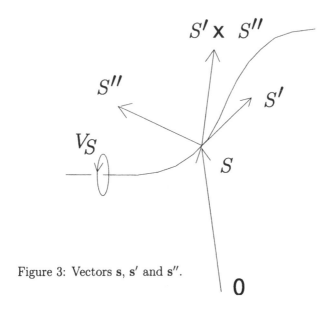

Figure 3: Vectors **s**, **s'** and **s''**.

The parameters α and α' describe the strength of the interaction between the normal fluid and the vortex line; they are often written in terms of *mutual friction coefficients*, B and B' defined by $\alpha = \rho_n B/(2\rho)$ and $\alpha' = \rho_n B'/(2\rho)$. The numerical values of B and B' are known from direct measurements. The calculation of these coefficients over the entire temperature range is still an open issue in superfluidity. Samuels and Donnelly (1990) showed that, in the high temperature range in which most experiments are performed, the friction arises from the scattering between rotons and the velocity field of a vortex line.

How does one detect superfluid vortex lines? The most widely used technique makes use of second sound. A second sound wave or pulse which travels inside liquid Helium II suffers an extra attenuation if it crosses a region of vortex lines. This extra attenuation can be related to the vortex line density. Note that second sound is attenuated by vortex lines only if the lines are perpendicular to the direction of sound propagation, so the second sound technique gives also information about the direction in which the vortex lines are aligned. It is interesting to notice that in the superfluid we can directly detect the vorticity but it is very difficult to visualize the velocity. In classical fluid dynamics the situation is the opposite: the laser Doppler method allows us to detect the velocity well, but it is difficult to reconstruct the vorticity.

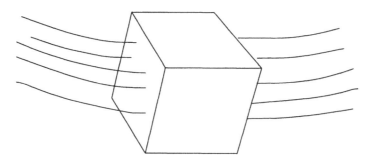

Figure 4: HVBK fluid particle.

6 Laminar Vortex Flows

In the previous section we saw that the simplest flow of Helium II that contains vortex lines is solid body rotation. This is an example of a laminar vortex flow. There are more laminar vortex flows that have been investigated: flow in an annulus, Taylor–Couette flow, flow in a driven cavity. These flows have been studied theoretically using a generalization of the two-fluid equations which takes into account the existence of vortex lines. These equations are called the Hall–Vinen–Bekharevich–Khalatnikov (HVBK) equations (Hall and Vinen 1956, Hall 1960, Khalatnikov 1965, Hills and Roberts 1977). The basic idea behind the HVBK model is to consider fluid particles large enough to be threaded by many vortex lines which are aligned in the same direction (see Figure 4). In this way the superfluid vortex lines are treated as a continuum and we can define a nonzero superfluid vorticity $\boldsymbol{\omega}_s$, in contrast to Landau's microscopic condition (3). Clearly the HVBK equations are valid only if the superfluid vorticity is organized spatially and not randomly oriented, and if the length scales of the flow under consideration are much bigger than the average separation between the vortex lines. In the previous example of Helium II inside a rotating cylinder we have $\boldsymbol{\omega}_s = 2\Omega\hat{\mathbf{z}}$.

In the HVBK theory equations (1) and (2) are replaced by

$$\frac{\partial \mathbf{v}_n}{\partial t} + (\mathbf{v}_n \cdot \nabla)\mathbf{v}_n = -\frac{1}{\rho}\nabla P - \frac{\rho_s}{\rho_n}S\nabla T + \nu_n \nabla^2 \mathbf{v}_n + \frac{\rho_s}{\rho}\mathbf{F}, \qquad (10)$$

$$\frac{\partial \mathbf{v}_s}{\partial t} + (\mathbf{v}_s \cdot \nabla)\mathbf{v}_s = -\frac{1}{\rho}\nabla P + S\nabla T + \mathbf{T} - \frac{\rho_n}{\rho}\mathbf{F}, \qquad (11)$$

where $\boldsymbol{\omega}_s = \nabla \times \mathbf{v}_s$, the mutual friction force is

$$\mathbf{F} = \frac{B}{2}\hat{\boldsymbol{\omega}}_s \times [\boldsymbol{\omega}_s \times (\mathbf{v}_n - \mathbf{v}_s - \nu_s\nabla \times \hat{\boldsymbol{\omega}}_s)] + \frac{B'}{2}\boldsymbol{\omega}_s \times (\mathbf{v}_n - \mathbf{v}_s - \nu_s\nabla \times \hat{\boldsymbol{\omega}}_s), \quad (12)$$

and $\hat{\omega}_s = \omega_s/\omega_s$ with $\omega_s = |\omega_s|$. Finally the *vortex tension* force $\mathbf{T}$ is $\mathbf{T} = -\nu_s \omega_s \times (\nabla \times \hat{\omega}_s)$ where $\nu_s = (\Gamma/4\pi)\log(b_0/a_0)$ is the vortex tension parameter and $b_0 = (2\omega_s/\Gamma)^{-1/2}$ is the intervortex spacing. Note that ν_s has the same dimension of a kinematic viscosity, but physically it is very different: it represents the ability of a superfluid fluid particle to oscillate because of the vortex waves which can be excited along the vortex lines threading the fluid particle itself.

The HVBK model has been used with success to study the transition from Couette flow to Taylor vortices. Barenghi's predictions (1992) of the critical Reynolds number of the transition and its temperature dependence have been confirmed by the experiments. It is interesting to note that the critical wavenumber of the transition has a strong temperature dependence which makes Helium II more stable than a classical fluid at higher temperatures and more unstable at lower temperatures. Taylor–Couette flow has been studied also in the nonlinear regime above the transition, and the calculations of torque (Henderson and Barenghi 1994) and second sound (Henderson, Barenghi and Jones 1995) are in agreement with the observations. More recently, the HVBK equations have been used to study the motion of Helium II in a simple rotating cavity (Henderson and Barenghi 2000). The theory predicts that the rotation of Helium II can be in the opposite direction to a classical fluid, but this flow has not been investigated experimentally yet.

All these results confirm the general trend that the *laminar* flow of Helium II, with or without vortices, is rather different from the flow of a classical fluid.

7 Turbulent Vortex Flows

Superfluid turbulence manifests itself as a *tangle* of vortex lines. Figure 5 shows a vortex tangle calculated in a periodic box.

Vortex tangles can be generated in many ways but it is useful to distinguish between turbulent *counterflow*, which has no classical analogy, and other forms of turbulence (*coflows*) which are more similar to the turbulence studied in classical fluid mechanics.

7.1 Counterflow turbulence

Because of its importance in heat transfer engineering, *counterflow turbulence* is the form of superfluid turbulence which is the most studied. As said in Section 4, laminar counterflow breaks down if the heat flux drives a counterflow velocity v_{ns} which exceeds a critical value v_{c1}. At this critical velocity a vortex tangle is created, as demostrated by Vinen (1957) in his pioneering

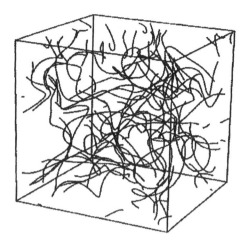

Figure 5: Vortex tangle.

investigations of the problem. The simplest quantity which characterizes the vortex tangle is the vortex line density L_0, which is the length of vortex line per unit volume. A great number of measurements were performed in pipes and channels and showed that $L_0 = \gamma v_{ns}^2$ for $v_{ns} > v_{c1}$ where γ is a temperature dependent parameter. Further measurements in circular pipes indicated that there exists a second critical velocity v_{c2} at which the vortex line density L_0 becomes suddenly larger (Tough 1987). The region of weak turbulence $v_{c1} < v_{ns} < v_{c2}$ and the region of strong turbulence $v > v_{c2}$ are called the T-1 and the T-2 *turbulent states* respectively. For many years the nature of the two turbulent states and the physical meaning of the transition at $v_{ns} = v_{c2}$ have been a mystery. In a recent paper Melotte and Barenghi (1998) showed that in the T-1 state the superfluid is turbulent (a vortex tangle is present) but the normal fluid is still laminar with Poiseuille-like profile; when $v = v_{c2}$ the normal fluid becomes unstable and turbulent. They calculated the critical velocity v_{c2} and the theory agrees well with the observations over a wide rage of experiments at different temperatures and different channel sizes.

7.2 Coflow turbulence

Vortex tangles can be created in many other ways besides counterflow. For example superfluid turbulence can be created using intense ultrasound: this technique is motivated by the desire to avoid boundary effects. Another example is the current attempt by McClintock and co-workers at the University of Lancaster to create a vortex tangle by suddenly crossing the lambda transition (Hendry *et al.* 1984). The motivation behind this work is the analogy between vortex lines as defects left over by the lambda transition and cosmic strings in the early Universe (Zurek 1985, Hendry 1994). So far there is no evidence to confirm the Kibble–Zurek mechanism (Dodd *et al.* 1998) but work is in progress.

We have seen in Sections 4 and 6 that the laminar motion of Helium II. with or without vortices, tends to be rather different from the motion of a classical fluid: the examples discussed include second sound, counterflow and Taylor–Couette flow. However, when Helium II is driven at high velocity and is made turbulent the situation becomes very different: turbulent Helium II seems to behave like classical turbulence. The evidence is the following:

1. Mass flow rates and pressure drops along pipes at $Re \approx 10^6$ can be well described by using classical relations for high Reynolds number classical flows (Walstrom, Weisend, Maddocks and VanSciver 1988).

2. Experiments on large scale turbulent vortex rings at $Re \approx 4 \times 10^4$ detect normal fluid vorticity and superfluid vorticity moving together as a single structure (Borner, Schmeling and Schmidt 1983; Borner and Schmidt 1985).

3. Experiments on turbulent Taylor–Couette flow at $Re \approx 4 \times 10^3$ show the typical structures of classical turbulent Taylor–Couette flow (Bielert and Stamm 1993).

4. Experiments on the decay of superfluid vorticity created by towing a grid show that the decay in time obeys the same laws as those of the decay of classical turbulence (Smith, Donnelly, Goldenfeld and Vinen 1993). Moreover the decay appears to be independent of temperatures in the explored range, from the lambda region down to 1.4K, where the normal fluid fraction is only 7.5% (Stalp 1998).

5. Experiments on turbulence created by rotating blades by Maurer and Tabeling (1998) show the classical Kolmogorov $-5/3$ power spectrum in the temperature range explored, from the lambda region down to $T = 1.4$K again.

6. Experiments on the drag on a moving sphere ($Re \approx 10^5$) show the same drag crisis observed in a classical fluid (Smith 1999).

The temperature independence of these observations is difficult to interpret. The normal fluid must be responsible for these classical aspects, but the dynamical importance of the normal fluid must be related to the fraction ρ_n/ρ, so it must be negligible at temperatures as low as 1.4K.

To explain the experiments it has been suggested (Donnelly 1991b) that the very high number of vortex lines which must be present in these flows can lock together the two fluid components of Helium II into a single fluid of density ρ which behaves like a classical turbulent fluid. This conjecture has not yet been proved, but it has lead to a number of investigations.

8 Vortex Dynamics

We have seen in Section 6 that the HVBK approach fails when the superfluid vorticity is spatially disordered. Without a model based on partial differential equations, progress can be achieved by direct numerical simulation of the motion of a large number of vortex lines. This approach was pioneered by Schwarz (1982, 1985, 1988). Consider a vortex line and call $\mathbf{s}(\xi, t)$ the position of a point on the line, where ξ is the arc length and t is time. The vectors $\mathbf{s}'$, $\mathbf{s}''$ and $\mathbf{s}' \times \mathbf{s}''$ are perpendicular to each others and point along the tangent, normal and binormal directions respectively, where a prime indicates the derivative with respect to ξ (see Figure 3).

The key observation to determine the equation of motion of $\mathbf{s}$ is that the vortex core is so small that it has negligible inertia. What determines the motion of the vortex is the condition that the total force acting upon it is zero. The forces acting on the unit length of vortex line are the drag force (9) and the Magnus force

$$\mathbf{f}_M = \rho_s \Gamma \mathbf{s}' \times (\mathbf{v}_l - \mathbf{v}_{sl}). \tag{13}$$

The Magnus force arises when a body with circulation about it moves in a flow: the circulation creates an increased total velocity of fluid on one side, which results in excess pressure from the other side. Since the key ingredient is the circulation rather than the details of the body, we apply the concept to a vortex line. Let $\mathbf{v}_l = d\mathbf{s}/dt$ be the velocity of the line in the laboratory frame and $\mathbf{v}_{sl}$ be the total velocity of the surrounding superfluid (in the laboratory frame). We have seen that $\mathbf{v}_{sl}$ consists of two parts, the superfluid velocity applied externally and the self-induced velocity of the vortex line: $\mathbf{v}_{sl} = \mathbf{v}_s + \mathbf{v}_i$. Since the total force acting upon the vortex line is zero we have

$$\mathbf{f}_D + \mathbf{f}_M = 0. \tag{14}$$

Solving for $d\mathbf{s}/dt$ we get

$$\frac{d\mathbf{s}}{dt} = \mathbf{v}_s + \mathbf{v}_i + \alpha \mathbf{s}' \times (\mathbf{v}_n - \mathbf{v}_s - \mathbf{v}_i) + \alpha'(\mathbf{v}_n - \mathbf{v}_s - \mathbf{v}_i). \tag{15}$$

The self-induced velocity $\mathbf{v}_i$ describes the motion which a vortex line induces onto itself because of its own curvature and is calculated via the Biot–

Savart integral

$$\mathbf{v}_i(\mathbf{s}) = \frac{\Gamma}{4\pi} \int \frac{(\mathbf{z} - \mathbf{s}) \times d\mathbf{z}}{|\mathbf{z} - \mathbf{s}|^3}. \tag{16}$$

The following Local Induction Approximation to the Biot–Savart law is often used in the literature:

$$\mathbf{v}_i = \beta \mathbf{s}' \times \mathbf{s}'',$$

where $\beta = \Gamma/(4\pi) \log(1/(|\mathbf{s}''|a_0)$. Since $|\mathbf{s}''| = 1/R$ where R is the local radius of curvature, we have

$$\mathbf{v}_i = \frac{\Gamma}{4\pi} \log(\frac{R}{a_0}) \hat{\mathbf{b}}, \tag{17}$$

where $\hat{\mathbf{b}}$ is the binormal.

A numerical simulation of the time evolution of any arbitrary configuration of vortex lines can be developed on the basis of equation (15). An initial vortex configuration is discretized into N points. The time evolution of each point is calculated using (15), given externally applied fields $\mathbf{v}_s$ and $\mathbf{v}_n$ and the temperature T, which determines the friction coefficients α and α'. The transverse part of the mutual friction, proportional to α', is smaller and is sometimes neglected. The number of points N and the time step must be allowed to vary during the evolution to take into account the appearance of regions of high or low curvature. Note that if one uses (16) the computational time is proportional to N^2, while if one uses (17) it is only proportional to N. However the use of (17) can give misleading results – see Ricca, Samuels and Barenghi (1999).

The computer code must also be able to perform *vortex reconnections* when two vortex lines come sufficiently close to each other. In the context of the dynamics of vortex filaments, this process is an arbitrary assumption, but it is justified by a calculation done by Koplik and Levine (1993) using the NLSE, a model which resolves what happens on the scale of the vortex core parameter a_0. Their calculation showed that vortex reconnections in fact take place.

9   Modelling the Vortex Tangle

A key observation to model turbulent Helium II was first made by Samuels (1992) that the presence of normal fluid vorticity has a significant effect on superfluid vortex lines. In most experiments under consideration the Reynolds number is high and the normal fluid must be turbulent. It is known from the observations and from the numerical simulations of classical turbulence (for example, She and Orszag 1990) that regions of intense concentrated vorticity appear in the flow, move about and disappear after a certain time. We expect that the normal fluid has similar regions of concentrated vorticity. To model

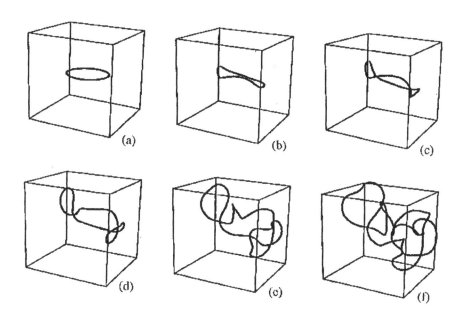

Figure 6: Time sequence of creation of vortex tangle.

these vortex tubes Barenghi, Samuels, Bauer and Donnelly (1997) chose an Arnold–Beltrami–Childress (ABC) flow. Using cartesian coordinates (x, y, z) the normal fluid velocity $\mathbf{v}_n$ has components given by

$$u_n = A \sin (2\pi z/\lambda) + C \cos (2\pi y/\lambda),$$

$$v_n = B \sin (2\pi x/\lambda) + A \cos (2\pi z/\lambda),$$

$$w_n = C \sin (2\pi y/\lambda) + B \cos (2\pi x/\lambda),$$

where λ is a length scale and A, B and C are parameters. ABC flows are solutions of the steady Euler equation and of the time dependent, forced Navier–Stokes equation. Despite the apparent simplicity, the streamlines have a complex Lagrangian pattern which includes chaotic particle path at certain values of parameters (Dombre *et al.* 1996). ABC flow have also been used to study turbulent processes of dynamo action in magnetohydrodynamics (Gilbert, Otani and Childress 1993, Galloway and Proctor 1992). Finally, ABC flows have nonzero helicity, a property which has been associated with turbulence structures both in experiments and numerical simulations (Moffat and Tsinober 1992).

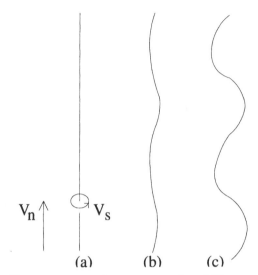

Figure 7: Time sequence of vortex wave instability.

The numerical simulation calculated the time evolution of an arbitrary initial superfluid vortex configuration in the presence of a driving normal ABC flow. The calculation was performed inside a three-dimensional periodic box of size λ. Typically the calculation started with an initial superfluid vortex ring. Under the influence of the normal flow, the ring became unstable and distorted, the total length of vortex line increased and a vortex tangle developed, as showed in the time sequence of Figure 6.

The physical mechanism at the base of this process is the instability of a superfluid vortex line to the growth of helical vortex waves (Kelvin waves), which is illustrated in Figure 7. The instability occurs if the component of the normal fluid velocity in the direction parallel to the vortex line exceeds a critical value (Cheng, Cromar and Donnelly 1973, Ostermeier and Glaberson 1975). As vortex waves become unstable and grow, more line length is created, hence more vortex line length undergoes the same instability, and so on, until nonlinear effects saturate the growth. Barenghi *et al.* noticed that the instability generates bundles of superfluid vortex lines which, driven by mutual friction, concentrate in the regions where the vorticity of the ABC normal fluid is high, see Figure 8.

Although the *microscopic* superfluid velocity pattern in the bundles is very complicated, its *macroscopic* average ω_s over a region larger than the inter-vortex separation is similar to the vorticity field ω_n of the normal fluid. This *vorticity matching* is consistent with the observations. Numerical investigation of the growth time scale for the vortex lines showed that it is of the same order of the ABC flow time scale; since the lifetime of the vortex tubes

Figure 8: Superfluid vortex bundles.

observed in turbulence is of the order of few turnover times, then there is enough time for the matching process to take place.

Although the ABC model is too simple to make direct quantitative comparison with the experiments, it confirms the locking mechanism which has been postulated to explain the experiments and provides a physical explanation for this mechanism.

10   Self-Consistent Superfluid Turbulence

The calculation described above determines the evolution of the vortex tangle for a *given* normal fluid, hence it is *kinematic* in character, because the back reaction of the tangle onto the normal fluid itself is not taken into account. This difficulty is present in *all* other previous calculations based on vortex dynamics (Schwarz 1982, 1985 and 1988; Samuels 1992 and 1993; Aarts and deWaele 1994; Penz, Aarts and deWaele 1995).

A new approach has been recently proposed (Barenghi and Samuels 1999) to study the evolution of superfluid vortices in a way which is dynamically *self-consistent*. In this approach the superfluid vortex tangle is calculated by moving each individual vortex line, as described in Section 8, but the normal fluid is not held fixed and evolves according to its own equation of motion. This equation of motion is equation (10) where at the place of the force (12) one has

$$\mathbf{F}_{ns} = \frac{B\rho_s \Gamma}{2\rho} \frac{1}{\mathcal{V}} \int_{\mathcal{V}} \mathbf{s}' \times [\mathbf{s}' \times (\mathbf{v}_n - \mathbf{v}_{sl})] d\xi. \tag{18}$$

In this way the friction force at the point (x, y, z) is obtained by performing an integral over the entire vortex tangle and averaging the result over a small fluid particle of size $\mathcal{V}$ centred around the point (x, y, z). The size of the fluid particle is the same as the mesh size used in the calculation of $\mathbf{v}_n$.

This new approach has been implemented in a self-consistent calculation of the decay of a three-dimensional vortex tangle interacting with a two-dimensional normal fluid eddy. The model is very idealized because of the reduced dimensionality, introduced to make the calculation less computationally expensive. Nevertheless the results are promising: it is found that the normal fluid energy decays independently of temperature, in agreement with the observations reported at the end of Section 7.

11   Vortex Tangles at Absolute Zero

At temperatures low enough that the normal fluid fraction ρ_n/ρ is negligible, a useful mathematical description of superfluidity is given by the NLSE (Section 5). Nore, Abid and Brachet (1997) studied the time evolution of a vortex tangle using this model. The initial condition was a Taylor - Green vortex. Expressed as a velocity $\mathbf{v}_s = (u_s, v_s, w_s)$ via the Madelung transformation, this initial condition is

$$u_s = \sin(x)\cos(y)\cos(z),$$

$$v_s = -\cos(x)\sin(y)\cos(z)$$

$$w_s = 0.$$

Nore, Abid and Brachet found that the solution ψ of the NLSE evolves into a vortex tangle, which then decays. By examining the energy power spectrum they discovered that the tangle has an inertial range compatible with Kolmogorov's $-5/3$ law. This surprising result is in agreement with the experiment of Tabeling and co-workers, who verified experimentally the existence of Kolmogorov's law in Helium II.

Nore, Abid and Brachet also found a significant transformation of the kinetic energy of the tangle into energy of sound waves. This may be the key mechanism of kinetic energy dissipation at low temperatures (Samuels and Barenghi 1998).

12   Discussion

We have seen that the dynamics of Helium II can be described as the interplay of a viscous Navier–Stokes fluid (the normal fluid component), an inviscid Euler fluid (the superfluid component) and superfluid vortex lines. In general when Helium II flows at small Reynolds numbers (with or without vortices) the motion tends to be very different from the motion of a classical fluid. But at high Reynolds numbers there are many vortex lines in the flow and the turbulent motion of Helium II has many features in common with classical turbulence.

There are many issues which are open to investigation. For example, is the similarity between classical and superfluid turbulence robust? Over which length scales and time scales does this similarity occur? More detailed investigations are needed because the evidence summarized in Section 7.2 deals only with average properties such as mean flows and energy spectra. The results of the studies based on vortex dynamics described in Section 9 suggest that spatial structures in the normal fluid affect the superfluid tangle, but we know very little about the opposite effect which the vortex tangle has on the normal fluid; the only study is the linear stability calculation of Melotte and Barenghi (1998) for counterflow turbulence. Can superfluid vortex lines generate vorticity structures in the normal fluid? Are region of concentrate normal fluid vorticity stable in the presence of superfluid vortices? The new self consistent approach described in Section 10 should give us more insight into the problem.

Besides the physical application to Helium II, the vortex tangle can be seen as a simplified, toy model of turbulence, perhaps a useful intermediate step toward understanding classical turbulence. First of all, the vortices have such a small core (10^{-8}cm) that the mathematical concept of vortex filament applies well; secondly the strength of each vortex line is the same because it is constrained by Planck's constant; thirdly the vortices consist of a fluid which has exactly zero viscosity; finally, dissipation takes the form of a simple friction, which is negligible if the temperature is low enough. Issues such as the topological complexity of turbulent flows could be investigated using this toy model.

There are also important questions about the NLSE and superfluids at zero temperature. What is the meaning of Kolmogorov's $-5/3$ law in this non-classical context? Can the NLSE be generalized to include the correct dispersion relation? What is the fundamental mechanism to dissipate kinetic energy at the temperature of absolute zero? What exactly happens during a vortex reconnection? Are reconnections responsible for most of the sound which is generated?

In conclusion, why is superfluid turbulence so classical? Attempts to answer this question may open new interesting issues about the nature of turbulent motion itself.

Acknowledgments

This research is funded by an equipment grant of the Royal Society of London.

References

Aarts, R.G.K.M. & deWaele, A.T.A.M. 'Numerical investigation of the flow properties of He II', *Phys. Rev. B* **50** 10069–10079 (1994).

Barenghi, C.F. 'Vortices and the Couette flow of helium II, *Phys. Rev. B* **45** 2290–2293 (1992).

Barenghi, C.F., Bauer, G., Samuels, D.S. & Donnelly, R.J., 'Superfluid vortex lines in a model of turbulent flow', *Phys. Fluids* **9** 2631–2643 (1997).

Barenghi, C.F., Donnelly, R.J & Vinen, W.F., 'Friction on quantised vortices in helium II', *J. Low Temp. Phys.* **52** 189–247 (1983).

Barenghi, C.F., Swanson, C.J. & Donnelly, R.J., 'Emerging issues in helium turbulence', *J. Low Temp. Phys.* **100** 385–413 (1995).

Barenghi, C.F. & Samuels, D.C., 'The self-consistent decay of superfluid turbulence', *Phys. Rev. B.* **60** 1252 (1999).

Bielert, F. & Stamm, G., 'Visualization of Taylor–Couette flow in superfluid helium', *Cryogenics* **33** 938–940 (1993).

Borner, H., Schmeling, T. & Schmidt, D.W. 'Experiments of the circulation and propagation of large scale vortex rings in He II', *Phys. Fluids* **26** 1410 (1983)

Borner, H. & Schmidt, D.W. 'Investigation of large scale vortex rings in He II by acoustic measurements of circulation', in *Flow of Real Fluids, Lecture Notes in Physics* **235** 135, Springer Verlag(1985).

Cheng, D.K., Cromar, M.W. & Donnelly, R.J. 'Influence of an axial heat current on negative ion trapping in rotating helium II', *Phys. Rev. Lett.* **31** 433 (1973)

Dodd,, M.E., Hendry, P.C., Lawson, N.S. & McClintock, P.V.E. 'Nonappearance of vortices in fast mechanical expansion of liquid He-4 through the lambda transition', *Phys. Rev. Lett.* **81** 3703–3706 (1998).

Dombre, T., Frisch, U., Greene, J.M., Henon, M., Mehr, A. & Soward, A.M. 'Chaotic streamlines in the ABC flows', *J. Fluid Mech.* **167** 353–391 (1986).

Donnelly, R.J., *Quantized Vortices in Helium II*, Cambridge University Press (1991a).

Donnelly, R.J., *High Reynolds Number Flows using Liquid and Gaseous Helium*, Springer Verlag (1991b).

Galloway, D.J. & Proctor, M.R.E. 'Numerical calculations of fast dynamos in smooth velocity fields with realistic diffusion', *Nature* **356** 691–693 (1992).

Gilbert, A.D., Otani, N.F. & Childress, S. 'Simple dynamical fast dynamos', in *Solar and Planetary Dynamos*, M.R.E. Proctor, P.C. Matthews & A.M. Rucklidge, (eds.) Cambridge University Press, 29–136 (1993).

Hall, H.E., 'The rotation of helium II', *Phil. Mag. Suppl.* **9** 89–146 (1960).

Hall, H.E. & Vinen, W.F., 'The rotation of helium II. (Part 2): the theory of mutual friction in uniformly rotating Helium II', *Proc. Roy. Soc. London A* **238** 215–234 (1956).

Henderson, K.L. & Barenghi, C.F., 'Calculation of torque in nonlinear Taylor vortex flow of Helium II', *Phys. Lett. A* **191** 438–442 (1994).

Henderson, K.L., Barenghi, C.F. & Jones, C.A., 'Nonlinear Taylor–Couette flow of Helium II', *J. Fluid Mech.* **283** 329–340 (1995).

Henderson, K.L. & Barenghi, C.F., 'The anomalous rotation of superfluid helium in rotating cavity', *J. Fluid Mech.* **406** 199 (2000).

Hendry, P.C., Lawson, N.S., Lee, R.A.M., McClintock, P.V.E. & Williams, C.D.H., 'Generation of defects in superfluid He-4 as an analog of the formation of cosmic strings', *Nature*, **368** 315–317 (1994).

Hills, R.N. & Roberts, P.H., 'Superfluid mechanics for a high density of vortex lines', *Arch. Rat. Mech. Anal.* **66** 43–71 (1977).

Khalatnikov, I.M., *An Introduction to the Theory Superfluidity*, Benjamin (1965).

Koplik, J. & Levine, H. 'Vortex reconnection in superfluid Helium', *Phys. Rev. Lett.* **71** 1375–1378 (1993).

Maurer, J. & Tabeling, P., 'Local investigation of superfluid turbulence', *Europhysics Lett.* **43** 29–34 (1998).

Melotte, D.J. & Barenghi, C.F., 'Transition to normal fluid turbulence in Helium II', *Phys. Rev. Lett.* **80** 4181–4184 (1998).

Moffatt, H.K. & Tsinober, A. 'Helicity in laminar and turbulent flow', *Ann. Rev. Fluid Mech.* **24** 281–312 (1992).

Nore, C., Abid, M. & Brachet, M.E. 'Kolmogorov turbulence in low temperature superflows', *Phys. Rev. Lett.* **78** 3896–3899 (1997).

Ostermeier, R.M. & Glaberson, W.I. 'Instability of vortex lines in the presence of an axial normal fluid flow', *J. Low Temp. Phys.* **21** 191 (1975).

Penz, H., Aarts, R. & deWaele, A.T.A.M. 'Numerical investigations of interactions between tangles of quantized vortices and second sound, *Phys. Rev. Lett.* **51** 11973–11976 (1995).

Ricca, R., Samuels, D.C. & Barenghi, C.F., 'The evolution of vortex knots', *J. Fluid Mech.* **391** 29 (1999).

Samuels, D.C. & Donnelly, R.J., 'Dynamics of the interaction of rotons with quantized vortices in Helium II', *Phys. Rev. Lett.* **65** 187–190 (1990).

Samuels, D.C., 'Velocity matching and Poiseuille pipe flow of Helium', *Phys. Rev. B* **46** 11714–11724 (1992).

Samuels, D.C. & Barenghi, C.F., 'Vortex heating in superfluid helium at low temperatures', *Phys. Rev. Lett.* **81** 4381–4383 (1998).

She, Z.S., Jackson, E. & Orszag, S.A. 'Intermittent vortex structures in homogeneous isotropic turbulence', *Nature* **344** 226–228 (1990).

Schwarz, K.W., Generating superfluid turbulence from simple dynamical rules', *Phys. Rev. Lett.* **49** 283–285 (1982); 'Three-dimensional vortex dynamics in superfluid⁴He. Line-line and line boundary interaction', *Phys. Rev. B* **31** 5782–5804 (1985); 'Three-dimensional vortex dynamics in superfluid ⁴He', *Phys. Rev. B* **38** 2398–2417 (1988).

Smith, M.R., Donnelly, R.J., Goldenfeld, N. & Vinen, W.F., 'Decay of vorticity in homogeneous turbulence', *Phys. Rev. Lett.* **71** 2583–2586 (1993).

Smith, M.R. private communication (1999).

Tough, J.T. 'Superfluid turbulence', in *Progress in Low Temperature Physics* vol. VIII, D. F. Brewer, (ed.), North Holland, page 133 (1987).

Vinen, V.F. 'Mutual friction in a heat current in liquid Helium II. I. Experiments on steady heat currents', *Proc. Roy. Soc. A* **240** 114 (1957); 'Mutual friction in a heat current in liquid Helium II. II. Experiments on transient effects', *ibidem,* **240** 128 (1957); 'Mutual friction in a heat current in liquid Helium II. III. Theory of the mutual friction', *ibidem* **242**, 493 (1957); 'Mutual friction in a heat current in liquid Helium II. IV. Critical heat currents in wide channels', *ibidem* **243** 400 (1957).

Walstrom, P.L., Weisend, J.G., Maddocks, J.R. & VanSciver, S.V. 'Turbulent flow pressure drop in various He II transfer system components', *Cryogenics* **28** 101 (1988).

Zurek, W.H. 'Cosmological experiments in superfluid helium?', *Nature* **317** 505–508 (1985).

Evolution of Localized Packets of Vorticity and Scalar in Turbulence

A. Leonard

1 Introduction

In this article we investigate the fate of small-scale elements or packets of vorticity as they evolve in an incompressible turbulent flow. Such objects are candidates for important structures in the inertial range and in the dissipation range of scales so that their geometry in physical space and their spectral properties in Fourier space (forward and backscatter of energy, energy spectrum, etc.) are of intrinsic interest. In addition the nature in which energy is exchanged between these packets and the larger scales could be of interest in sub-grid modelling for large-eddy simulations. Using the same techniques and similar assumptions, we also study the evolution of an initially small blob of scalar in turbulence.

Vortex structures of various types have long been considered as descriptions of turbulence, e.g. the study by Synge & Lin (1943) using Hill's spherical vortices and Townsend's (1951a) proposal that a random collection of stretched plane and axisymmetric Burgers' vortices is a useful representation of a turbulent vorticity field. In both cases above the proposed structures are, in isolation, solutions to the inviscid equation of motion (Hill's vortices) or to the viscous equations with simple, imposed external straining fields (Burgers' vortices). Perhaps the most celebrated vortex structure of this type is Lundgren's (1982) strained spiral vortex model, corresponding to an asymptotic solution to the two-dimensional viscous equations for axial vorticity with an imposed axisymmetric strain. Appropriate averaging over such structures yields Kolmogorov's $k^{-5/3}$ spectrum (Lundgren 1982). See the review by Pullin & Saffman (1998) for further discussion and examples.

Another approach, rather than using structures that, in isolation, are solutions to the fluid equations, is to consider the evolution of small-scale structures under the influence of strain-rate and rotation fields of the large scales. For example, Kida & Hunt (1989) take such an approach to study changes in small-scale (high wavenumber) turbulence spectra, anisotropy, etc. over short times due to distortions from random, but constant, straining and rotation by the large scales. By assuming self-induced motions are negligible they are able to use rapid distortion theory to obtain their results. Extension to nonlinear effects was studied by Kevlahan & Hunt (1997) assuming plane strain. In a similar approach, Reyl et al. (1998) emphasize the role of the unsteadiness of the velocity gradient tensor of the large scales, its effect on the small scales of turbulence, and the connection with Lagrangian chaos. Here, too, the analysis is done for the most part in Fourier space as the evolution wave packets of vorticity are followed as they progress (forward scatter) to high wavenumbers. In the present work we also wish to underscore the fact that a material element or a small-scale vortex structure in a turbulent flow is subjected to a random, time-dependent velocity gradient tensor. But we choose to consider the evolution of such structures in physical space rather than Fourier space, transforming to Fourier space when we wish to investigate spectra, etc. The underlying assumption of 'smallness' seems clearer. In any event small spatial structures have important contributions from both low wavenumbers and high wavenumbers. Thus we will be able to consider energy exchanges at low wavenumbers as well as forward scatter to high wavenumbers.

2   Vorticity packets – rapid distortion theory

It is imagined that these packets of vorticity are initially formed either from debris as a result of pairing, reconnection, or breakdown of vortices or from an outer region of vorticity that was shed by its parent structure. For early times in the evolution we will assume that the packet is small enough so that the velocity gradient tensor is spatially constant over the domain of the packet and that the vorticity field of the packet is weak so that self-induced motions are negligible. The velocity gradient tensor will, however, fluctuate in time, typical of that experienced by a material element in a velocity field producing chaotic advection. During this early stage we will use rapid distortion theory to determine the packet evolution. At later times the accumulated strain will intensify the packet vorticity so that self-induced motion will become important and, in addition, the packet will become elongated to the point at which the strain-rate will vary significantly over the length of the packet.

Let the total velocity and vorticity fields be given by

$$\hat{u} = \mathcal{U} + u$$
$$\hat{\omega} = \Omega + \omega, \tag{2.1}$$

where $\mathcal{U}$ and Ω are, respectively, the velocity and vorticity fields of the large-scale turbulence and u and ω the fields of the packet. We define $x_p(t)$ to be the location of the center of the packet. During the early stages $|u|$ is small compared to $|\mathcal{U}|$ so that its evolution is given by

$$\frac{dx_p}{dt} = \mathcal{U}(x_p(t), t). \tag{2.2}$$

In addition, during the early stages, the packet is assumed small relative to spatial variations in $\mathcal{U}$ so that the velocity field observed by the packet may be approximated by

$$\mathcal{U}_i(\boldsymbol{x}, t) = \mathcal{U}_i(\boldsymbol{x}_p(t), t) + U_{ij}(t)(x_j - x_{pj}), \qquad (2.3)$$

where $U_{ij}(t)$ is the velocity gradient tensor,

$$U_{ij}(t) = \frac{\partial \mathcal{U}_i}{\partial x_j}(\boldsymbol{x}_p(t), t), \qquad (2.4)$$

and x_{pj} is the j component of $\boldsymbol{x}_p(t)$.

Therefore, the vorticity field of the packet will, during the initial stage of development, satisfy the following linearized vorticity transport equation

$$\frac{\partial \omega_i}{\partial t} + U_{j\ell}(x_\ell - x_{p\ell})\frac{\partial \omega_i}{\partial x_j} = \Omega_j \frac{\partial u_i}{\partial x_j} + \omega_j U_{i,j} = \nu \nabla^2 \omega_i, \qquad (2.5)$$

where $U_{j\ell}$ and Ω_j are functions of time only.

As an initial condition for the packet we take a 'spherical' vortex ring with Gaussian profile,

$$\boldsymbol{\omega}(\boldsymbol{x}, 0) = \nabla \times \left[\boldsymbol{a}(0) \exp\left(-|\boldsymbol{x} - \boldsymbol{x}_p(0)|^2/\sigma^2\right)\right], \qquad (2.6)$$

where σ is a measure of the initial ring diameter and the magnitude of $\boldsymbol{a}$ is a velocity scale for the ring. The ring impulse, $\boldsymbol{I}$, is proportional to $\sigma^3 \boldsymbol{a}$. Specifically,

$$\begin{aligned} \boldsymbol{I}(0) &= \frac{1}{2} \int \boldsymbol{x} \times \boldsymbol{\omega}(x, 0) d\boldsymbol{x} \\ &= \pi^{3/2} \sigma^3 \boldsymbol{a}(0), \end{aligned} \qquad (2.7)$$

The initial circulation, $\Gamma(0)$, is

$$\Gamma(0) = \frac{\sqrt{\pi}}{2} \sigma |\boldsymbol{a}(0)|. \qquad (2.8)$$

The total energy of the flow is

$$\begin{aligned} \mathcal{E}_T(t) &= \frac{1}{2} \int |\mathcal{U} + \boldsymbol{u}|^2 d\boldsymbol{x} \\ &= \frac{1}{2} \int |\mathcal{U}|^2 d\boldsymbol{x} + \int \mathcal{U} \cdot \boldsymbol{u} \, d\boldsymbol{x} + \frac{1}{2} \int |\boldsymbol{u}|^2 d\boldsymbol{x}. \end{aligned} \qquad (2.9)$$

The last term on the RHS of (2.9) represents the self-energy, $\mathcal{E}_s$, of the packet

$$\mathcal{E}_s(t) = \frac{1}{2} \int |\boldsymbol{u}|^2 \, d\boldsymbol{x}. \qquad (2.10)$$

The self-energy of the initial ring is

$$\mathcal{E}_s(0) = \frac{\pi^{3/2}\sigma^3|a(0)|^2}{6\sqrt{2}}, \tag{2.11}$$

with a corresponding scalar energy spectrum $E_s(k,0)$ given by

$$E_s(k,0) = \frac{\pi\sigma^6|a(0)|^2}{6} k^2 \exp\left(-k^2\sigma^2/2\right). \tag{2.12}$$

More complex initial conditions may be formed by a superposition of such spherical rings each satisfying the linear transport equation during the initial stage before nonlinearity sets in and each with its own $x_p(t)$, and corresponding $U(t)$ and $\Omega(t)$. Of course the self-energy and self-spectrum would need to include cross-term contributions in addition to superpositions of (2.11) and (2.12), respectively.

For the general case, $\Omega \neq 0$, the first term on the RHS of (2.5) is nonlocal in space and is best handled by Fourier transform methods with a time-dependent wave vector to take care of the inhomogeneous convective velocity (second term on the LHS of (2.5)). This is the essence of rapid distortion theory (see, e.g., Cambon & Scott 1999) and leads to the following result

$$\boldsymbol{\omega}(\boldsymbol{x},t) = \frac{\sigma^3}{8\,\pi^{3/2}} \int e^{i\boldsymbol{k}(t)\cdot(\boldsymbol{x}-\boldsymbol{x}_p(t))} \left(i\boldsymbol{k}(t) \times \boldsymbol{a}(t,\boldsymbol{e}_{\boldsymbol{k}(0)}) \right)$$
$$\times \exp\left[-(\boldsymbol{F}(t)\boldsymbol{M}^{-1}(t)\boldsymbol{F}^T(t)):\boldsymbol{k}(t)\boldsymbol{k}(t)/4\right]d\boldsymbol{k}, \tag{2.13}$$

where $\boldsymbol{F}$ is the deformation tensor satisfying

$$\frac{d\boldsymbol{F}}{dt} = \boldsymbol{U}\boldsymbol{F} \qquad (\boldsymbol{F}(0) = \boldsymbol{I}) \tag{2.14}$$

The tensor $\boldsymbol{M}$ takes care of viscous effects by

$$\frac{d\boldsymbol{M}^{-1}}{dt} = 4\nu\boldsymbol{F}^{-1}\boldsymbol{F}^{T-1} \quad (\boldsymbol{M}(0) = \boldsymbol{I}/\sigma^2). \tag{2.15}$$

The time-dependent wave vector is given by

$$\boldsymbol{k}(t) = \boldsymbol{F}^{T-1}\boldsymbol{k}(0), \tag{2.16}$$

and $\boldsymbol{a}$ satisfies the evolution equation

$$\frac{d}{dt}\,\boldsymbol{a}\left(t,\boldsymbol{e}_{\boldsymbol{k}(0)}\right) = \frac{-(\boldsymbol{k}(t)\cdot\Omega)(\boldsymbol{k}(t)\times\boldsymbol{a})}{(\boldsymbol{k}(t)\cdot\boldsymbol{k}(t))} - \boldsymbol{U}^T\boldsymbol{a}. \tag{2.17}$$

The first term on the RHS above arises if Ω is nonzero and causes a complicated dependence of $\boldsymbol{a}$ on $\boldsymbol{e}_{\boldsymbol{k}(0)} = \boldsymbol{k}(0)/|\boldsymbol{k}(0)|$. This dependence is a manifestation of the nonlocal character of the $\Omega_j \partial u_i/\partial x_j$ term in the vorticity transport equation. However this nonlocal contribution to $\boldsymbol{\omega}$ violates our assumption that $\boldsymbol{\omega}$ is compact. Thus we will concentrate on the case of $\boldsymbol{\Omega}(t) = 0$, i.e., the packet evolves in fluctuating irrotational flow. We believe this to be a reasonable assumption anyway for high Reynolds number turbulence where the vorticity field is presumed highly intermittent.

If $\boldsymbol{\Omega}(t) = 0$, then (2.17) reduces to

$$\frac{d\boldsymbol{a}}{dt} = -\boldsymbol{U}^T \boldsymbol{a} \tag{2.18}$$

with the solution given in terms of the deformation tensor as

$$\boldsymbol{a}(t) = \boldsymbol{F}^{T-1} \boldsymbol{a}(0). \tag{2.19}$$

The vorticity field in physical space is now given by

$$\boldsymbol{\omega}(\mathbf{x}, t) = \tag{2.20}$$
$$\nabla \times \left\{ \mathbf{a}(t) \sigma^3 (det\boldsymbol{M})^{1/2} \exp\left[-(\boldsymbol{F}^{-1} \boldsymbol{M} \boldsymbol{F}^{T-1}) : (\boldsymbol{x} - \boldsymbol{x}_p(t))(\boldsymbol{x} - \boldsymbol{x}_p(t)) \right] \right\},$$

i.e., a compact object, ellipsoidal in shape. To proceed further we need to characterize $\boldsymbol{F}$.

3 Properties of Lagrangian chaos

For a velocity field that produces chaotic advection we expect exponential behavior in time for the eigenvalues of $\boldsymbol{F}$. In any case, because $\boldsymbol{F}\boldsymbol{F}^T$ is positive definite and symmetric it can be diagonalized by a rotation matrix $\boldsymbol{R}_1$:

$$\boldsymbol{R}_1^T [\boldsymbol{F}\boldsymbol{F}^T] \boldsymbol{R}_1 = e^{\boldsymbol{\lambda} t} = \begin{bmatrix} e^{2\lambda_1 t} & 0 & 0 \\ 0 & e^{2\lambda_2 t} & 0 \\ 0 & 0 & e^{2\lambda_3 t}, \end{bmatrix} \tag{3.1}$$

defining the finite-time Lyapunov exponents,

$$\lambda_i = \lambda_i(\mathbf{x}_p(0), t), \tag{3.2}$$

ordered so that

$$\lambda_1 \geq \lambda_2 \geq \lambda_3. \tag{3.3}$$

Incompressibility ($U_{ii} = 0$) yields the relation

$$\lambda_1 + \lambda_2 + \lambda_3 = 0. \tag{3.4}$$

For example, an infinitesmial line element satisfies

$$\frac{d\delta}{dt} = U\delta \tag{3.5}$$

and hence

$$\delta(t) = F\delta(0). \tag{3.6}$$

Therefore the length of $\delta(t)$ increases as

$$\frac{|\delta(t)|}{|\delta(0)|} \longrightarrow c_1 e^{\lambda_1 t}. \tag{3.7}$$

For an infinitesmial area element $A(t)$ ($|A|$ is the area, and A points in the direction normal to the surface) we have

$$\frac{dA}{dt} = -U^T A \tag{3.8}$$

and therefore

$$A(t) = F^{T-1} A(0). \tag{3.9}$$

so that

$$\frac{|A(t)|}{|A(0)|} \longrightarrow c_2 e^{-\lambda_3 t} = c_2 e^{(\lambda_1 + \lambda_2)t}. \tag{3.10}$$

Girimaji & Pope (1990) have obtained statistics for λ_i in forced, isotropic low Reynolds number turbulence using direct numerical simulation. Using 128^3 simulations they determined $F(t)$ (among other quantities) for 4,096 Lagrangian particles initially randomly placed. They found that

$$\begin{aligned}
\langle \lambda_1 \rangle &= 0.14\,\tau_\eta^{-1} \\
\langle \lambda_2 \rangle &= 0.02\,\tau_\eta^{-1} \\
\langle \lambda_3 \rangle &= -0.16\,\tau_\eta^{-1},
\end{aligned} \tag{3.11}$$

where the bracket, $\langle\ \rangle$, indicates average over particles and τ_η is the Kolmogorov timescale

$$\tau_\eta = \sqrt{\frac{\nu}{\epsilon}}, \tag{3.12}$$

with ϵ the dissipation rate per unit volume. The result was essentially independent of Taylor Scale Reynolds number Re_λ for the cases $Re_\lambda = 38$, 63, and 90.

Galluccio & Vulpiani (1994) have computed Lyapunov exponents for a flow map derived from ABC flow, a simple, steady three-dimensional flow known to produce chaotic advection. In this case they find

$$\langle \lambda_2 \rangle = 0 \tag{3.13}$$

so that

$$\langle \lambda_1 \rangle = -\langle \lambda_3 \rangle. \tag{3.14}$$

It is argued that the result (3.13) is related to the fact that the dynamics induced by ABC flow can be transformed into a system that is invariant under time reversal.

Of course (3.11) only gives the expected values for the λ_i for the cases studied by Girimaji & Pope. Each $\lambda_i(\boldsymbol{x}_p(0), t)$ can fluctuate significantly as a function of time and of initial location $\boldsymbol{x}_p(0)$. For example, the total stretch of a line element can be thought of as the accumulative effect of a large number of individual, essentially independent, stretching events. The expectation is then that $e^{\lambda_1 t}$ at large times would be approximately lognormally distributed or λ_1 would be normally distributed. Indeed it is found by Girimaji & Pope that

$$P(\lambda_1) \approx \frac{e^{-(\lambda_1 - \langle \lambda_1 \rangle)^2 t / 2\mu}}{\sqrt{2\pi\mu/t}} \tag{3.15}$$

is a good approximation for the distribution of $\lambda_1's$ over the particles for any fixed time t. The factor t in the exponent arises because the variance of $\ell n(|\boldsymbol{\delta}(t)|/|\boldsymbol{\delta}(0)|)$ and, therefore, the variance of $\lambda_1 t$ is linear in t for large times (Girimaji & Pope 1990). Similar results are found for the distribution of area growths (lognormal) or the distribution of $\lambda_1 + \lambda_2$ (normal) although the fit to, respectively, a lognormal or normal distribution is not as convincing in this case. In particular the fifth moment of the standardized log area is approximately 3 rather than 0.

4   Early time evolution of the vortex packet

We consider in more detail the vorticity field given by (2.20), valid for early times in the evolution, now with the additional assumption that viscous effects are also still negligible. In that case, (2.15) gives

$$\boldsymbol{M}(t) = \boldsymbol{I}/\sigma^2. \tag{4.1}$$

For convenience, we transform to the coordinate system

$$\tilde{\boldsymbol{x}} = \boldsymbol{R}_1^T(\boldsymbol{x} - \boldsymbol{x}_p), \tag{4.2}$$

where R_1 is the rotation matrix that diagonalizes FF^T (see (3.1)).

We find that

$$\boldsymbol{\omega}(\tilde{\boldsymbol{x}}, t) = \tilde{\nabla} \times \left\{ \hat{\boldsymbol{a}}(t) \exp\left[-\left(\frac{\tilde{x}_1^2}{\ell_1^2} + \frac{\tilde{x}_2^2}{\ell_2^2} + \frac{\tilde{x}_3^2}{\ell_3^2} \right) \right] \right\}, \tag{4.3}$$

where

$$\ell_i = \sigma e^{\lambda_i t} \tag{4.4}$$

and $\tilde{\boldsymbol{a}}(t)$ is given by

$$\tilde{\boldsymbol{a}}(t) = R_1^T \boldsymbol{a}(t) = \exp(-\boldsymbol{\lambda} t) R_1^T R_2 \boldsymbol{a}(0). \tag{4.5}$$

We have used (2.19), the fact that R_2 is the rotation matrix associated with the polar decomposition of F (see, e.g., Hunter(1983)),

$$F = V R_2, \tag{4.6}$$

and that V is the positive-definite and symmetric matrix related to F by

$$FF^T = V^2. \tag{4.7}$$

Although the circulation of the deformed ring has, of course, remained constant,

$$\Gamma(t) = \Gamma(0) = \frac{\sqrt{\pi}}{2} \sigma \, |\boldsymbol{a}(0)|, \tag{4.8}$$

the impulse is now

$$\boldsymbol{I}(t) = \pi^{3/2} \sigma^3 \tilde{\boldsymbol{a}}(t). \tag{4.9}$$

Referring to (4.6) we see that the impulse magnitude has increased significantly,

$$|\boldsymbol{I}(t)| \sim e^{-\lambda_3 t} |\boldsymbol{I}(0)| = e^{(\lambda_1 + \lambda_2)t} |\boldsymbol{I}(0)|, \tag{4.10}$$

i.e., an increase consistent with that of an area element.

To investigate the special properties of the packet we compute the Fourier transform of $\boldsymbol{w}(\tilde{\boldsymbol{x}}, t)$ given by (4.3) to find

$$\hat{\boldsymbol{w}}(\boldsymbol{k}, t) = \frac{1}{(2\pi)^{3/2}} \int \exp(-i\boldsymbol{k} \cdot \tilde{\boldsymbol{x}}) \boldsymbol{w}(\tilde{\boldsymbol{x}}, t) d\tilde{\boldsymbol{x}} = \frac{\sigma^3}{2\sqrt{2}} i \, \boldsymbol{k} \times \tilde{\boldsymbol{a}}(t) \, e^{-k_j^2 \ell_j^2/4} \tag{4.11}$$

The energy spectrum requires the following integration over spherical angles in $\boldsymbol{k}$-space

$$E_s(k, t) = \frac{1}{2} \int |\tilde{\boldsymbol{w}}|^2 d\Omega_k = \frac{1}{2} \int_0^{2\pi} d\psi \int_{-1}^{+1} d\mu \, |\hat{\boldsymbol{w}}|^2. \tag{4.12}$$

For convenience we chose (μ, ψ) as follows:

$$\boldsymbol{k} = (k_1, k_2, k_3) = k\left(\mu, \sqrt{1 - \mu^2}\sin\psi, \sqrt{1 - \mu^2}\cos\psi\right) \qquad (4.13)$$

so that

$$k_j^2 \ell_j^2 = k^2\sigma^2[\mu^2 e^{2\lambda_1 t} + (1 - \mu^2)\sin^2\psi e^{2\lambda_2 t} + (1 - \mu^2)\cos^2\psi e^{2\lambda_3 t}]. \qquad (4.14)$$

In addition, using (4.5) we can write $\tilde{\boldsymbol{a}}$ as

$$\tilde{\boldsymbol{a}}(t) = (\hat{a}_1 e^{-\lambda_1 t}, \hat{a}_2 e^{-\lambda_2 t}, \hat{a}_3 e^{-\lambda_3 t}) \qquad (4.15)$$

where $\hat{\boldsymbol{a}}$ has magnitude $|\boldsymbol{a}(0)|$ and is defined by

$$\tilde{\boldsymbol{a}} = \boldsymbol{R}_1^T \boldsymbol{R}_2 \boldsymbol{a}(0). \qquad (4.16)$$

Using simple asymptotics we find that, depending on the magnitude of $k\sigma$, $E_s(k, t)$ is given approximately as follows:

(i) small $k\sigma$ $(k\sigma \ll e^{-\lambda_1 t})$

$$E_s(k, t) \approx \frac{\pi\sigma^4 \hat{a}_3^2}{6} e^{2(\lambda_1 + \lambda_2)t}(k\sigma)^2 \qquad (4.17)$$

(ii) intermediate $k\sigma$ $(e^{-\lambda_1 t} \ll k\sigma \ll e^{-\lambda_2 t})$

$$E_s(k, t) \approx \frac{\sqrt{2}\pi^{3/2}\sigma^4 \hat{a}_3^2}{16} e^{(\lambda_1 + 2\lambda_2)t}(k\sigma) \qquad (4.18)$$

(iii) large $k\sigma$ $(e^{-\lambda_2 t} \ll k\sigma)$

$$E_s(k, t) \approx \frac{\pi\sigma^4}{4} \exp\left[-\frac{k^2\sigma^2}{2} e^{2\lambda_3 t}\right] e^{(\lambda_1 - \lambda_2)t}\left[\hat{a}_3^2 (k\sigma)^{-2} + \hat{a}_2^2 e^{-2(\lambda_1 + \lambda_2)t}\right]. \qquad (4.19)$$

The small $k\sigma$ result for E_s is seen to be proportional to $(k\sigma)^2$ and the square of the impulse as expected. For intermediate values E_s is proportional to $(k\sigma)$. This is the result one obtains for two vortex tubes that are nearly parallel to each other, separated by a small distance d ($kd \ll 1$), and with equal and opposite circulation. For large $k\sigma$ the sheetlike cross-section of the vortex tubes becomes apparent, hence the $(k\sigma)^{-2}$ behavior above and observed in the numerical simulations by Reyl *et al.* (1998). But also the fact that there are two opposing sheets is manifested by the constant component. The $(k\sigma)^{-2}$ component is dominant until $k\sigma \approx e^{-\lambda_3 t}$ where E_s begins to fall off exponentially. Integration of (4.17)–(4.19) over k shows that the self-energy is increasing as $\mathcal{E}_s(t) \sim \sigma^3 \hat{a}_3^2 e^{\lambda_1 t}$ with the increase being split roughly evenly between the intermediate and large k ranges.

5 Evolution of a scalar blob

In this section we determine the evolution of a small blob of scalar in a turbulent flow. Much of the analysis above will be applicable to the scalar case which is simpler and more straightforward. In addition there are no nonlinear effects to consider.

If it is assumed that the blob, centered at $x_p(t)$, is small relative to spatial variations in U, the advection-diffusion equation for the scalar density $\psi(x, t)$ is (see(2.5))

$$\frac{\partial \psi}{\partial t} + U_{j\ell}(x_\ell - x_{p\ell})\frac{\partial \psi}{\partial x_j} = \kappa \, \nabla^2 \psi, \tag{5.1}$$

where κ is the scalar diffusivity. If the initial distribution is spherical gaussian

$$\psi(x, 0) = \frac{C_0 \exp}{(\sqrt{\pi}\sigma)^3} \exp(-|x - x_p(0)|^2/\sigma^2), \tag{5.2}$$

where C_0 is the total amount of scalar

$$C_0 = \int \psi(x, 0)dx \,, \tag{5.3}$$

then the distribution at time t is given by

$$\psi = \frac{C_0(\det N)^{1/2}}{\pi^{3/2}} \exp[-(F^{-1}NF^{T-1}) : (x - x_p(t))(x - x_p(t)) \tag{5.4}$$

where F is the deformation tensor given by (2.14) and the tensor N, analogous to M, satisfies

$$\frac{dN^{-1}}{dt} = 4\kappa F^{-1}F^{T-1}, \quad N(0) = I/\sigma^2. \tag{5.5}$$

The above result is valid even if the vorticity $\Omega(t)$ is nonzero.

If diffusion is assumed negligible then

$$N(t) = I/\sigma^2 \tag{5.6}$$

and the coordinate transformation (4.2) gives ψ as

$$\psi(\tilde{x}, t) = \frac{C_0}{(\sqrt{\pi}\sigma)^3} \exp\left[-\left(\frac{\tilde{x}_1^2}{\ell_1^2} + \frac{\tilde{x}_2^2}{\ell_2^2} + \frac{\tilde{x}_3^2}{\ell_3^2}\right)\right], \tag{5.7}$$

reminiscent of Townsend's (1951b) result for a heat spot in a constant strain field. The Fourier transform of ψ is

$$\hat{\psi}(k, t) = \frac{C_0 \, e^{-k_j^2 \ell_j^2/4}}{(2\pi)^{3/2}} \,. \tag{5.8}$$

Consider now the scalar spectrum function

$$E_\psi(k, t) = k^2 \int |\hat{\psi}|^2 d\Omega_k .$$ (5.9)

As for the case of the energy spectrum of the packet of vorticity, simple asymptotics gives E_ψ for these ranges of $k\sigma$ as follows:

(i) small $k\sigma$ $(k\sigma \ll e^{-\lambda_1 t})$

$$E_\psi(k, t) \approx \frac{C_0^2 (k\sigma)^2}{2\pi^2 \sigma^2}$$ (5.10)

(ii) intermediate $k\sigma$ $(e^{-\lambda_1 t} \ll k\sigma \ll e^{-\lambda_2 t})$

$$E_\psi(k, t) \approx \frac{C_0^2 e^{-\lambda_1 t} (k\sigma)}{(2\pi)^{3/2} \sigma^2}$$ (5.11)

(iii) large $k\sigma$ $(e^{-\lambda_2 t} \ll k\sigma)$

$$E_\psi(k, t) \approx \frac{C_0^2 e^{-(\lambda_1 + \lambda_2)t}}{2\pi^2 \sigma^2} e^{-(k^2 \sigma^2)/2} e^{-2(\lambda_1 + \lambda_2)t}$$ (5.12)

For $\kappa \ll \nu$ and σ on the order of the Kolmogorov length or less we might expect these ellipsoidal distributions (5.7) to be candidates for structures in the viscous-convective subrange. But the result that the large k spectrum of a given blob is constant appears inconsistent with Batchelor's k^{-1} spectrum (Batchelor 1959). However one should consider a collection of these objects all at different stages of their evolution. The amplitudes of the older blobs are exponentially small in time but they extend to exponentially large values of k, $k\sigma \approx e^{\lambda_1 + \lambda_2)t}$. In fact if one assumes that $\lambda_1 + \lambda_2$ is independent of time, integration of (5.12) over time gives immediately a k^{-1} spectrum. More convincing is to take into account the distribution of $\lambda_1 + \lambda_2$ or $e^{(\lambda_1 + \lambda_2)t}$. Assuming the distribution of the latter to be lognormal as discussed above (Girimaji & Pope 1990) we have, for large $k\sigma$ (5.12),

$$\langle E_\psi(k, t) \rangle \approx \frac{C_0^2}{2\pi^2 \sigma^2} < e^{-(\lambda_1 + \lambda_2)t} e^{-(k^2 \sigma^2)/2 \exp[-2(\lambda_1 + \lambda_2)t]}$$

$$= \frac{C_0^2}{2\pi^2 \sigma^2} \int_{-\infty}^{\infty} e^{-p} e^{-(k^2 \sigma^2)/2 \exp(-2p)} \frac{e^{-[p - \lambda_1 + \lambda_2)t]^2/2\mu_A t}}{\sqrt{2\pi \mu_A t}} dp.$$ (5.13)

Now the integration over time gives

$$\int_0^{\infty} \langle E_\psi(k, t) \rangle \, dt \approx \frac{C_0^2}{(2\pi)^{3/2} \sigma^2} \frac{(k\sigma)^{-1}}{\langle \lambda_1 + \lambda_2 \rangle}$$ (5.14)

recovering the k^{-1} spectrum.

6   Discusssion

We have shown the early time development of a small localized packet of vorticity or scalar in turbulence to be ellipsoidal in shape with interesting spectral properties (see (4.17)–(4.19) and (5.10)–(5.12)). The self-energy of the vortex packet rises exponentially in proportion to the longest length of the ellipsoid, $\ell_1 = \sigma e^{\lambda_1 t}$, with roughly equal amounts of energy forward scattered from the large scales to wavenumbers $k > 1/\sigma$ and transferred or forward scattered to wavenumbers $k < 1/\sigma$.

The assumption that the velocity gradient tensor is constant over the packet is valid as long as ℓ_1 is small compared to the distance to the nearest large-scale vortex structure. Although, in general, we may have to restrict ℓ_1 to be on the order of the Kolmogorov length (Batchelor 1959). In any case, when the assumption is no longer valid the packet will become curved along its length and ultimately the packet will lie along a random, nonintersecting space curve. The total length of the curve will still be increasing exponentially but the net impulse will level off or only slowly increase because of cancellation due to the random orientation of $\boldsymbol{x} \times \boldsymbol{w}$ for each segment along the length. This leveling off might allow one to do a meaningful time (and $P(\lambda_1, \lambda_2)$) average of the energy spectrum for the packet although clearly more details need to be worked out.

As long as nonlinear effects remain small the packet will be ribbon-like along the space curve with a local width proportional to $\ell_2 = \sigma e^{\lambda_2 t}$ where λ_2 is the second finite-time Lyapunov exponent at the material point in question. Of course curvature along the length ℓ_2 could occur also because of nonuniform effects. But also the vorticity magnitude in the packet $|\boldsymbol{w}|$ is growing as $e^{\lambda_1 t}$ and when $|\boldsymbol{w}| \sim |\boldsymbol{U}|$ we would expect to see the onset of nonlinear effects so the initial sheetlike structures we have in cross-section would presumably become more tubelike. Appropriate two-dimensional computations with imposed three-dimensional straining fields would be useful and likely yield some insights.

Of course nonlinear effects are not present for the passive scalar case. Here also we can assume that ℓ_1 is smaller than the Kolmogorov length when considering $\kappa \ll \nu$ and the viscous-convective subrange. Individual packets have a constant high wavenumber component but averaging over time and the distribution λ_1 and λ_2 give the k^{-1} behavior of Batchelor (1959). Recovering the $k^{-5/3}$ scalar spectrum for the inertial-convective subrange is still open.

Finally, viscous or diffusive effects could be investigated by using (2.15) or (5.5) or by using

$$\frac{dE_s(k,t)}{dt} = T(k,t) - \nu k^2 E_s(k,t) \qquad (6.1)$$

for the viscous case and taking for $T(k,t)$ the time derivatives of the RHS's of (4.17)–(4.19).

Acknowledgments

Conversations with J.C.R. Hunt and J.C. Vassilicos are gratefully acknowledged.

References

Batchelor, G.K. 1959 Small-scale variation of convected quantities like temperature in turbulent fluid. Part I. General discussion and the case of small conductivity. *J. Fluid Mech.* 5:113–33

Cambon, C., Scott, J.F. 1999. Linear and nonlinear models of anisotropic turbulence. *Ann. Rev. of Fluid Mech.* 31:1–53

Galluccio, S., Vulpiani, A. 1994. Stretching of material lines and surfaces in systems with Lagrangian chaos. *Physica A* 212: (1–2) 75–98.

Girimaji, S.S., Pope, S.B. 1990. Material-element deformation in isotropic turbulence. *J. Fluid Mech.* 220:427–458

Hunter, S.C. 1983. *Mechanics of Continuous Media*, Second Edition, Chapter 5, Ellis Horwood Limited, Chichester, West Sussex.

Kevlahan, N.K.-R., Hunt, J.C.R. 1997. Nonlinear interactions in turbulence with strong irrotational straining. *J. Fluid Mech.* 337:333–364

Kida, S., Hunt, J.C.R. 1989. Interaction between different scales of turbulence over short times. *J. Fluid Mech.* 201:411–445

Lundgren, T.S. 1982. Strained spiral vortex model for turbulent fine structure. *Phys. Fluids* 25:2193–203

Lundgren, T.S. 1985. The concentration spectrum of the product of a fast bimolecular reaction. *Chem. Eng. Sci.* 40:1641–1652.

Pullin, D.I., Saffman, P.G. 1998. Vortex dynamics in turbulence. *Annu. Rev. Fluid Mech.* 30:31–51

Synge, J.L., Lin, C.C. 1943. On a statistical model of isotropic turbulence. *Trans. R. Soc. Canada* 37:45–79

Townsend, A.A. 1951(a). On the fine-scale structure of turbulence. *Proc. R. Soc. London Ser.* A 208:534–42

Townsend, A.A. 1951(b). The diffusion of heat spots in isotropic turbulence. *Proc. R. Soc. Lond.*, A 209:418–30

Vortical Structure and Modeling of Turbulence

E.A. Novikov

The dynamics and statistics of turbulent flows is better understood in terms of characteristics of motion which are local in physical space and have a mechanism of amplification. For the three-dimensional turbulence the primary local characteristic is the vorticity field. For the two-dimensional turbulence the corresponding local characteristic is the vorticity gradient. A description of turbulence in terms of the local characteristics naturally leads to the conditional averaging with a fixed local characteristic at a point. It also leads to a new scaling, which now has experimental support. Conditional averaging and the new scaling provide a basis for a statistical enslavement of the small-scale turbulence, which can be used for large-eddy simulations. Local characteristics also provide a natural way to measure the intermittency correction to the Kolmogorov's classical similarity.

The major goal of turbulence modeling is to slave the enormous number of degrees of freedom to a selected set of variables, manageable by computer simulations. We, probably, can not do it purely dynamically, because the dimension of corresponding inertial manifold (or approximate inertial manifold) is too high for practical purposes. So, we need specific information about the statistical structure of turbulent flows, which has independent interest.

One approach, which we are working on, is a conditional averaging of the Navier–Stokes equations (NSE), written in terms of local characteristics [1], which have a mechanism of amplification. For three-dimensional (3D) turbulence the primary local characteristic is the vorticity field and amplification is due to the vortex stretching [2, 3]. For 2D turbulence, the corresponding local characteristic is the vorticity gradient [4–6]. This approach allows, in particular, an analytical study of the conditionally-averaged 3D vorticity field [6]:

$$\overline{\Omega}_i(\mathbf{r}, \omega) = [f_1(\omega)]^{-1} \int \omega_i' f_2(\mathbf{r}, \omega, \omega') d^3\omega'. \tag{1}$$

Here $\mathbf{r}$ is distance from the point with fixed vorticity ω, f_1 and f_2 are the one-point and two-point probability density functions (pdf) for the vorticity field. We note that the conditionally averaged NSE with fixed vorticity in n points corresponds to a hierarchy of equations for the n-point pdf [3, 6]. The vorticity field (1) has a hyperboloidal component, corresponding to the effect of viscous smoothing, and a twisting component, corresponding to the effect of vortex stretching. For turbulent flows with high Reynolds number $Re = VL\nu^{-1}$ (where V is the characteristic velocity, L the external scale,

140

and ν the molecular viscosity), it was analytically predicted [6] that vortex stretching and twisting is statistically balanced with viscous smoothing on any level of fixed vorticity ω and other terms in the vorticity balance (particularly, the effects of inhomogeneity and nonstationarity) are $\sim Re^{-1/2}$. This prediction was confirmed by direct numerical simulations (DNS) [7], which also revealed that the conditionally-averaged rates of vortex stretching and dissipation increase exponentially with ω. This exponential growth of two opposing physical effects provides a statistical environment for the formation of strong localized vortices ('vortex strings'), followed by a quick breakdown and twisting of such vortices when they became unstable. Such a picture was observed experimentally [8].

The whole conditionally-averaged vorticity (CAV) field (1) with twisting and hyperboloidal components was obtained from DNS [9] and it is in qualitative agreement with a simple analytical model, suggested earlier [6]. However, certain details are different and in future we need a more accurate analytical description of the CAV field (1). Having this vorticity field, we can use it in a new relaxation subgrid-scale scheme [10] for the large-eddy simulations (LES). In such a scheme, it seems natural to choose the 'vortex strings' scale [11, 12] $l_s = LRe^{-3/10}$ as the grid scale. This scale is within the inertial range of scales between the Kolmogorov's internal scale $l_\nu \equiv LRe^{-3/4}$ and the external scale L. The scale l_s was obtained analytically [11, 12] from the correlation balance of vorticity, which is an exact consequence of NSE. This scaling for coherent vortex structures now has experimental support [13]. With such a scaling, we can potentially have a drastic reduction of the number of degrees of freedom: from the classical estimate [14] $N \sim (L/l_\nu)^3$ $\sim Re^{9/4}$ to $N_s \sim (L/l_s)^3 \sim Re^{9/10}$, [10].

As a complementary approach to turbulence modeling, we use general statistical schemes, like Markov processes (Lagrangian–Eulerian descriptions) [10, 15] and infinitely-divisible distributions (intermittency effects) [10, 16], making them statistically consistent with NSE and with experimental data. We are also working on some tools for turbulent flows with a free surface and complex geometry [17, 18].

These approaches are interconnected in a general strategy of the dynamical-statistical modeling of turbulence [17]. Let us indicate one novel connection [19]. Consider the deformation rates tensor $D_{ij} = \frac{1}{2}(\partial v_i/\partial x_j + \partial v_j/\partial x_i)$, where v_i is the velocity field at point $\mathbf{x}$. The correlation tensor of deformation rates is expressed in terms of spatial derivatives of the structural tensor of velocity $\langle u_i u_j \rangle$, where $\langle\ \rangle$ indicates the unconditional statistical averaging, $u_i = v_i' - v_i$ is the velocity increment, the prime indicates a field at point $\mathbf{x}' = \mathbf{x} + \mathbf{r}$. In the inertial range of locally isotropic turbulence $\langle u_r^2 \rangle \sim r^\alpha$, where $u_r = u_i r_i/r$ is the radial (longitudinal) component of velocity increment, $\alpha = 2/3 - \mu(2/3)$, and $\mu(2/3)$ is the intermittency deviation from the Kolmogorov '2/3 law' [14] expressed in terms of the breakdown coefficients [16]. It is known that $\mu(2/3)$ is negative [16] and also to be small experimentally. Consider the correlation for one component of deformation rates $(D_{11} = \partial v_1/\partial x_1)$: $\langle D_{11} D'_{11} \rangle \equiv D(r, \alpha, \phi)$, where ϕ is the angle between the direction of deformation and $\mathbf{r}$. Simple algebra gives [19]:

$$D(r, \alpha, \phi) = q(r) f(\alpha, \phi) \tag{2}$$

where $q(r) = \alpha \langle u_r^2 \rangle / 4r^2$, and

$$f(\alpha, \phi) = \alpha + (3 - \alpha)(2 - \alpha) \cos^2 \phi - (4 - \alpha)(2 - \alpha) \cos^4 \phi.$$

The remarkable property of this correlation is:

$$D(r, \alpha, 0) + D(r, \alpha, \pi/2) = -3\mu(2/3)q(r). \tag{3}$$

This means that the sum of longitudinal and transverse correlations is proportional to the intermittency correction. Thus, we have found a correlation characteristic of velocity gradients which is very sensitive to the effect of intermittency. Formula (3) presents a natural way to measure the intermittency correction, which is the subject of controversy in the literature.

In the derivation of (3) the power law $\langle u_r^2 \rangle \sim r^\alpha$ is assumed. On other hand, if we assume (3), then the power law follows [20]. This increases the importance of this relation and other statistical characteristics of the deformation rates for future studies of turbulence.

We note that the described results, including the new scaling and relation (3), are obtained by a transition to local characteristics. We can recommend this procedure for other areas of physics, where a local characteristic can be identified (a technique for such identification using conditional averaging is presented in [6]).

Acknowledgments

This work is supported by the U.S. Department of Energy under Grant No. DE-FG03-91-ER14188 and by the Office of Naval Research under Grant No. ONR-N00014-97-1-0186.

References

1. E.A. Novikov, *Fluid Dyn. Res.* **6**, 79 (1990).

2. G.I. Taylor, *Proc. Roy. Soc.* **A164**, 15 (1938).

3. E.A. Novikov, *Sov. Phys. Dokl.* **12**, 1006 (1968); *Phys. Rev.* **E52**(3), 2574 (1995).

4. E.A. Novikov, *Phys. Lett.* **A162**, 385 (1992).

5. E.A. Novikov, *J. Phys.* **A25**, L675 (1992).

6. E.A. Novikov, *Fluid Dyn. Res.* **12**, 107 (1993); *Phys. Lett.* **A236**, 65 (1997).

7. E.A. Novikov & D.G. Dommermuth, *Modern Phys. Lett.* **B8**(23), 1395 (1994).

8. S. Douady, Y. Couder & M.E. Brachet, *Phys. Rev. Lett.* **67**, 983 (1991), and video, produced by these authors.

9. R.C.Y. Mui, D.G. Dommermuth & E.A. Novikov, *Phys. Rev.* **E53** (3), 3255 (1996).

10. E.A. Novikov, 'Conditionally-averaged dynamics of turbulence, new scaling and stochastic modeling', in *Lévy Flights and Related Topics in Physics*, M. Shlesinger. G. Zaslavsky and U. Frisch (eds.), 35–50, Springer, 1995.

11. E.A. Novikov, *Phys. Rev. Lett.* **71**(17), 2718 (1993).

12. E.A. Novikov, *Phys. Rev.* **E49**(2), R975 (1994).

13. F. Belin, F. Moisy, P. Tabeling and H. Willaime (to be published); F. Moisy, P. Tabeling & H. Willaime, *Phys. Rev. Lett.* **82**(19), 3994 (1999).

14. L.D. Landau & E.M. Lifshitz, *Fluid Mechanics*, Pergamon Press, Oxford, 1987.

15. E.A. Novikov, *Phys. Fluids* **A1**, 326 (1989); *Phys. Fluids* **A2**, 814 (1990); *Phys. Rev.* **A46**(10), R6147 (1992); G. Pedrizzetti and E.A. Novikov, *J. Fluid Mech.* **280**, 69 (1994).

16. E.A. Novikov, *Appl. Math. Mech.* **35**, 231 (1971); *Phys. Fluids* **A2**, 814 (1990); *Phys. Rev.* **E50**(5), R3303 (1994); G. Pedrizzetti, E.A. Novikov & A.A. Praskovsky, *Phys. Rev.* **E51**(1), 475 (1996); E.A. Novikov & D.G. Dommermuth, *Phys. Rev.* **E56**(5), 5479 (1997).

17. E.A. Novikov, ' Dynamical-statistical modeling of turbulence', *Proc. 16th Sympos. Energy Engineering Science*, 128–137, Argonne Natl. Lab., Argonne, IL. 1998.

18. E.A. Novikov & D.G. Dommermuth, 'A variational approach to turbulent boundary layers', in *Trends in Mathematics, Fundamental Problematic Issues in Turbulence* , A. Gyr, W. Kinzelbach & A. Tsinober (eds.), Birkauser, Basel, 303–306 (1999).

19. E.A. Novikov, 'High order correlation tensors in turbulence', *Phys. Rev. Lett.* **82** (19), 3992 (1999).

20. E.A. Novikov,'Orthogonal antisymmetry of deformation rates and scaling in turbulence', *Phys. of Fluids* **12** 1605–1606 (2000)

The Issue of Local Isotropy of Velocity and Scalar Turbulent Fields

Z. Warhaft

1 Introduction

A preoccupation of many turbulence researchers is to describe the structures of typical turbulent eddies, for various kinds of flows. While flow visualization and Direct Numerical and Simulation (DNS) have indisputably shown that structures occur, not only in complex shear flows but in statistically isotropic flows such as grid turbulence, it is less clear that looking at isolated eddy structures will provide a proper description of a phenomenon that is fundamentally statistical. Appropriate statistical descriptors are needed to illuminate these structures. In this regard one-dimensional statistics such as spectra and correlation functions provided early insights into large-scale structures (Townsend 1956), but they have yielded relatively little insight into the more subtle structures that occur at the inertial and dissipation scales.

The generally accepted statistical view of turbulence in the inertial subrange is the Kolmogorov cascade description (Frisch 1995). Here, as the cascade proceeds (in wave number space), the turbulence is postulated to become more and more (statistically) isotropic, as the smaller scales loose information concerning the boundary conditions that affect the largest eddies. It is for the limit of infinite Reynolds number that the Kolmogorov theory of local isotropy is supposed to hold. The experimental evidence for this is not as strong as is commonly thought, and is the subject of this paper. On the other hand, experiment has clearly shown (e.g. Sreenivasan and Antonia 1997) that as the cascade proceeds, the local statistics (velocity differences) become more and more non-Gaussian (the velocity signal itself is close to Gaussian), reflecting the internal intermittency that becomes more and more pronounced at the small scales. And the internal intermittency also becomes stronger as the Reynolds number increases. The focus of much theoretical analysis (see reviews by Nelkin 1994, Frisch 1995) is to explain this intermittency. While presuming the input scales must play a role in determining the intermittency structure, there is very little discussion on whether the anisotropy of these eddies is in any way reflected at the small scales. Almost without exception the theorists assume that the small scales, as postulated by Kolmogorov, are isotropic.

What is the experimental evidence for local isotropy? While Kolmogorov scaling has well predicted the form of the turbulence spectrum (e.g. Chapter 5 of Frisch 1995), careful measurements of inertial subrange isotropy (or departures from it) are difficult. Measurements of the three components of velocity and its derivative must be done at very high Reynolds number since ample wave-number range is required for the Reynolds stress to diminish to sufficiently low levels so that the anisotropic large-scale stress does not have a direct effect on the small-scale structure. (Note the velocity spectrum has the wave-number dependence $E(k) \sim k^{-5/3}$ while for the Reynolds stress spectrum it is $R(k) \sim k^{-7/3}$ (Lumley 1967).) Atmospheric measurements are difficult and do not have sufficient stationarity and homogeneity to properly investigate the matter. The high Reynolds number boundary layer measurements of Saddoughi and Veeravalli (1994) were the first to convincingly show that when the Reynolds stress is negligible at high wave numbers, there is indeed a wave-number region in which $u/v = 3/4$, as required (kinematically) for local isotropy (where u and v are the longitudinal and transverse (mean shear direction) components of fluctuating velocity). These measurements have been considered to be an important benchmark and are thought by many to be a verification of local isotropy. Yet all statistical moments must be consistent with local isotropy, not just those at second order. We summarize here recent experimental evidence that shows that at third order there is significant anisotropy, at least to moderately high Reynolds numbers. If these measurements can be shown to hold at very high Reynolds number, then the local isotropy hypothesis is incorrect. But before discussing the evidence for local anisotropy of the velocity field we review new findings concerning the scalar field, which is easier to measure and interpret.

2   The Scalar Field

We have recently described a mapping of the three-point, two-dimensional correlation function of a passive scalar field, generated by means of a passive scalar gradient in decaying grid-generated turbulence (Mydlarski and Warhaft 1998a,b). The work, motivated by the theory of Pumir, Siggia and Shraiman (see Mydlarski *et al.* 1998), shows that the scalar ramp-cliff structures that occur in all turbulent flows with a mean scalar gradient – both with and without shear (Sreenivasan 1991, Tong and Warhaft 1994, Holzer and Siggia 1994) – have a unique signature. Moreover this anisotropic structure does not diminish with Reynolds number. The derivative skewness, $S_{\partial\theta/\partial y}$, is of order 1, independent of Reynolds number (Sreenivasan 1991, Mydlarski and Warhaft 1998a). The three-point correlation (Figure 1) has a characteristic shape that is due to the ramp-cliffs that are predominantly confined to the gradient direction, but fluctuate about it. If the ramp-cliffs were not present

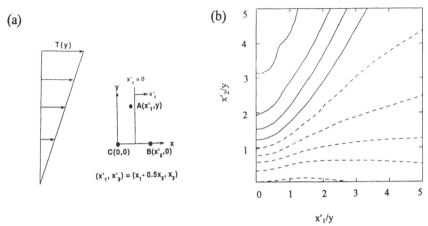

Figure 1. The three-point two-dimensional correlation for a passive scalar in grid generator turbulence. (a) The co-ordinate system. The flow is in the x-direction. (b) The contour plot of the triple correlation $\langle \theta_A \theta_B \theta_C \rangle$. The data are normalized by $-\langle \theta_A \theta_B \theta_C \rangle_{0,0}$ at the given transverse separation which is $y/l = 0.051$. Isocontours are separated by 0.2. Dashed lines are negative, full are positive (the -1 contour is the lower-most dashed line). From Mydlarski and Warhaft (1998b). Reproduced with permission from *Physics of Fluids* **10** p. 2885. (1998), American Institute of Physics.

and the scalar field were locally isotropic, the three-point correlation would be zero. Notice that this function provides salient information on structure. The two-point (one-dimensional) structure functions, while behaving anomalously (Mydlarski and Warhaft 1998a), do not provide insight into geometry (for discussion of the interpretation of Figure 1 see Mydlarski *et al.* 1998 and Mydlarski and Warhaft 1998b).

The ramp-cliff structures affect the higher-order odd moments also (Holzer and Siggia 1994, Mydlarski and Warhaft 1998a). However if the mean scalar gradient is removed (but the scalar fluctuations are still present, produced, for example, by means of a mandoline, (Warhaft and Lumley 1978), the odd moments will vanish and the scalar field will be statistically isotropic (Tong and Warhaft 1994). Large jumps (cliffs) in the scalar will still occur (Chen and Kraichnan 1998), but they will be randomly distributed, not having a preferred direction that is forced by the gradient. These jumps will produce intermittency, manifested as non-Gaussian even moments. Thus we argue that the scalar intermittency structure results from the same large scale eddy structure that produces anisotropy. These eddy structures form converging and diverging separatix (Holzer and Siggia 1994) and the temperature jump (or cliff) occurs at these separatrix, be there a mean gradient or not.

3   The Velocity Field

Here, detecting anisotropy is more difficult than for the scalar field. For the latter, in commonly studied shear flows, a single probe will measure a non-zero value for the scalar derivative third moment, and from this it directly follows that the small-scale scalar field is anisotropic (Sreenivasan 1991). It is over 30 years since Stewart (1969) brought this to the attention of the turbulence community. He reported high Reynolds number scalar atmospheric measurements that were in direct contradiction to local isotropy hypothesis.

On the other hand, for the velocity, measurements of the third moment of its fluctuating derivative are non-zero in the mean flow direction regardless of any anisotropy. This is due to the non-linear vortex interactions that are in balance with the dissipation. Thus the velocity derivative skewness

$$S_{\partial u/\partial x} \equiv \frac{\langle (\partial u/\partial x)^3 \rangle}{\langle (\partial u/\partial x)^2 \rangle^{3/2}}$$

is close to -0.4 (with a small Reynolds number dependence, e.g. Champagne 1978). In order to investigate small-scale anisotropy velocity, differences normal to the mean flow direction must be measured. Here, the test of local isotropy is whether

$$S_{\partial u/\partial y} \equiv \frac{\langle (\partial u/\partial y)^3 \rangle}{\langle (\partial u/\partial y)^2 \rangle^{3/2}}$$

is zero. Such a measurement is not routinely done, and until recently there have only been isolated (and unsystematic) results (e.g. Tavoularis and Corrsin 1984). It was Pumir and Shraiman (1995) and Pumir (1996) who, by analogy with the scalar anisotropy, looked at this issue in more serious light. In their numerical simulations they found that there was significant small-scale anisotropy in the vorticity and in the velocity difference at the third moment. Their Reynolds number was low and its range was small. Their work motivated the wind tunnel experiments of Garg and Warhaft (1998).

As mentioned in the Introduction, in order to investigate the issue of anisotropy in the inertial subrange, it is necessary to have a sufficient inertial subrange interval so that the Reynolds stress has decayed (in wave number), leaving an inertial subrange that is free of the large scale uv coherence that results from the imposed shear. Figure 2 shows the correlation spectrum (from Garg and Warhaft 1998) of uv at $R_\lambda = 156$ and 391. For the higher R_λ case there is negligible Reynolds stress for the dissipation subrange; however it is not negligible in the inertial range. These correlation spectra should be compared with the boundary layer correlation spectra (done at $R_\lambda = 1450$) of Saddoughi and Veeravalli (1994) (Figure 20 of their paper). In that experiment there was approximately one decade in the inertial subrange for which

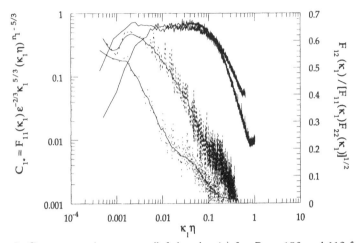

Figure 2. Compensated u spectra (left-hand axis) for $R_\lambda = 156$ and 112 for the wind-tunnel, approximately homogeneous shear experiment of Garg and Warhaft (1998). The right-hand axis shows the cross-correlation coefficient spectra for the two Reynolds numbers, the higher cross-correlation spectrum is for the lower R_λ case. Note that for the high R_λ case the cross-correlation spectrum is essentially zero at the beginning of the dissipation range ($\kappa_1\eta \sim 0.1$) but this is not the case for the low R_λ case. Thus for the high R_λ case, the dissipation scales are unaffected by the large scale Reynolds stress. From Garg and Warhaft (1998). Reproduced with permission from *Physics of Fluids* **10** p. 662. (1998), American Institute of Physics.

the coherence between u and v was negligible. Unfortunately $S_{\partial u/\partial y}$ was not measured in this well conditioned flow.

For the Garg and Warhaft (1998) experiment $S_{\partial u/\partial y}$ (Figure 3) was in the range 0.4 to 0.5, for R_λ in the range 300 to 400 (there is inevitable scatter in these measurements). This is a strong indicator that the flow is locally anisotropic the dissipation scales. Yet Garg and Warhaft (Figure 3) found that $S_{\partial u/\partial y}$ was decreasing with R_λ, from a value of approximately 0.8 at $R_\lambda = 156$. The empirical fit to their data is $S_{\partial u/\partial y} = 15.4R_\lambda^{-0.6}$; $150 \le R_\lambda \le 390$. Is $S_{\partial u/\partial y}$ decreasing to zero, albeit at very high Reynolds number, therefore confirming the local isotropy hypothesis, or is the value $S_{\partial u/\partial y} \sim 0.4$ to 0.5 the high Reynolds number limit, and the initial decrease is merely a low Reynolds number effect? We lean towards the latter. As shown above (see Figure 2), at low Reynolds numbers the dissipation scales are contaminated by the large scale anisotropy that is due to the Reynolds stress. This should enhance the derivative skewness. By $R_\lambda \sim 400$ this contamination is absent. Here the small scales are independent of the large scales. So we suggest the skewness has settled to its high Reynolds number limit. (We note that the *longitudinal* skewness $S_{\partial u/\partial x} \sim -0.4$ for high Reynolds number

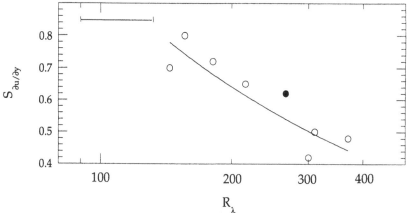

Figure 3. The transverse velocity derivative as a function of R_λ. Open symbols are from Garg and Warhaft (1998), the closed symbol is from Tavoulavis and Corrsin (1981b) and the bar-line is from the numerical results of Pumir (1996). From Garg and Warhaft (1998). Reproduced with permission from *Physics of Fluids* **10** p. 662. (1998), American Institute of Physics.

turbulence.) Nevertheless, more experiments at higher Reynolds number are needed. We are attempting to answer the above question by using an active grid in conjunction with a shear generator. We have achieved an R_λ of nearly 1000 in the presence of weak shear. Here the inertial range is significant – greater than two decades – and the Reynolds stress is essentially zero for a decade at the higher wave numbers of the inertial subrange i.e. the situation is similar to that of the (boundary layer) experiment of Saddoughi and Veeravalli (1994). Thus in this experiment we will be able to examine inertial as well as dissipation scale anisotropy.

If the result does indeed hold, the implications are important. From a conceptual viewpoint the return to isotropy hypothesis would have to be abandoned and the dynamics of the cascade would need to be re-interpreted. For the scalar situation, with the ramp cliff structure, there is a direct interaction between the large and small scales: the cliffs have the magnitude of the scalar variance, yet they occur over a dissipative length. The numerical simulations (Pumir and Shraiman 1995) indicate that the velocity field may be behaving in a similar manner, i.e. that there are sharp gradient sheets or fronts that may be characterized as ramp-cliffs. More insight into the velocity anisotropy may be achieved by determining the three-point correlation. Work towards this goal in progress.

We return to the connection between anisotropy and intermittency, and to the differences between the scalar and velocity fields. As noted above, for the scalar field the derivative skewness is of order 1, with no apparent decrease with R_λ (Sreenivasan 1991). It is also well known (Sreenivasan and Antonia 1997) that intermittency effects in the scalar structure function (manifest in its departure from K41) are first observed at the third order. On the other hand, for the velocity structure function, the data show (Sreenivasan and Antonia 1997) that there is no significant departure from K41 until the 5th or 6th order. It is quite feasible, then, that if the intermittency and anisotropy are connected, this will not become properly apparent for the velocity field until the 5th (or possibly even the 7th) moment. Thus even if $S_{\partial u/\partial y}$ does in fact continue to decrease with increasing R_λ (to zero at very high R_λ), the hyperskewness,

$$HS_{\partial u/\partial y} \equiv \frac{\langle (\partial u/\partial y)^5 \rangle}{\langle (\partial u/\partial y)^2 \rangle^{5/2}}$$

may be non-zero at high Reynolds numbers. This would constitute a violation of the return to isotropy hypothesis albeit at a higher order. Clearly the matter can only be resolved by further experiments.

Acknowledgements

I thank S. Garg, L. Mydlarski, and X. Shen. The work was funded by the US Department of Energy. I also thank Christos Vassilicos and members of the Issac Newton Institute for an enjoyable visit to the Workshop on Turbulence.

References

Champagne, F.H. (1978) 'The fine-scale structure of the turbulent velocity field', *J. Fluid Mech.* **86** 67–108.

Chen S., Kraichnan R.H. (1998) 'Simulations of a randomly advected passive scalar field', *Phys. Fluids* **10** 2867–2884.

Frisch U. (1995) *Turbulence: the Legacy of A. N. Kolmogorov*. Cambridge University Press.

Garg S., Warhaft Z. (1998) 'On small scale statistics in a simple shear flow', *Phys. Fluids* **10** 662–673.

Holzer M., Siggia E.D. (1994) 'Turbulent mixing of a passive scalar', *Phys. Fluids* **6** 1820–1837.

Lumley, J.L. (1967) 'Similarity and the turbulent energy spectrum', *Phys. Fluids* **10** 855.

Mydlarski L., Pumir A., Shraiman B.I., Siggia E.D., Warhaft Z. (1998) 'Structures and multipoint correlators for turbulent advection: predictions and experiments', *Phy. Rev. Lett.* **81** 4373–4376.

Mydlarski L., Warhaft Z. (1998a) 'Passive scalar statistics in high-Péclet–number grid turbulence', *J. Fluid Mech.* **358** 135–175.

Mydlarski L., Warhaft Z. (1998b) 'Three-point statistics and the anisotropy of a turbulent passive scalar', *Phys. Fluids* **10** 2885-2894.

Nelkin, M. (1994) 'Universality and scaling in fully developed turbulence', *Adv. Phys.* **43** 143-181.

Pumir, A. (1996) 'Turbulence in homogeneous shear flows', *Phys. Fluids* **8** 3112-3127.

Pumir, A., Shraiman, B.I. (1995) 'Persistent small scale anisotropy in homogeneous shear flows', *Phys. Rev. Lett.* **75** 3114-3117.

Saddoughi, S.G., Veeravalli, S.V. (1994) 'Local isotropy in turbulent boundary layers at high Reynolds number', *J. Fluid Mech.* **268** 333-372.

Sreenivasan, K.R. (1991) 'On local isotropy of passive scalars in turbulent shear flows', *Proc. Roy. Soc.* **A 434** 165-182.

Sreenivasan, K.R., Antonia, R.A. (1997) 'The phenomenology of small-scale turbulence', *Ann. Rev. Fluid Mech.* **29** 435-472.

Stewart, R.W. (1969) 'Turbulence and waves in stratified atmosphere', *Radio Science* **4** (12) 1269-1278.

Tavoularis, S., Corrsin, S. (1981a) 'Experiments in nearly homogeneous turbulent shear flow with a uniform mean temperature gradient. Part 1', *J. Fluid Mech.* **104** 11-347.

Tavoularis, S., Corrsin, S. (1981b) 'Experiments in nearly homogeneous turbulent shear flow with a uniform mean temperature gradient. Part 2. The fine structure', *J. Fluid Mech.* **104** 349-367.

Tong C., Warhaft Z. (1994) 'On passive scalar derivative statistics in grid turbulence', *Phys. Fluids* **6** 2165-2176.

Townsend, A.A. (1956) *The Structure of Turbulent Shear Flow*. Cambridge University Press.

Warhaft, Z., Lumley, J.L. (1978) 'An experimental study of the decay of temperature fluctuations in grid-generated turbulence', *J. Fluid Mech.* **88** 659-684.

Near-Singular Flow Structure: Dissipation and Eduction

J.C. Vassilicos

1 Introduction

Because turbulence is of a statistical nature, a lot of research on turbulent flows has concentrated on their statistics irrespective of the actual spatio-temporal flow structure of turbulence realisations. However the nature of this spatio-temporal flow structure is of central importance to the understanding of certain pivotal properties of the turbulence such as interscale energy transfer and the dependence of dissipation rates on molecular viscosity and diffusivities. It is not clear yet what the spatio-temporal structure of turbulence realisations is, and much research is and should be devoted to this question. But it can be argued that there must be near-singular flow structure in the turbulence. It is the object of this paper to briefly review the argument for near-singularities and then give a summary overview of some preliminary research carried out in the direction of understanding the dissipative properties of certain near-singular flow structures. Preliminary results relating to interscale transfer properties of near-singular flow structures may be found in Pedrizzetti & Vassilicos (2000). A discussion of the statistics of near-singularities in the turbulence requires the knowledge of what types of near-singularities actually exist in the turbulence and how the dynamics and spectral properties of specific near-singular flow structures relate to the average dynamical and spectral properties of the entire turbulence. To investigate the nature of near-singularities in turbulent flows new tools are needed for educing and characterising near-singular flow features in turbulent velocity fields and we present in some detail one such new tool which is suited to the post-processing analysis of 3-D Direct Numerical Simulation (DNS) data. The issue of the statistics of near-singularities is also briefly discussed by way of an example just to demonstrate the significant effects that different ways of calculating statistics can have on turbulence scalings.

It must be emphasised that the relevance to turbulence of the results summarised here rests ultimately on the outcome of future studies concerning the statistics of near-singularities and their collective dynamics in turbulent flows. Nevertheless, the results in this paper are interesting by themselves and form a significant part of two newly emerging fields: the physics of fractals and spirals, and the eduction of near-singularities.

2   The physics of fractals and spirals

Various universality and self-similarity arguments lead to $k^{-5/3}$ high-wave-number power spectra for turbulent scalar and velocity fields and experiments indeed confirm, at the very least, that these spectra decay with increasing wavenumber k at a rate strictly slower than k^{-2} (see, for example, Maurer *et al.* 1994 and Sreenivasan 1991). As noted by Hunt & Vassilicos (1991), such power spectra indicate the existence of a near-singular flow structure in the turbulence that is either cusp-like, spiral-like or fractal-like. The reason for such a near-singular flow structure is as follows. If all the realisations of the turbulence were analytic in the limit where the Reynolds number Re tends to infinity, and if we make the reasonable assumption that all these realisations have similar integral length-scales, then the energy spectrum of the turbulence would decay with increasing wavenumber at a rate faster than any power law in the limit $Re \to \infty$. However, universality and self-similarity arguments and experimental results indicate that the energy spectrum does not decay so fast in this limit, which therefore means that turbulence realisations have a non-zero probability to be non-analytic in the limit $Re \to \infty$. The additional fact that the energy spectrum decays at a rate slower than k^{-2} in the limit $Re \to \infty$ first and then $k \to \infty$ (in that order!) implies that the non-analyticity of turbulence realisations cannot be accounted for by isolated discontinuities (or near-discontinuities since we are concerned with the limit $Re \to \infty$) in the velocity field or its derivatives. This non-analyticity must therefore be the result of different near-singular flow structures which we classify in the following three broad categories.

i. *Isolated cusp-like near-singularities*, examples of which are the point vortex and the Burgers vortex and other cusp-like solutions of the Navier-Stokes equation (see Flohr & Vassilicos 1997).

ii. *Isolated spiral-like near-singularities*, examples of which are provided by the Lundgren family of asymptotic spiral vortex sheet solutions of the Navier-Stokes equation (see Lundgren 1982, Angilella & Vassilicos 1999).

iii. *Non-isolated near-singularities*, which may be cusp-like or spiral-like but what matters most is that they are non-isolated and therefore have a fractal-like property (see Vassilicos & Hunt 1991). Non-isolated means that there are singularities arbitrarily close to any singularity, and if the singularities are replaced by near-singularities because of finite Reynolds number effects, then every near-singularity is surrounded by other near-singularities over many length-scales ranging from a large integral scale to a minimal inner scale representative of viscous or diffusive smoothing. Whereas isolated singularities cannot have a purely random inner structure, non-isolated singularity structures can be either random or deterministic (or a mix of both). For example, a Cantor set of singularites forms a deterministic structure of non-isolated near-singularities. However, a spatially random distribution of singularities also forms a non-isolated singularity structure. Belcher & Vassilicos (1997) modelled the equilibrium range of wind-waves on the sea surface as a random spatial distribution of waves with discontinuous slopes. In their model the near-singular waves are characterised by an outer size or length-scale of their own, and the distribution of these outer sizes is given by a power law which, as it turns out, determines the wind-wave power spectrum.

By the same type of reasoning, black-or-white tracer (scalar) fields with a well-defined grey interface between black and white regions and a high-wavenumber power spectrum which decays slower than k^{-2} must have near-singularities in the geometry of the interface that are either isolated spiral-like or fractal-like (non-isolated). The dissipative properties of such scalar interfacial patterns are examples of dissipative properties of generic scalar near-singular flow structures that may be generated by either steady vortices in the case of spirals, or chaotic advection in the case of fractals (see Fung & Vassilicos 1991, Vassilicos & Fung 1995). Other examples are the dissipative properties of spiral vortex sheets which also have a well-defined interfacial structure corresponding to high gradients of velocity and can also have well-defined k^{-p} power-spectra over a large range of wavenumbers k. The study of the dissipative properties of spiral and fractal fields seems to be still in its infancy and may be a necessary prerequisite for the understanding of turbulence dissipative properties. Such a study is but one aspect of a broader field of study, the physics of fractals and spirals (Berry 1979, 1981, 1989, van den Berg 1994, Fleckinger et al. 1995, Gurbatov & Crighton 1995, Lapidus et al. 1996), also currently in its infancy.

Concerning the dissipative properties of spiral and fractal structures, the following broad principles can be drawn from the calculations of Vassilicos & Hunt (1991), Vassilicos (1995), Flohr & Vassilicos (1997), Angilella & Vassilicos (1998) and Angilella & Vassilicos (1999).

1. Dissipation is faster where the field is less autocorrelated in space. It is slower where the field is more autocorrelated in space.

2. On–off (black-or-white) fields with fractal or spiral interfacial structure are more autocorrelated in space for lower values of the fractal dimension (Kolmogorov capacity) and less autocorrelated in space for higher values of the fractal dimension of the interface. An increase in the degree in which an interface, whether spiral or fractal, is space-filling leads to a faster spatial decorrelation of the on–off scalar field.

3. A non-alternating field of delta functions on a fractal or spiral set of points on the 1-D axis is more autocorrelated in space for *higher* values of the fractal dimension of the set of points. It is less autocorrelated for *lower* values of this fractal dimension. This property of delta functions is in stark contrast with the corresponding property of on–off functions. The autocorrelation properties of non-alternating delta functions on a fractal or spiral curve in the 2-D plane or on a fractal or spiral curve or surface in 3-D space are not a trivial extension of the corresponding 1-D properties and remain the object of current research in the turbulence and mixing group in DAMTP.

4. From points 1 and 2 above it follows that on–off fields with well-defined fractal or spiral interfaces dissipate faster for more space-filling and slower for less space-filling interfaces. This qualitative conclusion can be made quantitative as follows: letting $\theta(\mathbf{x}, t)$ be the scalar field, initially sharply on–off (i.e. $\theta_0(\mathbf{x}) = \theta(\mathbf{x}, 0) = 0$ or 1) across a sharp fractal or spiral interface and taking averages over space (overbar),

$$\frac{\overline{\theta_0^2} - \overline{\theta^2}(t)}{\overline{\theta_0^2}} \sim (\eta(t)/L)^{E-D} \tag{1}$$

where $\eta(t)$ is a diffusive microscale at time t, L is an integral length-scale, E is the Euclidian dimension of the embedding space and D is the fractal dimension of the interface $(E - 1 \le D < E)$. The molecular diffusivity of the scalar θ being κ, the microscale $\eta(t)$ is equal to $\sqrt{\kappa t}$ when θ evolves under the action of diffusion alone,

$$\frac{\partial}{\partial t}\theta = \kappa \nabla^2 \theta, \tag{2}$$

but $\eta(t)$ can have a different dependence on time t when θ evolves under the combined action of both advection and diffusion,

$$\frac{\partial}{\partial t}\theta + \mathbf{u} \cdot \nabla \theta = \kappa \nabla^2 \theta. \tag{3}$$

In the case of equation (2) the interface is initially given a fractal or spiral structure with dimension D whereas in the case of equation (3) it is the velocity field $\mathbf{u}(\mathbf{x}, t)$ that generates a fractal or spiral interface from an initially regular interfacial geometry. When $\mathbf{u}(\mathbf{x}, t)$ is induced by a steady 2-D vortex and is purely azimuthal with a differential angular frequency $\Omega(r) = \Omega_0(r/L)^{-\alpha}$ where $\alpha > 1$, r is the radial distance to the centre of the vortex and Ω_0 is the angular frequency at $r = L$, then the interface adopts a spiral structure of dimension $D = 2\alpha/(1 + \alpha)$ at times $t \gg \Omega_0^{-1}$ and

$$\eta(t) = \Omega_0 t \sqrt{\kappa t} = \Omega_0 \kappa^{1/2} t^{3/2}. \tag{4}$$

In both cases of equations (2) and (3), the scaling (1) is valid as long as the fractal or spiral structure of the interface remains well-defined over a sufficient range of length-scales, that is for times t such that $\eta(t) \ll L$. When molecular diffusion acts alone without advection, $\eta(t) \ll L$ implies $t \ll L^2/\kappa$, but when molecular diffusion acts concurrently with the azimuthal advection

of the steady vortex, $\eta(t) \ll L$ implies $t \ll \Omega_0^{-1} Pe^{1/3}$ where $Pe \equiv (\Omega_0 L^2/\kappa$.
At this stage we should specify that (1) with (4) is valid for $Pe \gg 1$.

The dissipation of scalar variance in a steady vortex is therefore accelerated by two different effects, one local and one global. The local effect is the effect of the local shear caused by the differential rotation which acts locally on the interface to enhance gradients and thereby accelerate dissipation (Moffatt & Kamkar 1983, Rhines & Young 1983). This effect leads to an increase of the local diffusive microscale $\eta(t)$ from $\sqrt{\kappa t}$ to $\Omega_0 t \sqrt{\kappa t}$ and thereby to a reduction of the length of time for which the spiral interface is well defined from $\Omega_0^{-1} Pe$ to $\Omega_0^{-1} Pe^{1/3}$. The global effect is that of the space-filling geometry of the spiral interface and is manifest in (1). The more space-filling the spiral interface, the larger the value of D and the faster the decrease of the scalar variance in time. The dissipation rate of scalar variance is $\chi \equiv -\frac{d}{dt}\overline{\theta^2}(t)$ and (1) implies

$$\chi \sim \frac{\overline{\theta_0^2}}{t}(\eta(t)/L)^{E-D} \tag{5}$$

in the limit $Pe \gg 1$, which means that the dependence of χ on κ is progressively weaker as $D \to E$ and χ does not depend on κ in the limit $Pe \gg 1$ only if $D = E$, i.e. only if the spiral is completely space-filling.

5. From points 1 and 3 above, it follows that non-alternating delta functions on a fractal or spiral set of points on the 1-D axis evolving under the action of diffusion alone (equation (2)) dissipate *slower* for more space-filling geometries. Quantitatively, for an initial field $\theta_0(x) = \sum_i \delta(x - x_i)$ where the points x_i are those of a homogeneous fractal set with a fractal dimension D ($0 \le D < 1$) that is well-defined over a range of scales between η_0 and L ($\eta_0 \ll L$),

$$\frac{\overline{\theta^2}(t)}{\overline{\theta^2}(\eta_0^2/\nu)} \sim (\eta(t)/L)^{-1+D} \tag{6}$$

for $\eta_0^2/\nu \ll t \ll L^2/\nu$. The diffusive microscale $\eta(t)$ is equal to $\sqrt{\kappa t}$. (See Angilella & Vassilicos (1998) for a definition of 'homogeneous fractal' and for the 1-D spiral case which is qualitatively broadly similar but quantitatively different.) It is instructive to contrast (6) with the corresponding result for *alternating* delta functions on a fractal or spiral (e.g. $x_i \sim i^{-a}$ with $a > 0$) set, i.e. $\theta_0(x) = \sum_i (-1)^i \delta(x - x_i)$, where dissipation of scalar variance is *faster* for more space-filling geometries. Quantitatively,

$$\frac{\overline{\theta^2}(t)}{\overline{\theta^2}(\eta_0^2/\nu)} \sim (\eta(t)/L)^{-1-D} \tag{7}$$

for $\eta_0^2/\nu \ll t \ll L^2/\nu$.

6. However, the dissipation of 2-D spiral vortex sheets is different. Lundgren (1982) has shown that $\omega(r, \phi, t) = \gamma(r)\delta(\phi - \Omega(r)t)$ is a long-time asymptotic solution of the 2-D Helmholtz equation

$$\frac{\partial}{\partial t}\omega + \mathbf{u} \cdot \nabla \omega = \nu \nabla^2 \omega \qquad (8)$$

when $\nu = 0$ and provided that $\Omega(r)$ is a monotonic decreasing function of r and $r\gamma(r) = 2\pi \frac{d}{dr} r^2 \Omega(r)$. The latter condition is a result of the coupling $(0, 0, \omega) = \nabla \times \mathbf{u}$ which makes the difference between (8) and (3). Setting $\nu \neq 0$, and chosing $\Omega(r) = \Omega_0 (r/L)^{-\alpha}$ with $\alpha > 0$ so that the spiral's co-dimension is $D_0 = \alpha/(1 + \alpha)$, the average enstrophy of this solution decays as follows:

$$\frac{\overline{\omega^2}(t)}{\Omega_0^2} \sim (\eta(t)/L)^{2-4D_0} \qquad (9a)$$

for $D_0 \geq 3/4$ and

$$\frac{\overline{\omega^2}(t)}{\Omega_0^2} \sim (\eta(t)/L)^{-1} \qquad (9b)$$

for $D_0 \leq \frac{3}{4}$. The viscous microscale is given by (4) where κ should be replaced by ν, and equations (9) are valid for as long as $\eta(t) \ll L$, that is $t \ll \Omega_0^{-1} Re^{1/3}$ where $Re \equiv \Omega_0 L^2/\nu$. Lundgren's spiral vortex sheet solution of (8) is valid for large times in the sense that $\Omega_0^{-1} \ll t$, and our conclusions (9) are therefore valid in the limit $Re \gg 1$.

The dissipation of enstrophy is neither decelerated nor accelerated by the space-filling property of the spiral when $D_0 \leq 3/4$ and is in fact accelerated rather than decelerated by this space-filling property when $D_0 > 3/4$. This may be a surprising conclusion because the dissipation of non-alternating delta functions on a 1-D spiral or fractal set of points is decelerated by the space-filling property of the spiral (Angilella & Vassilicos 1998). For some understanding of this conclusion, it must be noted that the high-wavenumber energy spectrum of the Lundgren spiral vortex sheet considered here is $E(k) \sim k^{-2}$ for $D_0 \leq 3/4$ but $E(k) \sim k^{-5+4D_0}$ for $D_0 \geq 3/4$ (see Angilella & Vassilicos 1999). Hence, the autocorrelation property of the Lundgren spiral vortex sheet does not depend on D_0 if $D_0 \leq 3/4$ but does depend on D_0 otherwise, and in fact in a way that makes this spiral vortex sheet *less* autocorrelated as D_0 increases. Hence, conclusions (9) are consistent with point 1.

3 Eduction of near-singularities

Recent laboratory and numerical experiments have revealed the existence of vortex tubes in the small scales of the turbulence (see Cadot *et al.* 1995, Jiménez & Wray 1998, Flohr 1999 and references therein). Two questions arise. Firstly, do some of the near-singularities of the turbulence reside in these vortex tubes? And secondly, is it possible to define vortex tube flows topologically in such a way as to detect vortex tubes irrespective of their enstrophy level and independently of their enstrophy profile? If an entity such as a vortex tube exists in a turbulent flow, it is not necessarily characterised by high enstrophy throughout its extent, even if there is a near-singularity in the vortex tube. However, vortex tubes in real flows are generally characterized by spiral streamlines surrounding the core. Vassilicos & Brasseur (1996) have developed a systematic algorithm for the detection of spiral-like streamlines in Direct Numerical Simulations (DNS) of small-scale turbulence. This algorithm successfully identifies many vortex tubes in the small-scale turbulence irrespective of the enstrophy profile across and modulations along the vortex tube and gives direct access to the topology of the flow in and around these vortex tubes. An attempt to address the near-singularity question on the basis of this algorithm can be made by measuring the Kolmogorov capacity of streamlines around turbulent vortex tubes on the basis of the following theorem.

In an axisymmetric incompressible flow with bounded vorticity at infinity, when the Kolmogorov capacity D_K of a spiral-helical streamline is strictly larger than 1, there must exist a velocity singularity at the axis of symmetry of the flow.

The best way to explain this theorem is to sketch its proof. To prove it we prove the contrapositive statement: *in an axisymmetric, incompressible flow with bounded vorticity at infinity, if the velocity field is regular (no singularities) at the axis of symmetry, then the Kolmogorov capacity D_K of spiral-helical streamlines is necessarily equal to 1.*

A function $u(r)$ is regular around $r = 0$ if it can be Taylor expanded around $r = 0$, i.e. $u(r) = u(0) + ru'(0) + \frac{r^2}{2}u''(0) + \cdots$, where u' and u'' are, respectively, the first and second derivatives of u. In an axisymmetric flow, the velocity field is conveniently decomposed into azimuthal, radial and axial components u_ϕ, u_r and u_z and the spatial coordinates are r, the distance from the axis of symmetry, ϕ, the angle around the axis of symmetry, and z, the distance along the axis of symmetry. Our basic premises of axisymmetry and incompressibility imply the existence of a Stokes stream function $\psi(r, z)$, where

$$u_r(r, z) = -\frac{1}{r}\frac{\partial}{\partial z}\psi(r, z), \tag{10}$$

$$u_z(r, z) = \frac{1}{r}\frac{\partial}{\partial r}\psi(r, z). \tag{11}$$

The velocity field's regularity at the axis of symmetry means that, as $r \to 0$,

$$u_\phi(r, z) \approx r^m G(z), \tag{12}$$

$$\psi(r, z) \approx r^l F(z), \tag{13}$$

where $G(z)$ and $F(z)$ are regular functions of z for all values of z, m is an integer greater or equal to 1 for regularity and incompressibility, and l is an integer which must be greater or equal to 2 in order for u_z to be regular at the axis.

It turns out that regularity is too strong a requirement for the spiral-helical streamline's D_K to be equal to 1; *we prove that $D_K = 1$ provided that m and l are such that $m > -1$ and $l > 0$ and not necessarily integers*. Hence, there can exist singular flows where $D_K = 1$. But there can be no regular flow where $D_K > 1$. (Note that the existence of a spiral-helical streamline is assumed.)

The boundedness of vorticity at infinity and the regularity of $G(z)$ imply that $G(z)$ is a bounded function of z. By examining the spiral-helical streamline in the (r, z) plane we find that $F(z)$ must be a monotonically increasing function of z. Indeed, as $r \to 0$ *along* the spiral-helical streamline,

$$\frac{dr}{dz} = \frac{u_r}{u_z} \approx -\frac{r}{lF(z)}\frac{dF}{dz}(z) \tag{14}$$

which can be integrated to yield

$$r^l F(z) \approx \text{Const.} \tag{15}$$

as $r \to 0$ where r and z are streamline coordinates and one is therefore a function of the other (a more detailed discussion is given in Vassilicos & Brasseur 1996). It follows that $F(z) \to \infty$ as $r \to 0$ on the spiral-helical streamline, and since $F(z)$ is regular, $F(z)$ must be a monotonically increasing function of z.

Examining the spiral-helical streamline in the azimuthal plane (r, ϕ), we see that

$$\frac{d\phi}{dr} = \frac{u_\phi}{ru_r} \approx -r^{m-1}\frac{G(z)}{\frac{dF}{dz}(z)} \tag{16}$$

and

$$2\pi = \int_{n2\pi}^{(n+1)2\pi} d\phi \approx -\int_{r_n}^{r_{n+1}} dr\, r^{m-1}\frac{G[z(r)]}{\frac{dF}{dz}[z(r)]} \sim \int_{z_n}^{z_{n+1}} dz \frac{G(z)}{F^\beta(z)}, \tag{17}$$

where r_n and z_n are radial and axial coordinates of the spiral-helical streamline on the nth turn of the spiral, and $\beta = (m+1)/l$. Since β is strictly positive, $G(z)/F^\beta(z)$ is monotonically decreasing as $z \to \infty$, and therefore $\Delta z_n \equiv z_{n+1} - z_n \to \infty$ as $n \to \infty$.

The distance Δ_n between successive turns after the nth turn of the spiral streamline in 3-D space is given by $\Delta_n^2 = \Delta z_n^2 + \Delta r_n^2$ where $\Delta r_n \equiv r_n - r_{n+1}$. Because $\Delta z_n \to \infty$ as $n \to \infty$, $\Delta_n \to \infty$ as $n \to \infty$ and therefore, *no accumulation of spiral turns exists on the spiral streamline, which implies that* $D_K = 1$ (see Vassilicos & Hunt 1991). Indeed the Kolmogorov capacity D_K of streamlines is measured by covering the 3-D space with boxes, and even though the projection of the spiral-helical streamline on the azimuthal plane does accumulate ($\Delta r_n \to 0$ as $n \to \infty$), the spiral-helical streamline in 3-D space does not. It converges towards the central axis of symmetry, but as it does, the distance Δ_n between successive turns increases indefinitely. Hence, there is no accumulation that can be reflected in a non-integral D_K.

We have proved that *in an axisymmetric incompressible flow with bounded vorticity at infinity, if there is no singularity on the axis of symmetry of the flow or if there is a singularity of the type* (12) *and* (13) *where m and/or l are non-integral and* $m > -1$, $l > 0$, *then the Kolmogorov capacity* D_K *of a spiral-helical streamline is equal to 1.* The theorem of self-similar streamlines follows:

In an axisymmetric incompressible flow with bounded vorticity at infinity, when the Kolmogorov capacity D_K *of a spiral-helical streamline is strictly larger than 1, there must be a singularity on the axis of symmetry of the flow which is not a singularity of the type* (12) *and* (13) *where m and/or l are non-integral and such that* $m > -1$ *and* $l > 0$.

4   Application to DNS of homogeneous isotropic turbulence

Evidence has been obtained from a series of DNS of both decaying and forced homogeneous isotropic turbulence by Vassilicos & Brasseur (1996) and Flohr (1999), some of the low Reynolds number decaying simulations having exceptionally fine small-scale resolution ($k_{max}\eta \approx 6$ in 512^3 DNS at $Re_\lambda \approx 21$), all demonstrating that a significant number of streamlines around small-scale vortex tubes have a well-defined $D_K > 1$ in a range of length-scales between the Taylor microscale λ and the Kolmogorov length-scale η. In these simulations the Taylor-scale Reynolds number Re_λ was varied between about 20 and 130, and typically one decade of length-scales was found between λ and η even at the lower Reynolds numbers. The average value of D_K is between 1.4 and 1.5 irrespective of Reynolds number and resolution (as long as the DNS resolution is sufficiently good of course). Furthermore, evidence based on the spatial correlation of enstrophy with viscous force indicates that the spatial vorticity profile across the vortex tubes is not a well-resolved Gaussian even when the resolution of the DNS is half η. The flow is therefore not totally smoothed out by viscosity around scales of order η in some of the vortex tubes. Details can be found in Vassilicos & Brasseur (1996) and Flohr (1999).

These studies are an attempt to suggest that there exist near-singular vortex tubes at the small scales of homogeneous isotropic turbulence independently of Reynolds number. The new kinematics proposed to characterise these particular near-singularities are based on the Kolmogorov capacity of streamlines and the theorem of self-similar streamlines.

5 Statistics

Concerning assumptions on how local flow structures are related to global statistics, it may be interesting to close this paper by recalling Lundgren's (1982) assumption that a time-average over the history of one flow structure is equivalent to a spatial average over many such flow structures at different stages of their history. This assumption is not without consequences. A recent calculation by Khan & Vassilicos (1999) shows that for a scalar field advected in a 2-D sea of identical non-interacting axisymmetric vortices characterised by the same differential angular frequency $\Omega(r) = \Omega_0(r/L)^{-\alpha}$, the scaling of the structure functions $\langle \delta\theta^n(r) \rangle$ is

$$\langle \delta\theta^n(r) \rangle \sim r^{2(1-D_0)} \tag{18}$$

if the averaging operation represented by the brackets is taken over space and a homogeneous and isotropic distribution of vortices, but is

$$\langle \delta\theta^n(r) \rangle \sim \text{Const.} + \log r \tag{19}$$

if the average operation also includes an integration over time. These scaling laws hold in a range of scales r bounded from below by a microscale that depends on κ, and the co-dimension D_0 ($0 \leq D_0 < 1$) is that of the spiral structures generated by each one of these identical vortices. The scalings (18) and (19) are obtained by solving the advection-diffusion equation (3) and it is striking to see that they differ. An averaging assumption such as the one of Lundgren (1982) can therefore very significantly alter the predicted scalings, and even remove the dependence on the geometry (the co-dimension) of the spiral

However, the scalings (18) and (19) resemble in that the scaling exponents of these structure functions do not depend on the order n. This result is reminiscent of the structure function results obtained by Vassilicos (1992) without solving equation (3). Indeed, Vassilicos (1992) calculated the structure functions of on–off fields characterised by either spiral or fractal interfaces and found, as in (18) and (19), that the scaling exponents of these structure functions do not depend on the order n and are a function of D_0. The scalings (18) and (19) must therefore be seen as consequences of the spiral interfacial structure generated by the action of the vortices on the scalar field but also of the nature of the statistics involved.

Acknowledgements

I am grateful for financial support of the work described here to the Royal Society, EPSRC, NERC, European Commission and Hong Kong Research Grant Council. Also, I would like to thank the Isaac Newton Institute for its hospitality during the turbulence programme from 6 January to 2 July 1999.

References

Angilella, J.R. & Vassilicos, J.C. (1998) 'Spectral, diffusive and convective properties of fractal and spiral fields', *Physica D* **124** 23.

Angilella, J.R. & Vassilicos, J.C. (1999) 'Time-dependent geometry and energy distribution in a spiral vortex layer', *Phys. Rev. E* **59** (5), 5427.

Belcher, S.E. & Vassilicos, J.C. (1997) 'Breaking waves and the equilibrium range of wind-wave spectra', *J. Fluid Mech.* **342** 377.

van den Berg, M. (1994) 'Heat content and Brownian motion for some regions with a fractal boundary', *Probab. Theory Related Fields* **100** 439.

Berry, M.V. (1979) 'Diffractals', *Phys. A: Math. Gen.* **12** (6), 781

Berry, M.V. (1981) 'Diffractal echoes', *Phys. A: Math. Gen.* **14** 3101.

Berry, M.V. (1989) 'Falling fractal flakes', *Physica D* **38** 29.

Cadot, O., Douady, S. & Couder, Y. (1995) 'Characterisation of the low-pressure filaments in a three-dimensional turbulent shear flow', *Phys. Fluids* **7** 630.

Fleckinger, J., Levitin, M. & Vassiliev, D. (1995) 'Heat equation on the triadic von Koch snowflake: asymptotic and numerical analysis', *Proc. London Math. Soc. (3)* **71** 372.

Flohr, P. (1999) *Small-Scale Flow Structure in Turbulence: Fundamentals and Numerical Models.* PhD thesis, University of Cambridge.

Flohr, P. & Vassilicos, J.C. (1997) 'Accelerated scalar dissipation in a vortex', *J. Fluid Mech.* **348** 295.

Fung, J.C.H & Vassilicos, J.C. (1991) 'Fractal dimensions of lines in chaotic advection', *Phys. Fluids* **11** 2725.

Gurbatov, S.N. & Crighton, D.G. (1995) 'The nonlinear decay of complex signals in dissipative media', *Chaos* **5** (3), 524.

Hunt, J.C.R. & Vassilicos, J.C. (1991) 'Kolmogorov's contributions to the physical understanding of small-scale turbulence and recent developments', *Proc. R. Soc. Lond. A* **434** 183.

Jiménez, J. & Wray, A.A. (1998) 'On the characteristics of vortex filaments in isotropic turbulence', *J. Fluid Mech.* **373** 255.

Khan, M.A.I. & Vassilicos, J.C. (1999) 'The scalings of scalar structure functions in a velocity field with a coherent vortical structure', *Phys. Rev. E*, submitted.

Lapidus, M., Neuberger, J., Renka, R. & Griffith, C. (1996) 'Snowflake harmonics and computer graphics: numerical computation of spectra on fractal drums', *Int. J. Bifurcation Chaos* **6** 1185.

Lundgren. T.S. (1982) 'Strained spiral vortex model for turbulent fine structure', *Phys. Fluids* **25** 2193.

Maurer, J., Tabeling, P. & Zocchi, G. (1994) 'Statistics of turbulence between two counter-rotating disks in low temperature helium gas', *Europhys. Lett.* **26** 31.

Moffatt, H.K. & Kamkar, H. (1983) 'The time-scale associated with flux expulsion'. In *Stellar and Planetary Magnetism* (ed. A.M. Soward), pp. 91–97. Gordon & Breach.

Pedrizzetti, G. & Vassilicos, J.C. (2000) 'Interscale transfer in two-dimensional compact vortices', *J. Fluid Mech.* **406**, 109–129.

Rhines, P.B. & Young, W.R. (1983) 'How rapidly is a passive scalar mixed within closed streamlines?', *J. Fluid Mech.* **133** 133.

Sreenivasan, K.R. (1991) 'On local isotropy of passive scalars in turbulent shear flows', *Proc. R. Soc. Lond. A* **434** 165.

Vassilicos, J.C. (1992) 'The multispiral model of turbulence and intermittency'. In *Topological Aspects of the Dynamics of Fluids and Plasmas* (ed. H.K. Moffatt *et al.*), 427–442. Kluwer.

Vassilicos, J.C. (1995) 'Anomalous diffusion of isolated flow singularities and of fractal or spiral structures', *Phys. Rev. E* **52** R5753.

Vassilicos, J.C. & Brasseur, J.G. (1996) 'Self-similar spiral flow structure in low Reynolds number isotropic and decaying turbulence', *Phys. Rev. E* **54** (1), 467.

Vassilicos, J.C. & Fung, J.C.H. (1995) 'The self-similar topology of passive interfaces advected by two-dimensional turbulent-like flows', *Phys. Fluids* **7** (8), (1970.

Vassilicos, J.C. & Hunt, J.C.R. (1991) 'Fractal dimensions and spectra of interfaces with application to turbulence', *Proc. R. Soc. Lond. A* **435** 505.

Vortex Stretching versus Production of Strain/Dissipation

Arkady Tsinober

Abstract

A comparison between vortex stretching (VS) and production of strain/dissipation (PD) is made with the emphasis on the latter. These two processes are nonlocally interconnected, but weakly correlated. The energy cascade and its final result – dissipation are associated with the latter, i.e. with the quantity $-s_{ij}s_{jk}s_{ki}$ rather than with the enstrophy production $\omega_i\omega_j s_{ij}$. Moreover, vortex stretching suppresses the cascade and does not aid it, at least in a *direct* manner. On the contrary, it is the vortex *compression*, i.e $\omega_i\omega_j s_{ij} < 0$, that aids the production of strain/disspation and in this sense the 'cascade'.

Relation of VS and PD as well as of various alignments to the flow map of invariants of velocity derivatives tensor is given in qualitative terms.

1   Introductory notes, motivation

Velocity derivatives play an outstanding role in the dynamics of turbulence for a number of reasons. Their importance became especially clear since the papers by Taylor (1937, 1938)[1] and Kolmogorov (1941ab). Taylor emphasized the role of vorticity, whereas Kolmogorov stressed the importance of dissipation (strain).

Apart from vorticity and dissipation, looking at velocity derivatives is useful in a number of aspects as follows.

- The field of velocity derivatives is much more sensitive to the non-Gaussian nature of turbulence or more generally to its structure, and hence reflects more of its physics (Tsinober 1998c).

- In the Lagrangian description in a frame following a fluid particle, each point is a critical one, i.e. the direction of velocity is not determined. So

---

[1]Taylor (1937, 1938) was motivated by the assumption of von Karman (1937) *that the expression* $\sum_i \sum_k \omega_i\omega_k \frac{\partial u_i}{\partial u_k}$ (i.e. enstrophy production) *is zero* in the mean and that he (vK) *cannot see any physical reason for such a correlation.* Taylor (1937) has conjectured *that there is a strong correlation between* ω_3^2 *and* $\frac{\partial u_3}{\partial u_3}$ *so that* (the mean of) $\omega_3^2 \frac{\partial u_3}{\partial u_3}$ *is not equal to zero* (x_3 is directed along ω). He has shown that this is really the case (Taylor (1938)), and also expressed the view that *stretching of vortex filaments must be regarded as the principal mechanical cause of the the higher rate of disspation which is associated with turbulent motion.*

everything happening in its proximity is characterized by the velocity gradient tensor $A_{ij} = \partial u_i / \partial x_j$. For instance, local geometry/topology is naturally described in terms of critical points terminology (see Chertkov *et al.* (1999), Ooi *et al.* (1999) and references therein).

- There is a generic ambiguity in defining the meaning of the term *small scales* (or more generally scales) and consequently the meaning of the term *cascade* in turbulence research. The specific meaning of this term and associated interscale energy exchange/'cascade' (e.g. spectral energy transfer) is essentially decompostion/reperesentation dependent (for more details/discussion of this issue see Appendix I). Perhaps the only common factor in all decompostions/representations (D/R) is that the small scales are associated with the field of velocity derivatives. Therefore, it is natural to look at this field as the one *objectively* (i.e. D/R independent) representing the small scales. Indeed, the dissipation is associated precisely with the strain field s_{ij} both in Newtonian and non-Newtonian fluids.

The above mentioned reasons prompted us to study in some detail the processes associated with the field of velocity derivatives. In particular, along with vortex stretching and enstrophy production, of special interest is the production of strain. There are several reasons for this. First, though formally all the flow field is determined entirely by the field of vorticity the relation between the strain and vorticity is strongly nonlocal (e.g. Constantin (1994), Novikov (1968), Ohkitani (1994)); in many cases they are only weakly correlated. Second, energy dissipation is directly associated with strain and not with vorticity. Third, vortex stretching is essentially a proccess of interaction of vorticity and strain. Fourth, strain dominated regions appear to be the most active/nonlinear in a number of aspects (Tsinober (1998ab), Tsinober *et al.* (1999)). Finally, the energy cascade (whatever this means) and its final result – dissipation – are associated with predominant self-amplification of the rate of strain/production of dissipation and vortex compression rather than with vortex stretching. This last aspect is the main theme of this presentation.

2   Equations, notation, and some previous results

2.1   Equations and related things

We will need later the equations for vorticity, ω_i, and enstrophy, ω^2,

$$\frac{D\omega_i}{Dt} = \omega_j s_{ij} + \nu \nabla^2 \omega_i, \tag{1}$$

$$\frac{1}{2}\frac{D\omega^2}{Dt} = \omega_i\omega_j s_{ij} + \nu\omega_j\nabla^2\omega_j. \tag{2}$$

and the rate of strain tensor, s_{ik}, (Yanitsky (1982)) and the total strain, $s^2 \equiv s_{ij}s_{ij}$ (Brasseur and Lin (1995), Tsinober (1995))

$$\frac{Ds_{ij}}{Dt} = -s_{ik}s_{kj} - \frac{1}{4}(\omega_i\omega_j - \omega^2\delta_{ij}) - \frac{\partial^2 p}{\partial x_i\partial x_j} + \nu\nabla^2 s_{ij}, \tag{3}$$

$$\frac{1}{2}\frac{D2s^2}{Dt} = -2\left\{s_{ik}s_{kj}s_{ji} + \frac{1}{4}\omega_i\omega_j s_{ij} + s_{ij}\frac{\partial^2 p}{\partial x_i\partial x_j}\right\} + 2\nu s_{ij}\nabla^2 s_{ij}. \tag{4}$$

These equations clearly indicate that along with enstrophy, ω^2, and strain, $s^2 \equiv s_{ij}s_{ij}$, the third moments $\omega_i\omega_j s_{ij}$, $s_{ij}s_{jk}s_{ki}$, are the key quantities of turbulence dynamics. It is noteworthy that many aspects of the dynamics of velocity gradient tensor $\partial u_i/\partial x_j$ can be addressed via looking at its invariants: the second: $Q = 1/4(\omega^2 - 2s_{ij}s_{ij})$, and the third: $R = -1/3(s_{ij}s_{jk}s_{ki} + 3/4\omega_i\omega_j s_{ij})$, the first one: $P = \partial u_k/\partial x_k$ is vanishing due to incompressibility (see Chertkov et al. (1999), Ooi et al. (1999) and references therein). However it is not sufficient and, along with using the Q and R invariants, it is more transparent and physically meaningful in several respects to look directly at ω^2, s^2, $\omega_i\omega_j s_{ij}$, and $s_{ij}s_{jk}s_{ki}$. This is seen from the equations (3) and (4), which also show that the quantity $s_{ij}\frac{\partial^2 p}{\partial x_i\partial x_j}$, i.e. interaction of strain with pressure Hessian is of importance (there are two more: $\omega_i\omega_j\frac{\partial^2 p}{\partial x_i\partial x_j}$ and $s_{ik}s_{kj}\frac{\partial^2 p}{\partial x_i\partial x_j}$ in equations (5) and (6) below). Of course, formally the flow is determined entirely by the field of vorticity. However, due to the nonlocal relation between the rate of strain tensor and vorticity (e.g. Constantin (1994), Novikov (1968), Ohkitani (1994)) it is useful to look at the above mentioned quantities in parallel. Moreover, it appears that the dynamical equations for $\omega_i\omega_j s_{ij}$ and $s_{ij}s_{jk}s_{ki}$

$$\frac{D\omega_i\omega_j s_{ij}}{Dt} = \omega_j s_{ij}\omega_k s_{ik} - \omega_i\omega_j\frac{\partial^2 p}{\partial x_i\partial x_j} + \nu(2\omega_i s_{ij}\nabla^2\omega_j + \omega_i\omega_j\nabla^2 s_{ij}). \tag{5}$$

$$\frac{Ds_{ij}s_{jk}s_{ki}}{Dt} = 3\left\{-s_{ik}s_{kj}s_{il}s_{lj} + \frac{s_{ij}s_{ij}\omega^2 - \omega_j s_{ij}\omega_k s_{ik}}{4} - \frac{\partial^2 p}{\partial x_i\partial x_j} + \nu s_{ik}s_{kj}\nabla^2 s_{ij}\right\}. \tag{6}$$

are also instructive in several respects.

Since for homogeneous flows $\langle s_{ij}s_{ik}s_{kj}\rangle = -\frac{3}{4}\langle\omega_i\omega_j s_{ij}\rangle$ and $\langle s_{ij}\frac{\partial^2 p}{\partial x_i\partial x_j}\rangle = 0$ due to incompressibility it follows that the mean rate of production of strain/dissipation $-\langle 2(s_{ij}s_{ik}s_{kj} + \frac{1}{4}\omega_i\omega_j s_{ij} + s_{ij}\frac{\partial^2 p}{\partial x_i\partial x_j})\rangle = \langle\omega_i\omega_j s_{ij}\rangle$ is equal to that of enstrophy. Hence (and also due to $\langle\omega^2\rangle = 2\langle s_{ij}s_{ij}\rangle$) the choice of the coefficient 2 in (4), etc.

2.2 Vortex stretching and enstrophy production

We touch this aspect briefly. More details and references are given in Tsinober (1998ab) and Tsinober *et al.* (1999).

Ever since Taylor (1938) we have known that $\langle \omega_i \omega_j s_{ij} \rangle > 0$, i.e. vortex stretching prevails over vortex compressing (see also Betchov (1976), Tsinober (1998a) and references therein, and Appendix I). This basic phenomenon is closely associated with subtle geometrical relations such as strict alignment between vorticity ω_i and vortex streching vector $W_i \equiv \omega_j s_{ij}$, alignment between vorticity ω_i and the eigenvector, λ_2, of the rate of strain tensor, s_{ij}, corresponding to its intermediate eigenvalue Λ_2 (briefly intermediate eigenvector) and some others. However, enstrophy production is associated with two regions, characterised by alignment between between vorticity, ω_i, and the intermediate eigenvector, λ_2, and between vorticity, ω_i, and the largest eigenvector, λ_1. Moreover, the largest contribution to the enstrophy production comes from the regions with strong alignment between vorticity, ω_i, and the largest eigenvector, λ_1, and is associated with large curvature of vorticity lines and vorticity tilting, and large strain rather than with large enstrophy. The latter is true of all nonlinearities. It is noteworthy that the (approximate) balance between the *mean* enstrophy generation and its *mean* destruction via viscosity holds also in the enstrophy dominated regions. However, in the regions dominated by strain, the enstrophy generation is an order of magintute larger than both its destruction via viscosity and the enstrophy generation in the enstrophy dominated regions. Therefore most enstrophy generation occurs in the regions dominated by strain.

3 Generation of strain/dissipation

The appropriate level of dissipation moderating the growth of energy is achieved by the build up of strain of sufficient magnitude which is described by equation (4). It is seen from this equation that in the mean *the only term contributing positively to the production of strain/dissipation*, s^2, is the term $-s_{ij}s_{jk}s_{ki} = -(\Lambda_1^3 + \Lambda_2^3 + \Lambda_3^3) = -3\Lambda_1\Lambda_2\Lambda_3$, since $\langle s_{ij}s_{jk}s_{ki} \rangle = -3/4\langle \omega_i\omega_j s_{ij} \rangle$, and $\langle s_{ij} \frac{\partial^2 p}{\partial x_i \partial x_j} \rangle = 0$ due to homogeneity and incompressibilty. Moreoever, since, $\Lambda_1 > 0$ and Λ_2 is positively skewed, i.e. $\langle \Lambda_2^3 \rangle > 0$ the positivenes of $- \langle s_{ij}s_{jk}s_{ki} \rangle$ comes from the term $- \langle \Lambda_3^3 \rangle$. In other words, Λ_3 is doing most of the 'cascade', at least, one of the final results of the 'cascade' – dissipation of energy, which is directly associated with s_{ij} and not with ω_i. Hence, the cascade is directly associated with compressing/squeezing of fluid elements and not with (vortex) stretching. It is noteworthy that this idea is not entirely new: '*It is clear, therefore, that production of vorticity is associated essentially with* Λ_3 *and production of* ω_1 *and* ω_2. *This suggests that*

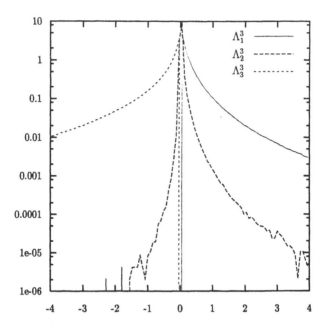

Figure 1. PDFs of cubed eigenvalues, Λ_i^3, of the rate of the strain
tensor s_{ij}, normalized on $\langle s^2 \rangle^{3/2}$.

the most important processes associated with production of vorticity and energy transfer resemble a jet collision and not the swirling of a contracting jet'
(Betchov (1956)). Betchov arrived at this conclusion by analysing the means
$\langle s_{ij} s_{jk} s_{ki} \rangle$ and $\langle \omega_i \omega_j s_{ij} \rangle$. Looking at the equation (4) it is seen that the above
conclusion is true of the production of strain, which is associated with Λ_3,
and with the 'jet collision' regions such as sheetlike structures as observed in
the laboratory (Frederiksen et al. (1996), Schwarz (1990)), and in numerical
experiments (Brachet et al. (1992), Boratav and Pelz (1997), Chen and Cao
(1997)). As for enstrophy production it is true in part: roughly two thirds
of its positive contribution occur in the 'jet collision' regions, the remaining
third happens in the 'swirling of a contracting jet' regions (Tsinober (1998)).
Also production of ω^2 requires s_{ij} and interaction between the two, but production of s_{ij} is in some sense less dependent on ω, though without vorticity
it is impossible.

   All the results shown in the sequel refer to a DNS simulation of NSE
decaying turbulence in a periodic box for the time moment(s) at which the
total enstrophy is (close to) maximal and $Re_\lambda \approx 80$. They are very similar

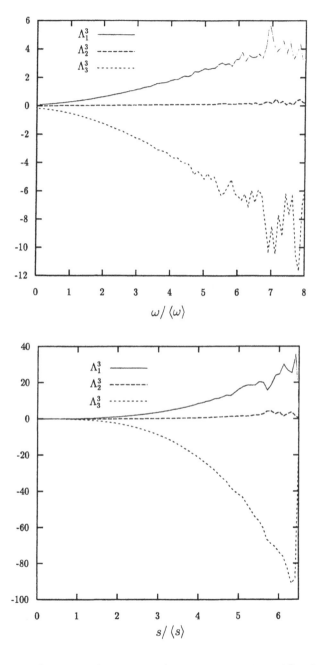

Figure 2. Conditional averages of cubed eigenvalues, Λ_i^3, of the rate of strain tensor, normalized on $\langle s^2 \rangle^{3/2}$, in slots of ω and strain s.

to those for a grid turbulent flow in which the Taylor hypothesis was used for computation of derivatives in the streamwise direction(Tsinober et $al.$ (1997)). This similarity indicates that the results below are not entitely local in time. Likewise, they also reflect some aspects of nonlocality in space which is mostly due to the use of conditional statistics and, of course, due to nonlocal relations between vorticity and strain and between the pressure Hessian $\frac{\partial^2 p}{\partial x_i \partial x_j}$ and velocity derivatives.

The ratios between the $\langle \Lambda_i^3 \rangle$ are as follows: $\langle \Lambda_1^3 \rangle : \langle \Lambda_2^3 \rangle : \langle \Lambda_3^3 \rangle$ =1.2 : 0.05 : -2.25. Their PDFs are shown in Figure 1 and their conditional averages in slots of ω and $s \equiv (s_{ij}s_{ij})^{1/2}$ are shown in Figure 2. One can see from the latter that Λ_i^3 are an order of magnitude larger in the strain dominated regions than in regions of strong vorticity.

A related phenomenon is shown in Figure 3. Namely, it is clearly seen that $s_{ij}s_{jk}s_{ki}$ is strongly correlated with the total strain $s_{ij}s_{ij}$, and is only weakly correlated with the enstrophy ω^2.

Similarly the production of strain $-2(s_{ij}s_{ik}s_{kj} + \frac{1}{4}\omega_i\omega_j s_{ij} + s_{ij}\frac{\partial^2 p}{\partial x_i \partial x_j})$ is correlated with the total strain $s_{ij}s_{ij}$, and is barely correlated with the enstrophy ω^2 (figure 4). It is noteworthy that the the production of strain $-2(s_{ij}s_{ik}s_{kj} + \frac{1}{4}\omega_i\omega_j s_{ij} + s_{ij}\frac{\partial^2 p}{\partial x_i \partial x_j})$ assumes its largest values in the regions with largest Λ_3^3 (see Section 4, Figure 9). On the other hand the enstrophy production is correlated with both the total strain $s_{ij}s_{ij}$ and with the enstrophy ω^2, but much more with the former (not shown here; they are similar to those obtained by Jimenez et $al.$ (1993)).

The behaviour of conditional averages of the total production of strain $-s_{ij}s_{jk}s_{ki} - \frac{1}{4}\omega_i\omega_j s_{ij} - s_{ij}\frac{\partial^2 p}{\partial x_i \partial x_j}$ and its separate terms is shown in Figure 5. The main feature is that the total inviscid rate of generation of strain/dissipation, $-2(s_{ij}s_{ik}s_{kj} + \frac{1}{4}\omega_i\omega_j s_{ij} + s_{ij}\frac{\partial^2 p}{\partial x_i \partial x_j})$, is more than an order of magnitude larger in the regions dominated by strain than in the enstrophy dominated regions. As seen from the Figure 6 the PDFs of $-2(s_{ij}s_{ik}s_{kj} + \frac{1}{4}\omega_i\omega_j s_{ij} + s_{ij}\frac{\partial^2 p}{\partial x_i \partial x_j})$ are fully consistent with the behaviour of their conditional averages in slots of ω and s. Again the main feature is the strong positive shift of the PDF of $-2(s_{ij}s_{ik}s_{kj} + \frac{1}{4}\omega_i\omega_j s_{ij} + s_{ij}\frac{\partial^2 p}{\partial x_i \partial x_j})$ in the regions dominated by strain $(s^2 > 2.5\langle s^2 \rangle)$.

It is noteworthy that though the mean $\langle s_{ij}\frac{\partial^2 p}{\partial x_i \partial x_j} \rangle = 0$, the PDF of $s_{ij}\frac{\partial^2 p}{\partial x_i \partial x_j}$ is positively skewed at $large$ strain (Figures 7 and 8), i.e. the interaction of strain and the pressure Hessian is such that it is opposing the production of strain when it becomes large. This is also seen from Figures 5 and 6 and from conditional averages of $s_{ij}\frac{\partial^2 p}{\partial x_i \partial x_j}$ in slots of ω and strain s (Tsinober et $al.$ (1999).

The next important point is that the enstrophy production $\omega_i\omega_j s_{ij}$ appears in the equation (4) with a negative sign, so that the vortex stretching is

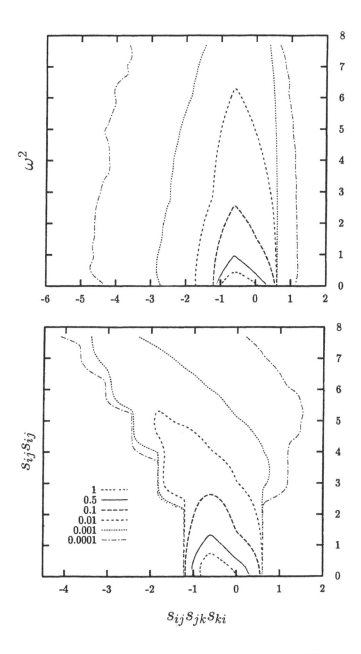

Figure 3. Joint PDF of $-s_{ij}s_{jk}s_{ki}$, normalized on $\langle s^2 \rangle^{3/2}$, versus total strain s^2 and enstrophy ω^2, both normalized on $\langle s^2 \rangle$.

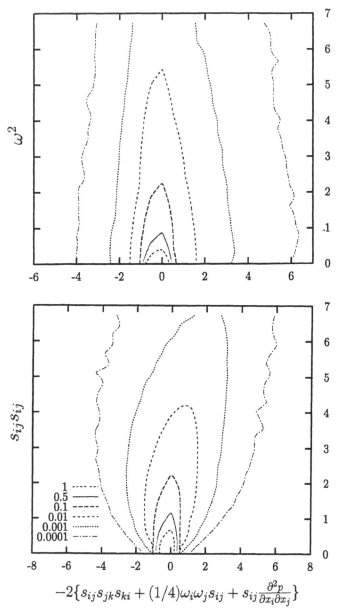

Figure 4.   Joint PDFs of production of strain $-s_{ij}s_{jk}s_{ki} -$ $\frac{1}{4}\omega_i\omega_j s_{ij} - s_{ij}\frac{\partial^2 p}{\partial x_i \partial x_j}$, normalized on $\langle s^2 \rangle^{3/2}$ versus s^2 and strain ω^2, both normalized on $\langle s^2 \rangle$.

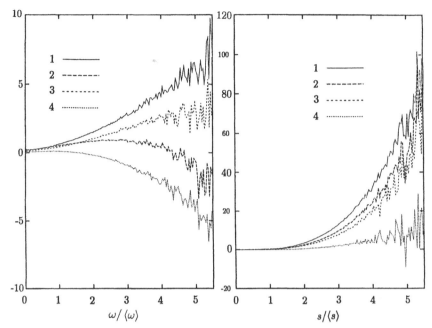

Figure 5. Conditional averages of production of strain and its separate terms in slots of ω and strain s. 1: $-s_{ij}s_{jk}s_{ki}$; 2: $-s_{ij}s_{jk}s_{ki} - \frac{1}{4}\omega_i\omega_j s_{ij}$; 3: $-s_{ij}s_{jk}s_{ki} - \frac{1}{4}\omega_i\omega_j s_{ij} - s_{ij}\frac{\partial^2 p}{\partial x_i \partial x_j}$; 4: $\frac{\partial^2 p}{\partial x_i \partial x_j}$. All are normalized on $\langle\omega_i\omega_j s_{ij}\rangle$.

opposing the production of dissipation/strain. Indeed, since $\omega_i\omega_j s_{ij}$ is essentially a positively skewed quantity all instantaneous positive values of $\omega_i\omega_j s_{ij}$ make a negative contribution to the right hand side of (4). In other words the energy cascade (whatever this means) is associated primarily with the quantity $-s_{ij}s_{jk}s_{ki}$ rather than with the enstrophy production $\omega_i\omega_j s_{ij}$ and that vortex stretching suppresses the cascade and does not aid it, at least in a *direct* manner (Tsinober, Ortenberg and Shtilman (1999)). On the contrary, it is the vortex *compression*, i.e. $\omega_i\omega_j s_{ij} < 0$, that aids the production of strain/disspation and in this sense the 'cascade'. Negative enstrophy production is associated with strong tilting of the vorticity vector and large curvature of vortex lines, which in turn are associated with large magnitudes of the negative eigenvalue, Λ_3, of the rate of strain tensor (Tsinober 1998ab, Tsinober *et al.* 1998). This is in full conformity with the above mentioned fact that Λ_3 is doing most of the 'cascade'.

One need not be confused by the equality $\langle s_{ij}s_{jk}s_{ki}\rangle = -3/4\langle\omega_i\omega_j s_{ij}\rangle$: though in the mean they are equal, their pointwise relation is strongly non-

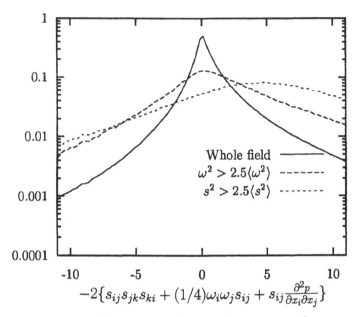

Figure 6. PDF of production of strain $(-s_{ij}s_{jk}s_{ki} - \frac{1}{4}\omega_i\omega_j s_{ij} - s_{ij}\frac{\partial^2 p}{\partial x_i\partial x_j})$ for the whole field and for $\omega^2 > 2.5\,\langle\omega^2\rangle$ and $s^2 > 2.5\,\langle s^2\rangle$.

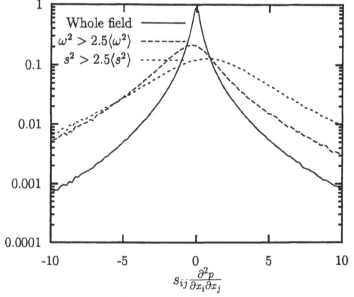

Figure 7. PDF of $s_{ij}\frac{\partial^2 p}{\partial x_i\partial x_j}$ for the whole field and for $\omega^2 > 2.5\,\langle\omega^2\rangle$ and $s^2 > 2.5\,\langle s^2\rangle$;

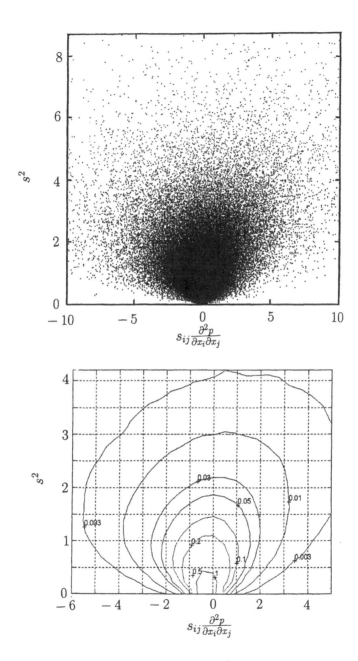

Figure 8. JPDF and scatter plot of $s_{ij} \frac{\partial^2 p}{\partial x_i \partial x_j}$ and s^2.

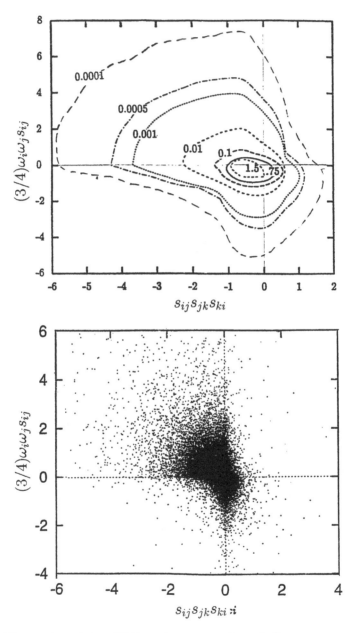

Figure 9.   Joint PDF and scatter plot of $\frac{3}{4}\omega_i\omega_j s_{ij}$ versus $-s_{ij}s_{jk}s_{ki}$.

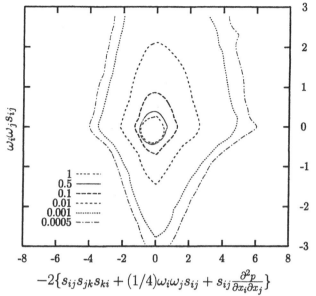

$$-2\{s_{ij}s_{jk}s_{ki} + (1/4)\omega_i\omega_j s_{ij} + s_{ij}\frac{\partial^2 p}{\partial x_i \partial x_j}\}$$

Figure 10. Joint PDF of of $\omega_i\omega_j s_{ij}$ versus $-s_{ij}s_{jk}s_{ki} - \frac{1}{4}\omega_i\omega_j s_{ij} - s_{ij}\frac{\partial^2 p}{\partial x_i \partial x_j}$.

local due to the nonlocal relation between vorticity and strain (Constantin (1994), Novikov (1968), Ohkitani (1994)). Consequently, locally they are very different as can be seen from their JPDF and scatter plots (Figure 9): they are only weakly correlated and there are great many points with small $\omega_i\omega_j s_{ij}$ and large $-s_{ij}s_{jk}s_{ki}$ and vice versa. The same is true of $-s_{ij}s_{jk}s_{ki} - \frac{1}{4}\omega_i\omega_j s_{ij} - s_{ij}\frac{\partial^2 p}{\partial x_i \partial x_j}$ and $\omega_i\omega_j s_{ij}$ (Figure 10).

4 Local flow properties in the R-Q plane

It is convenient to summarize the local flow propreties in the R-Q plane of the invariants of the velocity derivatives tensor $\partial u_i/\partial x_j$. In particular, this allows us to see the results given above from a different perspective. The results are given in three separate figures in order to avoid overloading a single figure with too much information. The main points are as follows.

Most of the (positive) enstrophy production occurs in the region $D > 0$, $R < 0$, whereas most of (positive) strain production occurs in the region $D > 0$, $R > 0$ (see Figure 11, where more details are given in the caption). Here $D = Q^3 + (27/4)R^2$ is the discriminant of the (cubic) equation defining the eigenvalues of $\partial u_i/\partial x_j$.

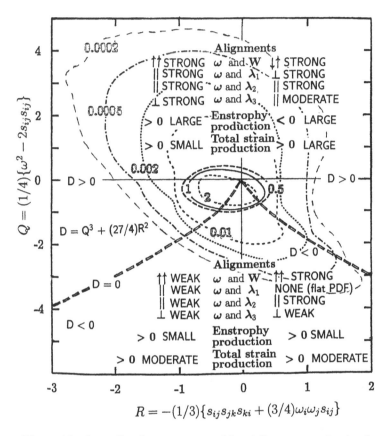

$$R = -(1/3)\{s_{ij}s_{jk}s_{ki} + (3/4)\omega_i\omega_j s_{ij}\}$$

Figure 11. A qualitative summary of local flow properties in the R-Q plane together with the joint PDF of R versus Q, corresponding to averaged quantitities over the four regions $D > 0$, $R > 0$; $D > 0$, $R < 0$; $D < 0$, $R > 0$ and $D < 0$, $R < 0$.
I. Alignments, enstrophy production and total strain production. Some additional features: *(i)*: the largest *rate* of enstrophy production, $\omega_i\omega_j s_{ij}/\omega^2$ occurs in the region $D > 0$, $R < 0$ and $Q < 0$; *(ii)*: the largest production of strain occurs in the proximity of the right branch $(R > 0)$ of the curve $D = 0$ (from both sides) and with the largest in magnitude negative eigenvalue, Λ_3, of the rate of strain; *(iii)*: the largest *rate* of production of strain, $(-s_{ij}s_{jk}s_{ki} - \frac{1}{4}\omega_i\omega_j s_{ij} - s_{ij}\frac{\partial^2 p}{\partial x_i \partial x_j}))/s^2$, occurs in the region $D > 0$, $R > 0$ and $Q < 0$; *(iv)*: the PDF of $\cos(\omega, \lambda_2)$ is flat in the region $D > 0$, $R < 0$ and $Q < 0$.

The region $D > 0$, $R > 0$ is characterized by large *negative* enstrophy production. Note that strong alignment between vorticity ω_i and the vortex stretching vector $W_i \equiv \omega_j s_{ij}$ occurs not only in the region with most of the (positive) enstrophy production $(D > 0, R < 0)$, but also in the region $D < 0$, $R > 0$, where the (positive) enstrophy production is relatively small. Naturally, the vorticity, ω_i, and the vortex stretching vector, $W_i \equiv \omega_j s_{ij}$, *antialign* in the region $D > 0$, $R > 0$ with large *negative* enstrophy production. Also noteworthy is that there is strong alignment between vorticity, ω_i, and *both* the largest, λ_1, and the intermediate, λ_2, eigenvectors of the rate of strain tensor s_{ij} in the region $D > 0$, $R < 0$. This is possible, since these alignments happen on *different* sets of points. Likewise strong alignment between vorticity ω_i and and the intermediate eigenvector, λ_2, occurs in three *qualitatively* different regions: $D > 0$, $R < 0$; $D > 0$, $R > 0$ and $D < 0$, $R > 0$. This shows that this most popular alignment is caused by different physical reasons in different flow regions (see also Figure 12). The above results are in conformity with those regarding curvature of vorticity lines and tilting of ω-vector as shown in Figure 13. An additional aspect shown in this figure concerns the interaction of the strain and pressure Hessian: it is large and positive in the region $D < 0$, $R > 0$, and it is large and negative in the region $D > 0$, $R < 0$.

It is noteworthy that some similar results in terms of energy flux using conditional averages in the plane of invariants of velocity derivatives, Q-R, were reported by Chertkov, Pumir and Shraiman (1999); see also Borue and Orszag (1998)[2].

5   Concluding remarks

• There exist two nonlocally interconnected weakly correlated processes: *(i)*: predominant vortex stretching/enstrophy production and *(ii)*: predominant self-amplification of the rate of strain/production of total strain. The energy cascade (whatever this means) and its final result – dissipation – are

---

[2]These authors used a parameterization of the SGS energy flux/dissipation in the form

$$-\langle s_{ij}\rangle_l \tau_{ij} \approx l^2\{-\langle s_{ij}\rangle_l \langle s_{jk}\rangle_l \langle s_{ki}\rangle_l + 1/4\langle\omega_i\rangle_l\langle\omega_j\rangle_l\langle s_{ij}\rangle\} \qquad (7)$$

They arrived at the conclusion that subgrid energy transfer over the scales (SGS-dissipation) takes place in regions with negative skewness of the filtered strain tensor, i.e. $\langle s_{ij}\rangle_l\langle s_{jk}\rangle_l\langle s_{ki}\rangle_l$) or where the vorticity stretching term is positive. The latter is due to the *positive* sign in front of $1/4\langle\omega_i\rangle_l\langle\omega_j\rangle_l\langle s_{ij}\rangle$ in equation (7), as opposed to the *negative* sign in equation (4). This shows how dangerous is drawing conclusions regarding the physics of turbulence from *models*. Also, though $\langle s_{ij}\frac{\partial^2 p}{\partial x_i \partial x_j}\rangle = 0$ because of homogeneity and incompressibility, it is unlikely that either $\langle s_{ij}\frac{\partial^2 p}{\partial x_i \partial x_j}\rangle_l$ or $\langle s_{ij}\rangle_l\langle\frac{\partial^2 p}{\partial x_i \partial x_j}\rangle_l$ or both will vanish.

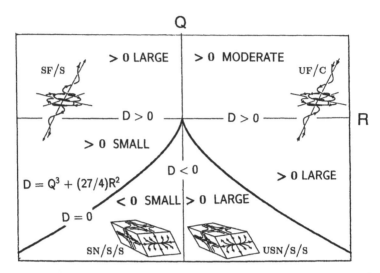

Figure 12. A qualitative summary of the behaviour of the second eigenvalue, Λ_2, of the rate of strain tensor in the R-Q plane corresponding to averaged quantitities over the six regions: $Q > 0$, $R > 0$; $Q > 0$, $R < 0$; $D < 0$, $R < 0$; $D < 0$, $R > 0$; $Q < 0$, $R > 0$ and $D > 0$; $Q < 0$, $R < 0$ and $D > 0$. Also shown are the schematic local flow fields (e.g. Ooi *et al.* (1999)): SF/S: stable focus/stretching; UF/C : stable focus/compressing; SN/S/S: stable node/saddle/saddle; USN/S/S: unstable node/saddle/saddle.

associated with the latter, i.e. with the quantity $-s_{ij}s_{jk}s_{ki}$ rather than with the enstrophy production $\omega_i\omega_j s_{ij}$. Moreover, vortex stretching suppresses the cascade and does not aid it, at least in a *direct* manner[3]. On the contrary it is the vortex *compression*, i.e. $\omega_i\omega_j s_{ij} < 0$, that aids the production of strain/disspation and in this sense the 'cascade'. The predominant vortex stretching/enstrophy production is associated mostly with the largest positive eigenvalue, Λ_1, of the rate of stain tensor and also with its intermediate eigenvector, Λ_2. The predominant self-amplification of the rate of strain/production of total strain is associated totally with the largest in magnitude negative eigenvalue, Λ_3.

• Nonlinearities such as enstrophy production, production of strain and many others (Tsinober (1998a,b), Tsinober *et al.* (1999)) are an order of magnitude larger in the regions dominated by strain than in the enstrophy dominated

---

[3]Contrary to the common belief: '*It seems that the stretching of vortex filaments must be regarded as the principal mechanical cause of the high rate of dissipation which is associated with turbulent motion*' (Taylor (1938)).

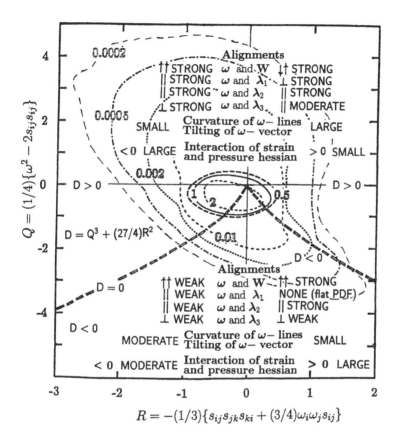

$$R = -(1/3)\{s_{ij}s_{jk}s_{ki} + (3/4)\omega_i\omega_j s_{ij}\}$$

Figure 13. A qualitative summary of local flow properties in the R-Q plane together with the joint PDF of R versus Q, corresponding to averaged quantitities over the four regions $D > 0$, $R > 0$; $D > 0$, $R < 0$; $D < 0$, $R > 0$ and $D < 0$, $R < 0$. II. Curvature of ω-lines, tilting of the ω-vector and interaction of strain and pressure Hessian $s_{ij}\frac{\partial^2 p}{\partial x_i \partial x_j}$.

Some additional features: *(i)* : in the region $Q < 0$, $R < 0$ and $D > 0$ the curvature of ω-lines and the tilting of the ω-vector are both large; they are both moderate in the region $Q < 0$, $R > 0$ and $D > 0$; *(ii)*: the interaction of strain and pressure Hessian $s_{ij}\frac{\partial^2 p}{\partial x_i \partial x_j}$ assumes its largest positive values in the proximity of the curve $D = 0$ (from both sides) and $R > 0$.

regions. In this sense the enstrophy dominated regions are characterized by reduced nonlinearities including the energy cascade whatever this means. In other words the most intense nonlinear processes occur in the strain dominated regions.[4] This supports the view that regions of concentrated vorticity in turbulent flows are not that important as previously thought (Jimenez *et al.*(1993), Dernoucourt *et al.* (1998), Roux *et al.* (1998), Tsinober (1998a)).

Appendix 1. Why are mean enstrophy and strain production positive?

So far there we have given no theoretical arguments in favour of the positiveness of $\langle \omega_i \omega_j s_{ij} \rangle$ (and also $\langle -s_{ij} s_{jk} s_{ki} \rangle$).[5]

The argument that the reason is the (approximate) balance between the enstrophy generation and its destruction via viscosity is misleading and puts the effects before the causes, since it is known that for the Euler equations the enstrophy generation increases very rapidly with time – apparently without limit (Yudovich (1974), Betchov (1976), Chorin (1982), Bell and Marcus (1992), Brachet *et al.* (1992), Fernandez *et al.* (1995), Grauer and Sideris (1995), Green and Boratav (1997), Kerr (1993)). It is noteworthy that there is another aspect in which the (approximate) balance between the *mean* enstrophy generation and its *mean* destruction via viscosity can be misleading as well. Namely, this balance holds also in the enstrophy dominated regions, but fails in the regions dominated by strain. The important point is that most of enstrophy generation occurs in the regions dominated by strain (see Section 2 and Tsinober (1998a,b), Tsinober *et al.* (1999)).

Another rather common view is that the prevalence of vortex stretching is due to the predominance of stretching of material lines is – at best – true in part only, since there exist several qualitative differences between the two processes. For example, for a Gaussian isotropic velocity field $\langle \omega_i \omega_j s_{ij} \rangle \equiv 0$, whereas the mean rate of stretching of material lines is essentially positive, i.e. the nature of vortex stretching process is to a large extent dynamical and not just a kinematic one (for more details on the differences between the two, see Tsinober (1998a,b)).

The following simple theoretical argument can be given for the case with a Gaussian velocity field [6] at the initial moment, $t = 0$. Let us look at the

---

[4]This reduction of nonlinearities is in some respect analogous to the processes occurring in the so called elliptical regions (corresponding to the enstrophy dominated regions) in two-dimensional turbulent flows (Weiss (1991)), and in a turbulent flow in the proximity of a large strained vortex (Andreotti *et al.* (1998)).

[5]Rigorous results on this issue would comprise a major contribution to the understanding of the *physics* of turbulence.

[6]It is sufficient that the velocity field satisfies the zero-fourth-cumulant relation, i.e. is

equation for the mean enstrophy production $\langle \omega_i \omega_j s_{ij} \rangle$ (dropping the viscous terms)

$$\frac{D}{Dt} \langle \omega_i \omega_j s_{ij} \rangle = \langle \omega_j s_{ij} \omega_k s_{ik} \rangle - \left\langle \omega_i \omega_j \frac{\partial^2 p}{\partial x_i \partial x_j} \right\rangle, \tag{8}$$

For a Gaussian velocity field $\langle \omega_i \omega_j s_{ij} \rangle_G = 0$, $\left\langle \omega_i \omega_j \frac{\partial^2 p}{\partial x_i \partial x_j} \right\rangle_G = 0$ and $\langle \omega_j s_{ij} \omega_k s_{ik} \rangle_G = \frac{1}{6} \langle \omega^2 \rangle^2 > 0$ (the quantity $\omega_j s_{ij} \omega_k s_{ik} \equiv W^2, W_i = \omega_j s_{ij}$, so it is positive pointwise for any vector field). Hence at $t = 0$

$$\left\{ \frac{D}{Dt} \langle \omega_i \omega_j s_{ij} \rangle \right\}_{t=0} = \{ \langle \omega_j s_{ij} \omega_k s_{ik} \rangle \}_{t=0} > 0, \tag{9}$$

It follows from equation (9) that, at least for a short time interval t, the mean enstrophy production will become positive. For later moments the vorticity–pressure Hessian correlation $\left\langle \omega_i \omega_j \frac{\partial^2 p}{\partial x_i \partial x_j} \right\rangle$ becomes finite, and nothing is known rigorously. As follows from DNS of NSE in a periodic box the correlation $\left\langle \omega_i \omega_j \frac{\partial^2 p}{\partial x_i \partial x_j} \right\rangle$ is positive, but is smaller than $\langle \omega_j s_{ij} \omega_k s_{ik} \rangle \equiv \langle W^2 \rangle$. Namely, $\left\langle \omega_i \omega_j \frac{\partial^2 p}{\partial x_i \partial x_j} \right\rangle \sim \frac{1}{3} \langle W^2 \rangle$, so that the RHS of (8) remains positive (Tsinober *et al.* (1995)). It is noteworthy that equation (8) with $\left\langle \omega_i \omega_j \frac{\partial^2 p}{\partial x_i \partial x_j} \right\rangle = 0$ is precisely the one arising using the quasi-normal approximation $\frac{D^2}{Dt} \langle \omega^2 \rangle = \frac{1}{3} \langle \omega^2 \rangle^2$ (Proudman and Reid (1954), see also Kaneda (1993)), since $\frac{D}{Dt} \langle \omega_i \omega_j s_{ij} \rangle = \frac{1}{2} \frac{D^2}{Dt} \langle \omega^2 \rangle$ and under quasi-normal approximation $\langle \omega_j s_{ij} \omega_k s_{ik} \rangle = \frac{1}{6} \langle \omega^2 \rangle^2$ and $\left\langle \omega_i \omega_j \frac{\partial^2 p}{\partial x_i \partial x_j} \right\rangle = 0$. The essential point is that at $t = 0$ the relation (9) is precise due to the freedom of the choice of the intitial condition.

In a similar way one can see from equation (6) that the mean rate of production of strain, $-2 \left\{ \langle s_{ij} s_{jk} s_{ki} \rangle - \frac{1}{4} \langle \omega_i \omega_j s_{ij} \rangle \right\}$, becomes positive at small times (at $t = 0$ it is vanishing) for an initially Gaussian velocity field (note that $\left\langle s_{ij} \frac{\partial^2 p}{\partial x_i \partial x_j} \right\rangle = 0$ due to incompressibility) . It is seen also from the equation for $\langle \omega_i \omega_j s_{ij} \rangle$ that an initially Gaussian and nearly potential velocity field with small seeding of vorticity will produce, at least for a short time, an essentially positive enstrophy generation as well. This process seems to be of importance in the phenomenon of entrainment of nonturbulent fluid into the turbulent region in the proximity of the region separating turbulent and nonturbulent fluid.

It is noteworthy that equations (8) and (9) and similar ones for $\langle s_{ij} s_{jk} s_{ki} \rangle$ is one of the manifestations of the statistical irreversibililty[7] of turbulent flows (Betchov (1974), Novikov (1974)). There exist, at least, two different aspects of this problem. The first one is related to purely inertial behaviour

---

quasi-normal.

[7]The corresponding dynamical *instantaneous* (inviscid) equations are reversible. Hence, the term *statistical*.

governed by the Euler equations as mentioned above. It is closely related to the (possible) formation of singularities in 3D Euler flows in finite or infinite time. The above example (equations (8)-(9) and similar ones for $\langle s_{ij} s_{jk} s_{ki} \rangle$) is closely related to this aspect. The second aspect is associated with the dissipative nature of turbulent flows. Viscosity provides a sink of energy, enstrophy, etc. moderating their unbounded growth in the inviscid case.

Appendix 2. On difficulties in defining scales, cascades and related matters.

As mentioned above, there is a generic ambiguity in defining the meaning of the term *small scales* (or more generally scales) and consequently the meaning of the term *cascade* in turbulence research. The specific meaning of this term and associated interscale energy exchange/'cascade' (e. g. spectral energy transfer) is essentially decompostion/reperesentation dependent (see, for example, Borue & Orszag (1998), Frick and Zimin (1993), Fuehrer and Friehe (1999), Germano(1999), Holmes et al (1996), Mahrt & Howell (1994), Meneveau (1991), Pullin and Saffman (1997), Sirovich (1997), Tsuge (1984), Waleffe (1993) and references therein)[8]. Perhaps the only common factor in all decompositions/representations is that the small scales are *always* associated with the field of velocity derivatives. Therefore, it is natural to look at this field as the one *objectively* (i.e. decomposition/represenation independent) representing the small scales. Indeed, the dissipation is associated precisely with the strain field s_{ij}. The advantage of this 'definition' of small scales can be seen from the following example. It is well known that there is no contribution from the nonlinear term in the total energy balance equation (in a homogeneous/periodic flows its contribution is null as well in the mean) since the nonlinear term has the form of a spatial flux, $\partial \{...\}/\partial x_j$. In other words, the nonlinear term redistributes the energy in physical space, but does it do more than that?[9] The usual claim is that *the nonlinear term redistributes the energy among the scales of motion* (Frisch (1995), p.22), whereas in reality the nonlinear term redistributes the energy, e.g. in the Fourier space between the Fourier components of the turbulent field. However, the non-

---

[8]Indeed, the meaning of *scales* is different for different representations: it is not the same for *Fourier* ('regular' and helical) and similar (Fourier, Weierstrass, Gabor, Littlewood–Paley) decompositions; *Wavelets* (wavepackets, solitons); *POD; LES*. It is also different in various heuristic representations, e.g. 'two-fluid' (organized/incoherent, deterministic/random and some other two-fluid models); intermittency-prompted (breakdown coefficients/multipliers, (multi)fractals); Moffatt's 'smart decomposition' and the 'punctuated' conservative dynamics.

[9]It is straightforward to see that in a homogeneous turbulent flow the mean energy of volume of any scale (Lagrangian and/or Eulerian) is changing due to viscous dissipation and external forcing only.

linear term does it in a *different* way between the components of *different* decompositions, such as the Fourier – or wavelet – representations, the POD, and so on (Frick and Zimin (1993), Holmes *et al.* (1996), Mahrt & Howell (1994), Meneveau (1991), Sirovich (1997), Waleffe (1993) and references therein). In other words, the term 'cascade' corresponds to a process of interaction/exchange of (not necessarily only) energy between components of some *particular* decomposition/representation of a turbulent field associated with the nonlinearity and the nonlocality of the turbulence phenomenon (the two n's out of three: nonlinearity, nonlocality and nonintegrability, which make the problem so impossibly difficult). On the other hand, the energy transfer, like any physical process, should be invariant of particular the decompositions/representations of a turbulent field.[10] Taking the velocity derivatives as a basic notion allows one to resolve/clarify (to some extent) the ambiguities associated with the terms 'energy transfer', 'scale', and so on in the following way. While the mean contribution of the nonlinearity in the energy balance is vanishing, the nonlinear term definitely *creates* vorticity (and strain, see Sections 2 and 3) in *physical space*, since the enstrophy production $\langle \omega_i \omega_j s_{ij} \rangle > 0$ is strictly positive as well as the corresponding term for the production of strain. As mentioned above it is naturally and justified from the *physical* point of view to associate the field of velocity derivatives with *small scales.* It is immediately seen that 3D turbulent flows have a natural tendency to create small scales. Namely, the velocity field (and its energy) arising in the process of (self-) production of the field of velocity derivatives is the one which is associated with the small scales. This process is what can be called as energy transfer from large to small scales in physical space. The latter are not necessarily created via a stepwise turbulent 'cascade': it can be bypassed (and most probably this is the case in turbulent flows), e.g.

---

[10]The difficulty is not a trivial one and seems to be 'generic'. Under turbulent motion/dynamics the interaction of 'modes' (whatever they are) is strong. The resulting structure(s) is (are) not represented by the modes (of whatever decomposition), e.g. Fourier decomposition of a flow in a periodic box. Hence the ambiguity (for another aspect of the Fourier transform ambiguity see Tennekes (1976)). Recall the suggestion by Dryden (1948): '*It is necessary to separate the random processes from the nonrandom processes*'. The implication is that such a separation is possible. But this is not obvious at all, as is seen from the futility of the enormous efforts to do so. For example, it is not even known how to separate random gravity-wave motion (which does not produce vertical transport) and genuine turbulence (which does) in a stably stratified fluid (Stewart (1959)). All the attempts to find a 'good' decomposition are related to what R. Betchov (1993) called the '*dream of linearized physicists*', i.e. a *superposition* of some (desirably simple) elements. The dream is, of course, to find sets (consisting of small number) of *simple weakly interacting* elements/objects *adequately* representing the turbulent field. Those are known so far as strongly interacting (most of them nonlocally) – a fact reflecting one of the central difficulties in 'solving the turbulence problem' as a whole, in general, and the 'closure problem' (such as LES and other reduced descriptions of turbulence (Kraichnan (1988)), in particular, as well as in construction of a kind of statistical mechanics of turbulence (Kraichnan and Chen (1989)).

via broad-band instabilities with highest growth rate at short wavelengths (e. g. Pierrehumbert & Widnall 1982, Smith & Wei 1994) or some other approximately single step process (Betchov (1976), Douady, Couder & Brachet (1991), Vincent & Meneguzzi (1994), Garg & Warhaft (1998); the problem goes back to Townsend (1951): ... *the postulated process differs from the ordinary type of turbulent energy transfer being fundamentally a single process.*; see also Corrsin (1962), Tennekes (1968)). Indeed two large neighbouring eddies can dissipate energy directly by rubbing each other on a very small scale. Note that the process of vorticity production is not just creation of the field of velocity derivatives. It literally involves creation of small scale structure in the following sense. Namely, an inevitable concomittant process to vortex stretching is tilting and folding of vorticity due to the energy constraint (Chorin (1982), Tsinober (1998a,b)). Simultaneously, the strain field is built up at the same rate (see Section 3 and Tsinober *et al.* (1999)). This, together with limitations on the volume scale, leads to formation of fine small scale structure. Since the flow field (including velocity, which is mostly a large scale object) is determined entirely by the field of vorticity, i.e. the velocity field is a functional of vorticity $\mathbf{v} = F\{\omega(\mathbf{x}, t)\}$), the production of vorticity 'reacts back' in creating the corresponding velocity field.[11] It is noteworthy that due to the nonlocality of the relation $\mathbf{v} = F\{\omega(\mathbf{x}, t)\}$ mostly small scale vorticity is, generally, also creating some large scale velocity. Therefore from the *physical point* it seems incorrect to treat small scales as a kind of 'passive' object and impossible to 'eliminate' them (as is done in many theories) reducing their reaction back only to some eddy viscosity or similar thing. In view of the above arguments it seems that in *physical space* the energy is dissipated not necessarily via a multistep cascade-like process. Instead, there is an exchange of energy (and everything else) in both directions, whereas the dissipation occurs in 'small scales'. So it is quite possible that in physical space the famous verse by Richardson (1922, p.66)

Big whirls have little whorls,
Which feed on their velocity.
And little whorls have lesser whorls
And so on to viscosity
(In the molecular sense)

should be replaced by the one by Betchov (1976, p. 845)

Big whirls lack smaller whirls,
To feed on their velocity.
They crash and form the finest curls
Permitted by viscosity.

---

[11]This essential process is fully ignored in approaches like RDT.

References

Andreotti, B., Douady, S. and Couder, Y. (1998) 'Experimental investigation of the turbulence near a large-scale vortex', *Eur. J. Mech.*, B/*Fluids*, **17**, (4), 451–470.

Bell, J.B. and Marcus, D.L. (1992) 'Vorticity intensification and transition to turbulence in the three-dimensional Euler equations', *Commun. Math. Phys.*, **147**, 371–394.

Betchov, R. (1956) 'An inequality concerning the production of vorticity in isotropic turbulence', *J. Fluid Mech.*, **1**, 497–504.

Betchov, R. (1974) 'Non-Gausian and irreversible events in isotropic turbulence', *Phys. Fluids*, **17**, 1509–1512.

Betchov, R. (1976) 'On the non-Gaussian aspects of turbulence', *Archives of Mechanics* (Archiwum Mechaniki Stosowanej), **28**, (5-6), 837–845.

Betchov, R. (1993) in *New Approaches in Turbulence*, T. Dracos and A. Tsinober, eds., Birkäuser, p. 155.

Boratav, O.N. and Pelz, R.B. (1997) 'Structures and structure functions in the inertial range of turbulence', *Phys. Fluids*, **9**, 1400–1415.

Borue, V. and Orszag, S.A. (1998) 'Local energy flux and subgrid scale statistics in three-dimensional turbulence', *J. Fluid Mech.*, **366**, 1–31.

Brachet, M.E., Meneguzzi, M. Vincent, A., Politano, H. and Sulem, P.-L. (1992) 'Numerical evidence of smooth self-similar dynamics and possibility of subsequent collapse for three-dimensional ideal flows', *Phys. Fluids*, **A4**, 2845–2854.

Brasseur, J.G. and Lin, W.-Q. (1995) 'Dynamics of small-scale vorticity and strain-rate structures in homogeneous shear turbulence', *Proc. Tenth Symposium on Turbulent Shear Flows*, 3.19–3.24.

Cadot, O., Douady, S. and Couder, Y. (1995) 'Characterization of the low-presure filaments in three-dimensional turbulent shear flow', *Phys. Fluids*, **7**, 630–646.

Chen, S. and Cao, N. (1997) 'Anomalous scaling and structure instability in three-dimensional passive scalar turbulence', *Phys. Rev. Lett.*, **78**, 3459–3461.

Chertkov, M., Pumir, A. and Shraiman, B.I., (1999) 'Lagrangian tetrad dynamics and the phenomenology of turbulence', *Phys. Fluids* **11**, 2394–2410.

Chorin, A.J. (1982) 'The evolution of a turbulent vortex', *Comm. Math. Phys.*, **83**, 517–535.

Constantin, P. (1994) 'Geometrical statistics in turbulence', *SIAM Rev.*, **36**, 73–98.

Corrsin, S. (1962) 'Turbulent dissipation fluctuations', *Phys. Fluids*, **12**, 1301–1302.

Dernoncourt, B., Pinton, J.-F. and Fauve, S. (1998) 'Experimental study of vorticity filaments in a turbulent swirling flow', *Physica D*, **117**, 181-190.

Dryden, H. (1948) 'Recent Advances in Boundary Layer Flow', *Adv. Appl. Mech*, **1**, 1–40.

Fernandez, V.M., Zabusky, N.J. and Gryanik, V.M. (1995) 'Vortex intensification and collapse of the Lissajous-elliptic ring: single- and multi- filament Biot–Savart simulations and visiometrics', *J. Fluid Mech.*, **299**, 289–331.

Frederiksen, R.D., Dahm, W.J.A. and Dowling, D.R. (1996) 'Experimental assessment of fractal scale similarity in turbulent lows. Part 2', *J. Fluid Mech.*, **338**, 89–126.

Frick, P. and Zimin, V. (1993) 'Hierarchical models of turbulence', in *Wavelets, Fractals, and Fourier Transforms*, M. Farge, J.C.R. Hunt and J.C. Vassilicos, eds., Clarendon Press, pp. 265–283.

Frisch, U. (1995) *Turbulence: the Legacy of A.N. Kolmogorov*, Cambridge University Press.

Fuehrer, P.L. and Friehe, C.A. (1999) 'A physically-based turbulent velocity time series decomposition', *Boundary Layer Meteorology*, **90**, 241–295.

Garg, S. and Warhaft, Z. (1998) 'On the small scale structure of simple shear flow', *Phys. Fluids*, **10**, 662–673.

Germano, M. (1999) 'Basic issues of turbulence modelling', in *Fundamental Problematic Issues in Turbulence*, A. Gyr, W. Kinzelbach and A. Tsinober, eds., Birkhauser, pp. 325–336.

Grauer, R. and Sideris, T. (1995) 'Finite time singularities in ideal fluids with swirl', *Physica D*, **88**, 116-132.

Greene, J.M. and Boratav, O. (1997) 'Evidence for the development of singularities in Euler flow', *Physica D*, **107**, 57-68.

Holmes, P.J., Berkooz, G. and Lumley, J.L. (1996) *Turbulence, Coherent Structures, Dynamical Systems and Symmetry*, Cambridge University Press.

Jimenez, J., Wray, A.A., Saffman, P.G. and Rogallo, R.S. (1993) 'The structure of intense vorticity in homogeneous isotropic turbulence', *J. Fluid Mech.*, **255**, 65–90.

von Karman, Th. (1937) 'The fundamentals of the statistical theory of turbulence', *J. Aeronaut. Sci.*, **4**, 131–138.

Kaneda, Y. (1993) 'Lagrangian and Eulerian time correlations in turbulence', *Phys Fluids*, **A5**, 2835–2843.

Kerr, R.M. (1993) 'Evidence for a singularity of the three-dimensional incompressible Euler equations', *Phys Fluids*, **A5**, 1725–1748.

Kolmogorov, A.N. (1941a) 'The local structure of turbulence in incompressible viscous fluid for very large Reynolds numbers', *Dokl. Akad. Nauk SSSR*, **30**, 299–303.

Kolmogorov, A.N. (1941b) 'Dissipation of energy in locally isotropic turbulence', *Dokl. Akad. Nauk SSSR*, **32**, 19–21.

Kraichnan, R.H. (1988) 'Reduced descriptions of hydrodynamic turbulence', *J. Stat. Phys.*, **51**, 949–963.

Kraichnan, R.H. and Chen, S. (1989) 'Is there a statistical mechanics of turbulence?', *Physica D*, **72**, 160–172.

Mahrt, L. and Howell, J.F. (1994) 'The influence of coherent structures and microfonts on scaling laws using global and local transforms', *J. Fluid Mech.*, **260**, 247–270.

McWilliams, J.C. (1990) 'A demonstration of the suppression of turbulent cascades by coherent vortices in two-dimensional turbulence', *Phys. Fluids*, **A2**, 547–552.

Meneveau, C. (1991) 'Analysis of turbulence in the orthonormal wavelet representation', *J. Fluid Mech.*, **232**, 469–520.

Meneveau, C. and Lund, T.S. (1994) 'On the Lagrangian nature of the turbulence energy cascade', *Phys. Fluids*, **6**, 669–671.

Novikov, E. (1974) 'Statistical irreversibility of turbulence', *Archives of Mechanics* (Archiwum Mechaniki Stosowanej), **4**, 741–745.

Novikov, E.A. (1968) 'Kinetic equations for a vortex field ', *Soviet Physics Doklady*, **12**, 1006–1008.

Ohkitani, K. (1994) 'Kinematics of vorticity: Vorticity-strain conjugation in incompressible fluid flows', *Phys. Rev.*, **E 50**, 5107–5110.

Ooi, A., Martin, J., Soria, J. and Chong, M. S. (1999) 'A study of the evolution and characteristics of the invariants of the velocity-gradient tensor in isotrpic turbulence', *J. Fluid Mech.*, **381**, 141-174.

Pierrehumbert, R.T. and Widnall, S.E. (1982) 'The two- and three-dimensional instabilities of a spatially periodic shear layer', *J. Fluid Mech.*, **114**, 59–82.

Proudman, I. and Reid, W.H. (1954) 'On the decay of a normally distributed and homogeneous turbulent velocity field', *Phil. Trans. Roy. Soc.*, **A247**, 163–189.

Pullin, D.I. and Saffman, P.S. (1997) 'Vortex dynamics and turbulence', *Ann. Rev. Fluid Mech.*, **30**, 31–51.

Richardson, L.F. (1922) *Weather Prediction by Numerical Process*, Cambridge University Press.

Roux, S., Muzy, J.F. and Arneodo, A., (1999) 'Detecting vorticity filaments using wavelet analysis: about the statistical contribution of vorticity filaments to intermittency in swirling turbulent flows', *Eur. Phys. J.*, **B** (Condensed Matter Physics), in press.

Schwarz, K.W. (1990) 'Evidence for organized small-scale structure in fully developed turbulence', *Phys. Rev. Lett.*, **64**, 415–418.

Sirovich. L. (1997) 'Dynamics of coherent structures in wall bounded turbulence', in *Self-Sustaining Mechanisms of Wall Turbulence*, R.L. Panton, ed., Comp. Mech. Publ., 333–364.

Stewart, R.W. (1959) 'The problem of diffusion in a stratified fluid', in *Atmospheric Diffusion and Air Pollution*, F.N. Frenkiel and P.A. Sheppard, eds., Academic Press, 303-311.

Taylor, G.I. (1937) 'The statistical theory of isotropic turbulence', *J. Aeronaut. Sci.*, **4**, 311–315.

Taylor, G.I. (1938) 'Production and dissipation of vorticity in a turbulent fluid', *Proc. Roy. Soc.*, **A164**, 15-23.

Tennekes, H. (1968) 'Simple model for the small-scale structure of turbulence', *Phys. Fluids*, **11**, 669–671.

Tennekes, H. (1976) 'Fourier-transform ambiguity in turbulence dynamics', *J. Atmosp. Sci.*, **33**, 1660–1163.

Townsend, A.A. (1951) 'On the fine-scale structure of turbulence', *Proc. Roy. Soc. Lond.*, **A208**, 534–542.

Tsinober, A. (1995) 'On geometrical invariants of the field of velocity derivatives in turbulent flows', *Actes 12me Congrès Francais de Mécanique*, **3**, 409–412.

Tsinober, A. (1998a) 'Is concentrated vorticity that important?', *Eur. J. Mech., B/Fluids*, **17**, 421–449.

Tsinober, A. (1998b) 'Turbulence – Beyond Phenomenology', in *Chaos, Kinetics and Nonlinear Dynamics in Fluids and Plasmas*, S. Benkadda and G.M. Zaslavsky, eds., *Lect. Notes in Phys.*, **511**, Springer, 85–143.

Tsinober, A. (1998c) 'On statistics and structure(s) in turbulence', IUTAM-IUGG Symposium *Developments in Geophysical Turbulence*, NCAR, Boulder, Colorado, USA, June 16–19, 1998.

Tsinober, A., Shtilman, L., Sinyavsky, A. and Vaisburd, H. (1995) 'Vortex stretching and enstrophy generation in numerical and laboratory turbulence', *Lecture Notes in Physics*, **462**, 9-16. M. Meneguzzi, A. Pouquet, and P.-L. Sulem, eds., Springer.

Tsinober, A., Shtilman, L. and Vaisburd, H. (1997) 'A study of properties of vortex stretching and enstrophy generation in numerical and laboratory turbulence', *Fluid. Dyn. Res.*, **21**, 477-494.

Tsinober, A., Ortenberg, M. and Shtilman, L. (1998) 'On geometrical properties of velocity derivatives in numerical turbulence', *Advances in Turbulence*, **VII**, 335-338.

Tsinober, A., Ortenberg, M. and Shtilman, L. (1999) 'On depression of nonlinearity', *Phys. Fluids* **11**, 2291-2297.

Tsugé, S. (1984) 'Separability into coherent and chaotic time dependencies of turbulent fluctuations', *Phys. Fluids*, **27**, 1370-1376.

Vincent, A. and Meneguzzi M., (1994) 'The dynamics of vorticity tubes in homogeneous turbulence', *J. Fluid Mech.*, **258**, 245-254

Waleffe, F. (1993) 'Inertial transfers in the helical decomposition', *Phys. Fluids*, **A5**, 677-685.

Yanitskii, V. E. (1982) 'Transport equation for the deformation-rate tensor and description of an ideal incompressible fluid by a system of equations of the dynamical type', *Sov. Phys. Dokl*, **27**, 701-703.

Yudovich, V., (1974) 'On loss of smoothness for the Euler equations', *Dinamika sploshnoi sredy*, **16**, 71-78 (in Russian).

Weiss, J. (1991) 'The dynamics of enstrophy transfer in two-dimensional hydrodynamics', *Physica D*, **48**, 273-294.

Dynamics and Statistics of Vortical Eddies in Turbulence

J.C.R. Hunt

Abstract

This article analyses the vortical eddies of turbulence in terms of their kinematics and dynamics and in terms of their individual and collective statistical properties. This leads to insights and quantitative estimates of the formation, mutual interaction and destruction of eddies and their relation to both larger scale straining motions and smaller scale background fluctuations. Analysis of the overall dynamics and movement of eddies leads to estimates for their contributions to mean momentum and scalar fluxes in homogeneous turbulence and in inhomogeneous shear flows. An overview emerges of the scale-dependent, visco-elastic like properties of turbulent flow fields depending on the space and time scales on which they are distorted, their non-uniformity and the effects of boundary conditions and body forces.

1 Introduction

The early statements about the nature of turbulence theory, for example those of Landau (1959) and Batchelor (1953) emphasised the statistical nature of turbulence. Later work on the coherent structures of turbulence and linear distortion of random fields showed how deterministic calculations can describe significant local features of the flow and explain their underlying dynamics (e.g. Betchov 1956, Hussain 1986). In fact, as in statistical physics, most of our conceptual understanding and quantitative models of turbulence are actually based on a combination of statistical and deterministic analyses, even if they are labelled as rather more one than the other. Just as the kinetic theory of gases requires statistical models of distributions and dynamical models of collisions, so Kolmogorov's (1941) statistical theory is based on concepts of eddy motion and dynamics, which he pointed out in 1961 were based on those of L.F. Richardson (1922). Conversely deterministic analysis of individual eddy motion and distorted flows are only good approximations for a certain period, because there are unpredictable motions within and between eddies, both of which have to be modelled statistically.

In most branches of natural science, there have been fruitful combinations of deterministic and statistical approaches; one of the reasons has been because in all these complex nonlinear systems the fields (whether of material or kinematic properties) have a finite number of characteristic localised forms (faults, clouds, eddy motions etc.) (Hunt 1999). In the case of turbulence many of the key processes can be understood and modelled quantitatively by local analyses of the characteristic eddy structures – as Prandtl (1925) demonstrated in his studies of shear flows structural studies. This has the scientific merit of generating particular hypotheses that can be tested numerically and experimentally, such as those about the role of eddy motions in determining the statistical properties of the turbulence, notably the variations in the energy spectrum $E(k)$ when it decreases in proportion to k^{-2p} and $2p$ changes from one flow to another, (Lundgren 1982, Perry *et al.* 1986, Vassilicos & Hunt 1991), or the invariants of the velocity gradients and how they relate to each other and vary over time (Tsinober 1998) and why extremely large (or even singular) velocities do or do not occur at particular points in the flow (Moffatt 2000).

Recent studies of the statistical measures of turbulence is helping to answer the major dynamical questions about turbulence, (Constantin 1999; Hunt *et al.* 2000, Warhaft 2000). For example what causes the nonlinear, but self similar mechanisms of the cascade process and how do these produce a wide spectrum of the small scale eddy motion on the time scale of the large eddies L/u_0? Why is this approximately universal in certain respects (e.g. second order two point spectra), but in others (e.g. higher order spectra) depends significantly on the largest scales, a property that invalidates the assumption that there can be a universal model of 'subgrid' scale motions in large eddy simulation (Hunt *et al.*, 2000)? A related question, of some practical importance, is to explain and predict the relaxation time scale $\tau(k)$ for estimating how the different eddy scales in the spectrum adjust when turbulence is strained by large scale eddies or a mean straining motion. This is the crucial parameter for changes in estimating the spectra of the small scale turbulence (Townsend 1976) or of sheared turbulence (Mann 1994).

The combined dynamical and statistical study of the eddy structure of turbulence has been particularly valuable for explaining and predicting how heavy and light particles have characteristic non-uniform distributions in these flows and (e.g. Squires & Eaton, 1990; Sene *et al.* 1993) settle at average speeds that differ significantly from their values in still fluid (Maxey 1987).

This article first reviews the deterministic analysis of small eddies in large scale straining flows away from boundaries showing how characteristic eddy structures emerge; some new concepts are presented about how this development also depends both on the instabilities within these distorted flows and on the interaction with uncorrelated background motions outside of the eddy structure. These build on earlier work of Betchov 1956, Ruetsch & Maxey 1992, Kida & Tanaka 1994, the results of flow structure visualisation from (a) numerical simulations, (b) computation of the invariants of the velocity field, and (c) dynamical arguments. The analysis concludes that the small scale eddies form and reform in a repetitive cycle of stages, each of which corresponds to structures of characteristic shape (or none). Similar cycles, but driven by different mechanisms, have already been demonstrated for the elongated eddy structures near the wall in turbulent boundary layers, at moderate values of the turbulent Reynolds number Re, (i.e. $\mathrm{Re} \lesssim 10^4$), (e.g. Sandham & Kleiser 1992; Holmes *et al.* 1997).

In section 3 we consider how, in fully developed turbulent shear flows, the large scale eddies that are not attached to any fixed boundary tend to be in the form of coherent vortical structures which move relative to the flow and, as they do so, interact with the mean shear or any mean straining motion (Prandtl 1925). The structures also interact directly with rigid or flexible boundaries, or the convoluted vortex sheets that bound any finite volume of turbulent motion. Analysing the vortex dynamics of large scale eddies in shear and straining flows, including the effects of boundaries, leads to concepts and quantitative models more directly than by calculating the vorticity of infinitesimal fluid elements using formulae for the motion of fluid lumps (Hunt 1987). This approach which can be applied in quite complex flows, might well be further developed in future, as Spalding (1987) has also suggested.

Section 4 is devoted to the construction and analysis of spectra of velocity fields consisting of random ensembles of different types of eddy structure. Currently there is no consensus or systematic algorithm about whether or when the spectra are determined on the one hand by a random distribution of independent structures or fluctuations (e.g. as explained by Tennekes & Lumley 1971), such as those constituting the near 'wall' structures in the high Reynolds number turbulent boundary layer by Perry *et al.* (1986), or, on the other hand, by the complex structure of particular types of eddy such as the vortex sheet spiral (Lundgren 1982, Hunt & Vassilicos 1991). The analysis given here, based on the results of the first part of the paper indicates under what conditions either one or other factor is dominant. A novel proposal is that the relevant structure model varies with the time scale on which the eddy structures are being analysed, for example depending on whether the flow distortion on the turbulence process is rapid or quite slow. This problem is analogous to that in analysing the spectrum of water waves, where there is a similar issue of whether it is determined by distribution of waves or by the forms of their individual shapes. The recent work in this field by Belcher & Vassilicos (1997) is applied here to the spectrum of turbulence.

The aim of this article is to develop a fairly complete 'story' of these processes, based on mechanisms each of which can be, and to some extent have been, tested theoretically and experimentally.

2   Isolated Eddies

2.1   Dynamics and invariants

We analyse the interaction starting at $t = 0$ between a uniform steady straining field or $U_i = \alpha_{ij} x_j$, whose scalar magnitude is the straining rate $S = \|\alpha_{ij}\|$, and a small scale 'eddy' field with velocity $\mathbf{u}(\mathbf{x}, t)$ and vorticity $\boldsymbol{\omega} = \nabla \mathbf{u}$, the total velocity. In this section we first consider the case where the large scale is uniform over a distance Λ and the small scale motion is confined within a distance of order ℓ, with characteristic velocity u_0 – for example that of a vortical eddy. The spectrum energy tensor $\Phi_{ij}^{(e)}(\mathbf{k})$ for wave number $\mathbf{k}$, energy spectrum $E^e(k)$, where $k = |\mathbf{k}|$, or one-dimensional spectrum $E_{ij}(k)$ for a one-dimensional wave number can be defined most generally if these eddies are considered to be distributed over space (Townsend 1976, Leonard 2000). The full spectrum for broad distributions of eddies, Φ_{ij}, $E(k)$, $E_{ij}(k)$ are analysed in Section 4.

For certain types of eddy and at certain stages of the development of most eddies, the development of low-amplitude random 'incoherent' motions $\mathbf{u}'(\mathbf{x}, \mathbf{t})$ within or outside the eddies have to be considered from their initial state $\mathbf{u}_0'(\mathbf{x}, \mathbf{t} = 0)$, with r.m.s. value u_0'. In Section 2.4 an ensemble of large scale motions is considered in order to estimate the interactions between large and small scale and incoherent motions in turbulence and thence how each contribute to the overall statistics, such as the spectrum of the total velocity $\mathbf{u}^* = \mathbf{U} + \mathbf{u} + \mathbf{u}'$ shown in Figure 1. (Townsend 1976 p. 100, Kida & Hunt 1989). In the analysis of Section 3 the large scale straining motion corresponds to the mean flow and the small scale is the full turbulent velocity field.

The equations for the small scale velocity field is expressed in coordinates following the large scale flow as

$$\frac{D\boldsymbol{\omega}}{Dt} \equiv \frac{\partial \boldsymbol{\omega}}{\partial t} + (\mathbf{U} \cdot \nabla)\boldsymbol{\omega} = (\boldsymbol{\omega} \cdot \nabla)\mathbf{U} + (\boldsymbol{\Omega} \cdot \nabla)\mathbf{u} + NL, \qquad (2.1)$$

where the nonlinear term

$$NL \equiv \nabla \wedge (\mathbf{u} \wedge \boldsymbol{\omega}) = -(\mathbf{u} \cdot \nabla)\boldsymbol{\omega} + (\boldsymbol{\omega} \cdot \nabla)\mathbf{u}. \qquad (2.2a)$$

Their order of magnitude is μ times the linear terms where

$$\mu = \hat{\mu}\mu_0, \quad \mu_0 = (u_0/\ell)/S \qquad (2.2b)$$

and $\hat{\mu}(t)$ is a variable coefficient which depends on the forms and orientations of the eddy in relation to that of the large scale strain. Initially $\hat{\mu}(t = 0)$ is of order one. However during the strain, since $|\boldsymbol{\omega}|$ and $|\mathbf{u}|$ usually increase by a factor A, where A is of order St in shear flows or $\exp St_*$, in straining flows, it follows that $\hat{\mu}(t \sim S^{-1})$ is of order A and increases until the nonlinear terms

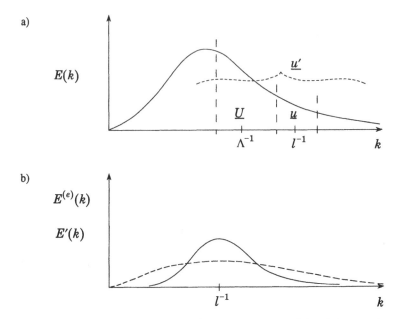

Figure 1. Schematic diagram of energy spectra: (a) of the total turbulent velocity field $E(k)$ showing the length scale ranges of typical 'large scale' straining fields $\mathbf{U}$, the small motions of scale ℓ and of the incoherent motions u' (that span these scales); (b) of the eddy motions of scale ℓ, $E^{(e)}(k)$ and of the incoherent motions.

(for certain components and for motions of length scale ℓ) become comparable with the linear terms. This occurs when $t \sim \tau_{NL}$ where $\tau_{NL} \sim (\ell/u_0)$, in shear or with pure strain $\tau_{NL} \sim S^{-1} \ln \left(\frac{1}{\mu_0} \right)$. Note that for the very smallest (micro) length scales of the distorted motion, the nonlinear effects are dominant at a smaller time, comparable to the time scale of these motions. However this is not the relevant time scale for estimating the changes to $\mathbf{u}$ and the overall structure of the distorted motion. (Contrast Batchelor & Proudman's (1954), criteria with that of Kevlahan & Hunt (1997).)

The aim of this section is to provide a geometrical and dynamical description of the stages through which small scale vortical structures pass as they are distorted by the large scale motion and deformed by nonlinear and instability mechanisms. See Figures 2 to 5. Section 2.2 shows how these stages correspond to the statistical measurements and to certain statistical properties of the turbulence especially the spectra and transfer of energy. The nature of the large scale straining is characterised by the second and third invariants

II, III of the velocity gradient tensor, the first of which can be described in terms of the magnitudes of the symmetrical or 'pure' strain component of the tensor Σ and vorticity Ω

$$\mathrm{II} = \partial U_i/\partial x_j \, \partial U_j/\partial x_i = \Sigma^2 - \frac{1}{2}\Omega^2, \qquad (2.3a)$$

where

$$\Sigma^2 = \Sigma_{ij}\Sigma_{ji}, \quad \Sigma_{ij} = \frac{1}{2}\left(\frac{\partial U_i}{\partial x_j} + \frac{\partial U_j}{\partial x_i}\right),$$

$$\Omega^2 = \Omega_i\Omega_i \text{ and } \Omega_i = \epsilon_{ijk}\left(\frac{\partial U_k}{\partial x_j} - \frac{\partial U_j}{\partial x_k}\right),$$

and

$$\mathrm{III} = \frac{\partial U_i}{\partial x_j}\frac{\partial U_j}{\partial x_k}\frac{\partial U_k}{\partial x_i}. \qquad (2.3b)$$

In normalised form $\widetilde{\mathrm{II}} = \mathrm{II}/\left(\Sigma^2 + \frac{1}{2}\Omega^2\right)$ and $\widetilde{\mathrm{III}} = \mathrm{III}/\left(\Sigma^2 + \frac{1}{2}\Omega^2\right)^{3/2}$. If the straining has a significant component of 'pure' strain, i.e. $\widetilde{\mathrm{II}} > -1$ then the vortex lines of **u** are stretched. If $\widetilde{\mathrm{III}} > 0$, typically large extensional strains in one direction are associated with weaker compressional strain in the other two. However in most turbulent flows the average value of $\widetilde{\mathrm{III}} < 0$, which is associated with two extensional and one compressive component of strain. In two-dimensional flows $\widetilde{\mathrm{III}} = 0$ (Betchov 1956).

In order to calculate the nonlinear development of the velocity field it is necessary to consider its initial distribution. In this second phase the distortion of any one vortex element depends on that of others (which it does not in the linear phase), and the distortion of the small scale velocity field depends on the spatial distribution of vorticity, as will emerge from the later calculations. What are the typical distributions? Since small scale vorticity is continually stretched and rotated by larger scale motions, it tends to be concentrated within elongated 'vortex structures' (VS) whose envelopes are parallel to vortex lines (Figure 2). The vectors normal to these surfaces are parallel to the gradient of the modulus of vorticity $\mathbf{g}^{(\omega)} = \nabla\omega$ where $\omega = |\boldsymbol{\omega}|$ (cf. Ruetsch & Maxey 1992).

As the small scale elongated structures undergo different types of linear and nonlinear distortion and then break up, their geometrical form can be described in terms of 'tubes', 'sheets','spheres' etc. depending on the 'aspect ratios' of the cross sectional slope of these surfaces. Then as nonlinear effects develop, the structures tend to deform, for example into rolled up sheets and 'sausage' instabilities. To avoid such imprecise geometrical descriptions, it is proposed to characterise the shapes of the scalar contours in terms of the invariants of the local average, taken over a surface S_ω of constant $|\boldsymbol{\omega}|$, of the anisotropy tensor $\langle g_{ij}^{(\omega)}\rangle$ of the vector $\mathbf{g}^{(\omega)}$.

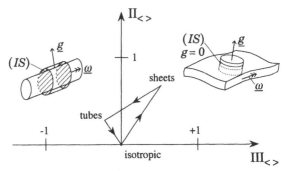

Figure 2. Schematic diagram to show how objective, frame-invariant measures $\text{II}_{()}$, $\text{III}_{()}$ for the geometrical forms of vortical eddies, vary over time during the cycle of deformation and break-up.

Thus if $\langle g_{ij}^{(\omega)} \rangle = \left(\frac{\langle g_i^{(\omega)} g_j^{(\omega)} \rangle}{\langle g_k^{(\omega)} g_k^{(\omega)} \rangle} - \frac{1}{3}\delta ij \right)$, where the average $\langle \rangle$ is taken over the surface of the S_ω. (For surfaces that are very elongated in one or more directions (e.g. a sheet), for S_ω to have a finite size it is constructed so that part of it is outside and tangential to VS and part of it consists of one (for a flat shaped VS) or two (for a cylindrical shape) intersecting surfaces, IS, on which $g_i = 0$. Thus if the contours are spherical, $\langle g_{ij}^{(\omega)} \rangle = 0$, or aligned with a cylinder, (whose axis is in the x_3 direction) $\langle g_{\alpha_i}^{(\omega)} \rangle = (1/6, 1/6, -1/3)\,\delta_{\alpha_3}$ for $\alpha = 1, 2, 3,$. The second and third invariants of $\langle g_{ij}^{(\omega)} \rangle$ for spheres, cylinders, sheets are

$$\text{II}_{()}^{(\omega)} = 0,\ 1/6,\ 2/3, \text{ and } \text{III}_{()}^{(\omega)} = 0,\ -1/36,\ 2/9.$$

Note firstly that the averages of the invariants are not the same as the invariants of the averages, and secondly the pointwise values of $g_{ij}^{(\omega)}$ do not have much significance. This notation provides a straightforward correlation between the invariants of the large scale straining flow field and those of the distorted flow. Thus in the linear phase when $\widetilde{\text{III}} > 0$ the vortex lines form into tubes (which are not necessarily circular) and $\text{III}_{()}^{(\omega)} < 0$, while when $\widetilde{\text{III}} < 0$, the vortex lines form into sheets and $\text{III}_{()}^{(\omega)} > 0$. Surface irregularities on VS are not described by this measure. To describe the internal structure of the motion within VS local measures such as spectra or wavelet distributions are required, as shown later.

2.2 How eddy motion is distorted

We now consider how VS evolve in different kinds of large scale distortions. In the special case of irrotational large scale strain (i.e. $\tilde{\mathrm{II}} = 1$), it follows that, whatever the initial orientation of the vorticity, $\boldsymbol{\omega}_0$, since $\frac{D\boldsymbol{\omega}}{Dt} = (\boldsymbol{\omega} \cdot \nabla)\,\mathbf{U}$, the change of vorticity is given by $\omega_i(t) = \omega_{oj}\, e^{\alpha_{ij} t}$. So that as $St \to \infty$, $\boldsymbol{\omega}$ becomes aligned parallel to the major extensional axis $\boldsymbol{\alpha}^{(E)}$ of α_{ij} (Betchov 1956, Kida & Hunt 1989) and because of the exponential growth $\boldsymbol{\omega}(t)$ is largest where $|\boldsymbol{\omega}_0|$ is largest. Secondly the shape of the resulting vortex structures is determined by how the large scale motions deform the contours of $|\boldsymbol{\omega}|$ and $\mathbf{g}^{(\omega)}$ through the term $(\boldsymbol{U} \cdot \nabla)\,\boldsymbol{\omega}$. If α_{11} is the principal eigenvector, then $dg_2^{(\omega)}/dt = -\alpha_{22}\, g_2^{(\omega)}$, $dg_3/dt = -\alpha_{33} g_3^{(\omega)}$ which shows that tubes form if $\alpha_{22} < 0$, $\alpha_{33} < 0$; and sheets if $\alpha_{22} \gtrsim 0$, $\alpha_{33} < 0$. As explained in Section 2.2, in any three-dimensional multi-scale vortical flow, $\overline{(\partial u_1/\partial x_1)}^3 < 0$. Since the straining is never perfectly axisymmetric about the principal strain, in general $\tilde{\mathrm{III}} < 0$, which is why most small scale vortical structures as they undergo significant large scale stresses are formed into sheets (where $\mathrm{III}_{()}^{(\omega)} < 0$) (Betchov 1956). Even if the large scale strain is not irrotational, for example if it is sheared, there is a similar trend to form flattened elongated vortex structures (Kida & Tanaka 1994, Carruthers *et al.* 1990). The important difference between purely irrotational strains and those with a rotational strain component is that in the latter case the deformed vortex structures take on a characteristic form that is quite insensitive to their initial orientation and form (Hunt & Carruthers 1990).

During this linear deformation stage, the NL term in (2.2a) generally increases at a rate depending on the orientation and slope of the vortex structure in relation to the large scale strain. Since the VS is stretched the gradients of velocity in the direction $\boldsymbol{\alpha}^{(E)}$ decrease and therefore, paradoxically, the nonlinear vortex stretching term $(\boldsymbol{\omega} \cdot \nabla)\,\mathbf{u}$ does not increase as fast as the linear term. The most significant nonlinear effect is the advection of the distorted vorticity by the velocity induced in the vortex structure, defined by the term $(\mathbf{u} \cdot \nabla)\,\boldsymbol{\omega}$. This term can be zero in certain circumstances, generally where VS is located symmetrically in relation to α_{ij}; for example in a plane irrotational straining flow where $\boldsymbol{\alpha}^{(E)} \cdot \boldsymbol{\omega}^{(0)} = 0$ the shape of the VS is flattened and stretched parallel to $\boldsymbol{\alpha}^{(E)}$ and perpendicular to $\boldsymbol{\omega}$. But the actual vorticity is not stretched. This leads to a velocity component parallel to $\boldsymbol{\alpha}^{(E)}$, and a steadily decreasing value of $(\mathbf{u} \cdot \nabla)\,\boldsymbol{\omega}$ (Kevlahan & Hunt 1997). This discussion is consistent with that of other investigators who have focused on the relation between the ω and the eigenvectors of the total strain $(\alpha_{ij} + \partial u_i/\partial x_j)$, (e.g. Vincent & Meneguzzi, 1991, 1994).

If α_{ij} is extensional with axisymmetric symmetry (with $\tilde{\mathrm{III}} = 1/36$) the VS becomes an axisymmetric vortex tube, so that in the resulting small scale azimuthal velocity fields the nonlinear advective term is zero.

Because of the rarity of such conditions, and because most large scale straining motions are partly rotational, the significant vortex structures become extended sheets of thickness $\delta \sim \ell e^{-St}$ (see Figure 3). Because of their finite width of order ℓ the vorticity in these strips induces a velocity u_3 at the edges with magnitude of order $u_2 \sim Stu_0 e^{St}$. This nonlinear velocity, through the term $(\mathbf{u} \cdot \nabla)\boldsymbol{\omega}$, leads to the advection of $\boldsymbol{\omega}$ normal to the sheet, with a speed that is largest at the edges. Therefore as the strip rolls up into a spiral its diameter D_S and the number of turns (N) in the spiral grow. D_S becomes comparable to the width of the strip when $t \sim \tau_{sp} = \ell/(Stu_0 e^{St})$ or $t \sim S^{-1} \log n (\ell S/u_0)$ (Kevlahan & Hunt 1997). For theoretical studies and experiments of vortex sheet rollup in high Reynolds number flows both (see Batchelor 1967 on the basic aspects) and (Werlé 1973) for applications such as the flow over delta wings and the vortex wakes of aircraft.

Note that as the total length of the vortex sheet is increased $\ell_{sp} \sim D_{sp}N_{sp}$ by the self-induced rollup, its thickness $\delta(t)$ and strength $\Delta u \sim \omega\delta(t)$ are reduced in proportion to ℓ/ℓ_{sp} relative to its value across the sheets without nonlinear effects.

In the special case of a purely rotational large scale strain (i.e. $\tilde{II} = -1$) its linear effects are firstly to rotate vortex lines of the small scale motion and not to stretch them, and secondly to generate small scale vortical fluctuations through the bending of the large scale vorticity $\boldsymbol{\Omega}$, (through the vortex distortion term $(\boldsymbol{\Omega} \cdot \Delta)\mathbf{u}$). The kinetic energy $\frac{1}{2}\langle u_i \cdot u_i \rangle$ of the small scale motion in this case does not increase with time. However, in this flow (unlike the previous one) only when nonlinear effects are eddy strong enough, are characteristic structures formed. As a result of the distance $\Delta = |\Delta|$ between pairs of particles increasing with time in any locally homogeneous motion (which has been proved rigorously by Etemadi (1990) and confirmed numerically and experimentally) it follows that $d\langle\langle \boldsymbol{\omega} \cdot \boldsymbol{\Omega} \rangle\rangle/dt > 0$ where $\langle\langle \ \ \rangle\rangle$ denotes a space–time average on a time scale of order ℓ/u_0. This leads to initially selective stretching of the vorticity $\boldsymbol{\omega}$ parallel to the vorticity of $\boldsymbol{\Omega}$, and to the generation of persistent VS aligned parallel to $\boldsymbol{\Omega}$ with circulation $\gamma \sim \Omega$ (Hopfinger et al. 1982; Cambon & Scott 1999). As in two-dimensional turbulence the larger of these structures tend to wrap weaker structures around them (Kevlahan & Farge 1997). Also they tend to elongate by increasing longitudinal convergence at each end of the vortex structure.

In many complex rotating flows, including turbulent flows, large scale flows are largely rotational but contain some straining. This means that the large scale streamlines are elliptical i.e. $0 > \tilde{II} > \mu^{-1}$. VS develop in the form of localised regions of small scale fluctuations which are amplified rapidly on a timescale Ω^{-1} (Leweke & Williamson 1998) and VS tend to grow parallel to the rotational axis. This is one of the mechanisms seen in the growth of turbulent spots within recirculating streamlines (Malkus & Waleffe, 1991).

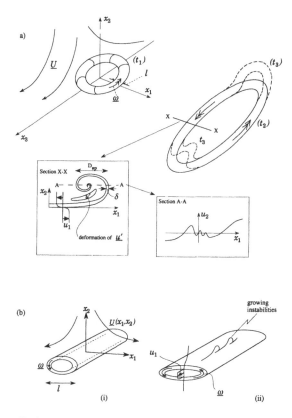

Figure 3. Deformation of typical eddy field **u** by typical large scale strain **U** with one compressional component and two extensional (so that $\widetilde{III} < 0$). (a) Shows how a ring eddy at time (t_1) becomes flattened and elongated by time (t_2), so that at time (t_3) the cross section of the flattened vortex ring and its whole shape are deformed by self induction (a non-linear process) and the tendency to form an accumulation point on AA. Note how the 'incoherent' motion is distorted by this process. (b) Shows how the different orientation of the ring eddy can be distorted, but in such a way that the direct non-linear effects are small. But in this case the resulting eddy motion is highly unstable to Kelvin-Helmholtz billows, which become non-linear.

2.3 Role of instabilities and external fluctuations on eddy rollup and breakup

For some types of eddy with particular orientation when they are distorted by the large scale strain, the nonlinear term NL in the vorticity equation tends to zero. In other words $\nabla \wedge (\boldsymbol{u} \wedge \boldsymbol{\omega}) \to 0$, as in (2.2a). Usually the VS become flattened vortex sheets (or flattened pairs of sheets) (when III < 0, and III$^{(\omega)} > 0$) or cylindrical vortex tubes (when III > 0 and III$^{(\omega)} < 0$). However this does not mean that μ simply evolves according to the linear theory until viscous processes intervene (assuming the spatial and time scale of the straining motions Λ and τ are large enough). The reason is that these flows with reduced nonlinearity become progressively more unstable to small disturbances as they are distorted. When the VS is formed into a sheet like structure with a velocity jump Δu a fluctuating velocity field $\boldsymbol{u}'(x,t)$ grows exponentially through Kelvin–Helmholtz billows, i.e.

$$|\boldsymbol{u}'(x,t)| \sim u_0'(x, t = 0) \exp\left(k\Delta u\, t_{NL}\right), \tag{2.4}$$

where the most unstable wave number is of the order of the thickness δ^{-1} of the VS, where $\delta^{-1} \sim \ell_0^{-1} \exp(St)$ or $\ell_0^{-1}(St)$ depending on whether the large scale straining is more irrotational or more sheared. Note that under the action of the large scale strain $\mathbf{U}$, the velocity jump Δu is increasing at similar rates. Since the instability in (2.4) can only develop if the large scale velocity jump is changing more slowly than the growth rate of the disturbance, they are more significant on the slowly growing sheared small scale eddies than those being strained irrotationally (which can be understood by considering the changed pressure distribution either side of the deformed velocity sheet, or, by analogy, the stabilising of an accelerating pendulum). It follows from (2.4) that provided the conditions are suitable for the growth of instabilities, when $t \geq S^{-1}$ these amplitudes increase so as to be comparable with the amplitude u_0 of the linearly distorted eddy, for example with small scale vorticity or billows, as can be seen in the laboratory or natural turbulence. Note that Ruetsch & Maxey (1992) and Kida & Tanaka (1994) focused on the quasi-steady Kelvin–Helmholtz mechanism, whereas their results were are also consistent with the vortex rollup mechanisms described in Section 2.2.

Another kind of instability occurs when the elongated vortex structures VS are in the form of circular or near circular tubes, whether generated linearly by axisymmetric large scale straining or nonlinearly in a large scale rotation. Typically, waves travel along the vortex, and grow rather slowly. They do not necessarily lead to destruction of the vortex (Hopfinger & Browand 1982), without the assistance of mechanisms, associated with the surrounding turbulence and other vortices, which we consider in Section 2.4.

So far it has been assumed that the large scale motions applied over a length scale Λ are unaffected by the small scale strain. In fact the nonlinear terms of (2.2) i.e. $\langle NL \rangle = \langle \nabla \wedge (\mathbf{u} \wedge \boldsymbol{\omega}) \rangle$ are on average non-zero and lead to changes in the mean vorticity $\langle \boldsymbol{\omega} \rangle$ over the scale Λ and thence to changes in $\boldsymbol{\Omega}$ and $\mathbf{U}$. These effects do not simply act like an isotropic eddy viscosity by reducing the gradients in $\mathbf{U}$; rather the anisotropic form of NL (and anisotropic

Reynolds stress gradients $\nabla \langle u_i u_j \rangle$) can lead to a significant change in **U**, for example by setting up large scale vortices on the scale Λ. Such motions, which have been analysed in the special case of stagnation flows by Kerr & Dold (1994) and confirmed experimentally by Andreotti *et al.* (1997), tend to limit the anisotropy and thence the extent of the stretching of the small scale vorticity. This is a particular example of the general tendency of large scale to be driven by Reynolds stresses set up in anisotropic small scale turbulence. (Sulem *et al.* 1999). To summarise, both nonlinear and instability mechanisms acting on small scale turbulence change significantly the distortions produced in the linear mechanisms over a time scale of order S^{-1}. The main effect is to amplify those velocities and vorticity components that tend to be diminished by the linear distortion process while not appreciably reducing those components with a high rate of growth. (Kevlahan & Hunt 1997). Thus the nonlinear mechanisms do not act isotropically, especially at the larger, energy containing scales of turbulence. (See for example the discussion by Townsend (1976) pp. 66–72 on this point.)

2.3.1 Interactions between distorted vortical eddies and external fluctuations

We now consider how external fluctuations with r.m.s. velocity u_0' interact with a typical vortical eddy in a large straining flow that has been flattened and stretched, and when either the nonlinear induction or the instability mechanism has begun to cause rolled-up vortices to form at the edges or over the central region of the strip. If the length scale, ℓ', of the external fluctuations **u**′ is less than the scale of the rolling up vortex, its vorticity and velocity are amplified by the straining of the vortical eddy which is of the order (u/ℓ). Furthermore, as the rolling up continues, the velocity jump Δu across the sheet weakens until it is of the same order as the amplified external velocity fluctuations u_0'. This leads to the random distortion of the vortex sheet and to the merging of the rolled up sheets into a coherent vortex.

Fluctuations and instability within the sheet also contribute to this diffusive process which, of course, is a widespread phenomena in high Reynolds flows wherever vortex sheets are generated in wakes or impulsively started in jets etc. (e.g. Batchelor 1967 p. 590). The time scale for this nonlinear rollup and diffusion is of the order of $\ell/u_0 \sim S^{-1}$. In turbulence at Reynolds numbers of order 300 or less, because of the relatively high viscous diffusion vortex sheets are quite thick and $\left(\delta \sim \ell R_\ell^{-1/2} \right)$ where R_ℓ is the Reynolds number for the small scale motion (i.e. $R_\ell \sim u\ell/v$). Also the tendency of the sheet to roll up is reduced by the enhanced viscous diffusion, caused by the spiral geometry (Angilella & Vassilicos 1999). Nevertheless even in the range of Re for the characteristic type of homogeneous turbulence generated from an initial set of vortices (the 'Taylor–Green' problem), the rolled-up sheets diffuse into elongated vortices (Ohkitani 1999). Flow visualisation studies (e.g. Sreenivasan 1991) show that typically the number of turns is about 2 to 3.

Note that large scale external motions have quite a small effect on these internal dynamics within the rolled up vortex, because the outer vortex sheets shelter the interior motions. (Kevlahan & Farge 1997; Hunt & Durbin 1999). Typically elongated vortices are formed, whether by this diffusive rollup process or by an axisymmetric strain on the time scale S^{-1}, which is that of the lifetime of the large scale strains (for reasons discussed below). Therefore some vortices continue to interact strongly with the strain, extracting energy from it while also dissipating it on a smaller scale (Moffatt, Kida & Ohkitani 1993). Subsequently, since vortices exist on their own without any external strain, they can persist for long after the strain has decreased, or more likely they have moved out of the local region of significant large scale strain (Jimenez et al. 1993). They are not particularly significant for the overall dynamics of the turbulence in terms of energy transfer, but they are of interest for many transfer processes.

These vortices' subsequent transformations are determined either by self-induced velocity fields which can rapidly deform a vortex on a time scale ℓ/u_0 (Hussain & Husain 1989), or by instabilities (for nearly straight or circular vortices) (e.g. Widnall 1975) on a similar time scale, or thirdly by interactions with the external velocity field $\mathbf{u}'$. This latter process may be quite rapid because the velocity field of the vortices distorts the external vorticity ω', so that it forms external helical or even ring like vortices; as these move along the vortices, they trigger waves and break down on the vortex on a time scale of order (ℓ/u_0'); see Figure 4. Specific numerical and laboratory experiments verify this basic process, which suggests that even if coherent vortices are generated in turbulent flows, they break up (Melander & Hussain 1996; Miyazaki & Hunt 2000). Only in the special situation of a rotating flow do strong vortices persist and extend over the whole length of the flow (Hopfinger et al. 1982); generally they extend over a limited distance, at maximum of the order of the scales Λ of the large scale turbulence.

2.3.2 The cycle of eddy dynamics within a cascade

The concept of Kida & Tanaka (1994), further developed here, is of a deterministic model for the characteristic eddy time scale in terms of the cyclical generation, transformation and breakup of particular types of vortex structure. It is now based on well defined studies of each stage of the cycle. However evidence for the existence of the complete cycle is still fragmentary. If this occurs within a turbulent flow, it involves the assumption that there are, at least initially, distinct large and small scale motions $\mathbf{U}, \mathbf{u}$, the latter being defined within a vortex structure. Also there is a weak background velocity field $\mathbf{u}'$ with scales that span the range of those of $\mathbf{U}, \mathbf{u}$, but only the length

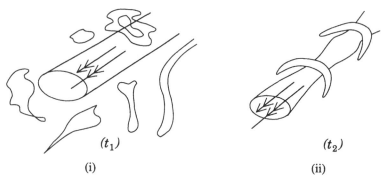

(t_1) (t_2)

(i) (ii)

Figure 4. Later stages of an eddy when it becomes an elongated vortex tube interacting with and distorting the surrounding incoherent turbulence. This may lead to the eventual breakdown of the vortex, in some cases via the generation of waves on the vortex.

scales less than that of the eddy are effective.

The next question is whether the cycle at the scale ℓ repeats itself at smaller scales and whether such motions are linked or are effectively independent. As shown in Figure 3, the nonlinear rolling up motion in the eddies induces a largely planar strain (so that the mean normalised third invariant for the strain tensor is negative $\overline{\widetilde{III}}^{(u)} < 0$). This leads to a distortion of the small scales of $\mathbf{u}'$, in a very similar manner to the way that $\mathbf{U}$ distorts $\mathbf{u}$. Note that the principal axis of the straining by $\mathbf{u}$ on $\mathbf{u}'$ is perpendicular to that of $\mathbf{U}$ acting on $\mathbf{u}$, which contributes a tendency towards isotropy and to independence between motions at different scales. At the same time, as we have already noted, the background field $\mathbf{u}'$ reacts back on the velocity field $\mathbf{u}$ by hastening the tendency of the rolling up sheets to diffuse into vortex tubes, and later by hastening the growth of longitudinal waves along the tubes and their eventual breakup. This process is quite analogous to the Kerr & Dold (1994) upscale effect of the eddy motions $\mathbf{u}$ tending to break up large scale stagnation flows $\mathbf{U}$ into cellular motions. The interaction between $\mathbf{u}$ and $\mathbf{u}'$ also demonstrates why most of the energy transferred to $\mathbf{u}'$ is at scales smaller than ℓ, typically of the order of the $\ell/3$.

Because the scales are close together the total strain acting on $\mathbf{u}'$ is caused by a combination of $\nabla\mathbf{U}$ and $\nabla\mathbf{u}$. Since $(\partial u_2/\partial x_2)^2$ increases where $(-\partial U_2/\partial x_2)$ is large, it follows that the average skewness of the total strain rate $\left(\frac{\partial}{\partial x_2}(U_2 + u_2)\right)^3$ is negative, and therefore the mean of the third invariant $\overline{\widetilde{III}}^{(U+u)}$ of the large scale motions acting on $\mathbf{u}'$ is negative, just as we assumed in the analysis of $\mathbf{U}$ acting on $\mathbf{u}$. For the same reason this tends to cause sheets to form on the small scale background field in the initial stage of distortion. This picture (based on earlier studies by several authors) of a semi-deterministic cascade includes more stages and more processes than that proposed by Melander & Hussain (1996) who focused on the interaction between $\mathbf{u}$ and $\mathbf{u}'$ at the later stage when $\mathbf{u}$ is in the form of tubes and causes tubes to form in the background field.

2.4   Statistical descriptions of eddy motions

In the complex cycle of the eddy structure the set of mechanisms and the characteristic kinematical features of the motion at each stage can be described in terms of statistical properties of the eddies' flow fields. The influence of the eddies on the whole turbulent flow and on its overall statistics are analysed later when we consider the statistics of the movements and large scale interactions within the eddies, the flow and their size and strength distributions.

2.4.1   Invariants of the velocity gradients

Tsinober (1998), Ooi *et al.* (1999), and others have found that in different types of turbulent flow the second and third invariants $II(t), III(t)$ (introduced in Section 2.1) of the gradients of the total velocity field $u_i^* (\mathbf{x}, t)$ have similar distributions when plotted against each other. In these studies, the signs of the invariants have been changed. Also they are rescaled and renamed as Q, R,

$$Q = -\frac{1}{2}II^* = \frac{1}{4}\omega_i^* \omega_i^* - \frac{1}{2}s_{jk}^* s_{kj}^* \qquad (2.6a)$$

and

$$R = -\frac{1}{3}III^* = -\frac{1}{3}s_{ik}^* s_{km}^* s_{mi}^* - \frac{1}{4}\omega_i^* \omega_k^* s_{ik}^*. \qquad (2.6b)$$

The variations of $Q(t)$ and $R(t)$ in an ensemble of strained eddies can be estimated using the analysis of the eddy cycle in Section 2.3 by the three components of the velocity field produced by strong large scale, initially weak small scale and weaker background motions so that

$$\mathbf{u}^* = \mathbf{U} + \mathbf{u} + \mathbf{u}' \qquad (2.7a)$$

and

$$s_{ij}^* = \alpha_{ij} + s_{ij} + s_{ij}'. \qquad (2.7b)$$

This approach explains the variations observed in different types of fully developed turbulence. For the first stages of the cycle the effects of the background turbulence $\mathbf{u}'$ can (as we showed in Section 2.3) be ignored in this analysis, and the effect of the large scale straining motion is dominated by its straining effect; thence it can be assumed to be approximately irrotational. Chertkov,

Pumir & Shraiman (1999) explained this cycle more in terms of the local dynamics of a vortical structure than the interactive dynamics with a large scale strain.

When the 'cycle' begins at $t = 0$, the small scale strain is weak, so that $||s_{ij}|| \ll ||\alpha_{ij}||$ and $Q \sim -||\alpha_{ij}|| = -S^2 < 0$. The average value of $\text{III}^{(U)}$ depends on the previous history of straining; as explained in sec 2.3, $\widetilde{\text{III}} < 0$, so that initially $R \simeq -\frac{1}{3}\alpha_{ik}\alpha_{km}\alpha_{mi} > 0$.

As the eddies begin to be strained, so that $0 < St \lesssim 1$, $|\omega|$ grows by a factor A, leading to an increase in Q; the vortex stretching term $\omega_i\omega_k s_k$ is the dominant term in (2.1b), so that R becomes negative. Progressively the vorticity $|\omega|$ and strain rate $||s_{ij}||$ of the eddies grow until they exceed the imposed strain, i.e. $|\omega_i| \sim ||s_{ij}|| \sim A\omega_0 \sim ||\alpha_{ij}|| = S$ (Kevlahan & Hunt 1997). Eventually the increase in Q, which is now positive, is limited by the nonlinear processes (such as rollup and instability) to being of order ω_0^2 or S^2,. At this stage the dominant term in R is the product of the large scale strain and the square of the small scale strain, so that $R \approx -\frac{1}{3}S_{ik}S_{km}\alpha_{mi} - \frac{1}{4}\overline{\omega}_i^2 Q_{ii}$. As shown by linear theory for typical cases (e.g. $-\frac{\partial U_2}{\partial x_2}\left(\frac{\partial u_2}{\partial x_2}\right)^2 > 0$ in a plane strain) R becomes positive.

In the final stage when the nonlinear mechanisms acting on the distorted vorticity field of the small eddies causes them to form vortex tubes, much of the velocity field on the scale of the small eddies is now the strain field outside the vortical region, where $\omega_i = 0$ and therefore $Q \sim || s_{ij} ||^2 \sim -A\omega_0^2$. Thus Q is dominated by the strained eddy field.

Previous investigators have pointed out how this clockwise cycle starts and ends in the quadrant where $Q < 0$ and $R > 0$. This pattern at a given instant in a flow with eddies at all stages is reflected by points at every point in the cycle, lying in a 'teardrop' shape (Figure 5).

The final stage of the cycle involves the interaction between these rolled up tube-like structures and the growth of the background turbulence, leading to the destruction of vorticity and strain rate (i.e. decrease of $|\omega|^2$ and $||s_{ij}||$) on the eddy scale ℓ. Therefore the values of Q and R approximately return to their initial values. The stages of the cycle have been most clearly delineated in numerical simulations over a range of moderate Reynolds numbers ($Re \lesssim 10^3$) (e.g. Ooi *et al.* 1999). In these flows there were no fully developed inertial range spectra. However, as proposed in Section 2.3, these mechanisms can operate at different length scales within the same flow (as one observes in clouds), in which case the cycle might be detectable if Q and R are defined in relation to a range of scales, say of the order of ℓ.

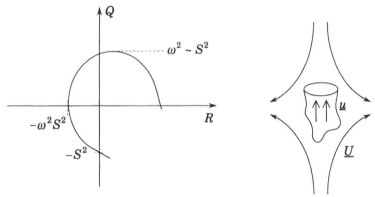

Figure 5. Schematic diagram (following Tsinober, Ooi *et al.*) of the
variation of the second and third strain invariants $Q - R$ for the
combined flow field of a small scale eddy and a large scale straining
motion, during the cycle of deformation and break-up. Note the
similarity with the $Q - R$ plots for fully developed turbulence
– probably because of the dominant non-linear interactions are
between eddies having rather similar length scales.

2.4.2   Spectra

Some aspects of the distortion of the eddies' velocity field (**u**) under the influence of the large scale strain **U** are described by the changes to the spectra
$\Phi_{ij}^{(e)}$ (**k**), $E^{(e)}$ (k) as (defined in Section 2.1). Firstly the dynamical effects of
the strain change the amplitudes and phases of the Fourier coefficients and
thence the spectra for each wave number component **k** and for each direction
i (and combination i, j). Secondly there is the kinematic effect on the wave
number **k**, in which they are increased/decreased/rotated depending on the
relative magnitude of the compression or dilatation of elements of the symmetric part of the strain tensor α_{ij}, and of the vorticity Ω_i (the antisymmetric
part). Thirdly these spectra, especially for the larger scales, depend on the
geometrical forms of the eddies and the internal velocity distribution within
them. At high Reynolds numbers these may be topologically quite complex.
For example the energy spectra depend on how the velocity is induced by
vortex lines distributed within the confines of an eddy, whose shape changes
significantly around the 'cycle'. Furthermore the spectra of these highly inhomogeneous eddy velocity fields may differ significantly from that of the overall
flow which is more homogeneous (see Section 4).

The distorted vorticity tends to become concentrated in elongated structures with thickness $\delta\,(t)$ and length ℓ in particular planar sheets, rolled up
sheets, and then tubes. Therefore the velocity spectra for intermediate wave
numbers $\ell^{-1} \leq k \leq \delta^{-1}$ are determined by the forms of the irrotational

velocity field just outside the vortical surfaces within which the vorticity is significant. So, as the forms of these vortical surfaces are distorted by the action of the larger scale straining flows, the velocity spectra of the eddies change, and in a way that differs from the changes to the spectra of small scale homogeneous turbulence under the influence of a uniform large scale strain. This difference affects appreciably how the spectra of the eddy motions contribute to the spectra of the whole velocity field.

Consider a vortical eddy (e.g. as in Figure 3a) which is a ring vortex whose axis is parallel to x_2 direction. The initial vorticity profile in the plane $x_3 = 0$ (shown in Figure 3a) is

$$\omega_3 (t = 0) = \omega_0 \, e^{-\left(x_1^2 + x_2^2\right)/\delta_0^2}.$$

When this is distorted by a plane strain (e.g. $\underline{U} = (0, -Sx_2, Sx_3)$) the exponential form of the **vorticity** spectrum is unchanged, though it is extended to higher wave numbers where $k \sim \delta^{-1} \rightarrow \infty$, since $E_\omega^{(e)}(k) \propto \exp\left(-k^2\delta^2(t)/4\right)$. By contrast the 'intermediate' velocity spectrum $E_{11}(k)$ over the wave number range $\ell^{-1} < k < \delta_0^{-1}$ now varies as a power law, i.e. $E_{11} \propto k^{-2}$. This significant change results from the 'jump' in velocity u_1, across the vortex sheet that forms when the vortex ring has been flattened by the strain (i.e. $\delta t \gg 1$) (see Section 4).

This result cannot be applied generally to small scale homogeneous turbulence being distorted by any large scale strain field. In fact such a qualitative change in the form of the total spectra for the whole velocity field $E_{ij}(k)$ only occurs if α_{ij} is rotational. For example in the case of large scale shearing e.g. $\mathbf{U} (= Sx_2, 0, 0)$, $E(k)$ for the distorted spectrum also decays in proportion to k^{-2} because of spanwise velocity discontinuities; (in the x_3 direction in other words the eddy structures are elongated high-speed and low-speed streaks (Lee, Moin, & Kim 1989, Hunt & Carruthers 1990). If α_{ij} is irrotational, the form of the high wave number spectrum (e.g. algebraic decay or exponential) does not change during the distortion.

In the second stage of the cycle of distortion, when the vortex sheets tend to roll up, this changes the nature of the mathematical 'singularity' which determines the exponent $2p$ of the decay of the spectrum (Hunt & Vassilicos 1991). From a simple discontinuity across the vortex sheet, there is now an 'accumulation' of such discontinuities at decreasing intervals towards the centre of the spiral rollup. This causes a decrease in the exponent $2p$, depending on the largely inviscid dynamics of the rollup. In the model of Lundgren (1982) $2p = 5/3$. If the diffusive effects dominate so that the rollup becomes a vortex tube of finite length, this implies that $2p > 1$ (Townsend 1951). Note that there is some ambiguity about this result, depending on the elongation of the vortex (Leonard 2000). The physical significance of these exponents when $1 < 2p < 2$, is that they indicate to what extent the intense vorticity occupies an increasing region of space, because the latter measure, using the 'box-counting' fractal dimensions D_k (cf. Sreenivasan (1991), is closely connected to the decrease in the exponent $2p$, (Vassilicos & Hunt 1991). However when the rolled up vorticity diffuses into a tube, the minimum length scale of ω immediately **increases**. Also the rate of production of small scale vorticity, which is a maximum in the first two stretching and straining phases, decreases.

A more detailed study of the transfer of energy from larger scales to smaller scales using the analysis of triadic interactions $\mathbf{k}_1$, $\mathbf{k}_2$, $\mathbf{k}_3$, where $\mathbf{k}_1 + \mathbf{k}_2 + \mathbf{k}_3 = 0$ shows that these nonlinear interactions are generally most intense where one of the wave numbers is much less than the other two, i.e. $|\mathbf{k}_1| \ll |\mathbf{k}_2|, |\mathbf{k}_3|$ (Domaradskii & Rogallo 1990, Brasseur & Lee 1990). This corresponds to the picture of strained eddies presented in Section 2.3, and to the weakly nonlinear computations of Kevlahan & Hunt (1997). In the second and third stages of vortex tube formation and breakup, Ohkitani (1999) shows that some energy continues to be transferred, but at a lower rate than in the first stage; this weaker mechanism involves non circular motions around the tubes interacting with the large scale straining field $\mathbf{U}$ (Moffatt *et al.* 1994) Also energy is transferred to small scale background turbulence during these final stages (Miyazaki & Hunt 2000). Analyses of inter-wave number energy transfer are needed for better understanding of this stage.

2.4.3 Energy transfer

The average rate of energy transfer $\bar{\epsilon}(\ell)$, to motions of scale ℓ, and quantitative estimates for the spectra $E(k)$ (and correlation functions) can be estimated by considering the distorted eddies. It is assumed that the primary interaction for a typical eddy scale ℓ is linear for a certain period $\tau(\ell)$ after which the interactions become nonlinear and the primary interactions cease. The detailed analysis in Sections 2.2, 2.3 of the different stages in the cycle of period T_c now enable this hypothesis to be examined more deterministically for each type of interaction or quantity being estimated. Note that only in the exceptional case of pure rotational strain shown in Section 3, is the nonlinear phase the most significant. If $\bar{\epsilon}(\ell, t)$ is the time dependent energy transfer rate for scales of order ℓ during this cycle, then $\bar{\epsilon}(\ell) = \frac{1}{T_c} \int_0^{T_c} \bar{\epsilon}(\ell)\, dt \sim \left[S u_0^2(\ell)\, \tau(\ell) \right]/T_c$ where $\tau(\ell) \sim \lambda S^{-1} \sim T_c$, the factor λ being a weak function of $(S \ell / u_0)$ (as discussed in Section 2.2). (This result for $\tau(\ell)$ is not obvious; it is sometimes assumed axiomatically that $\tau \sim \ell/u_0(\ell)$). This concept may be applied to a full energy spectrum $E(k)$ of approximately homogeneous, isotropic turbulence as in Tennekes & Lumley (1971) and Townsend (1976 pp. 99–100).

To estimate the transfer of energy $\bar{\epsilon}(\ell)$ to all eddies with length scale less than k^{-1} from eddy scales larger than k^{-1}, it is assumed that, as a result of nonlinear dynamics at high enough values of Re (Kevlahan & Hunt 1997), the strain rate of the smaller scales, except in the dissipation ranges, is greater than the strain rate of the largest scales of order $u_0 (L) / L$. This implies that $k^2 E(k)$ increases with k (or $2p < 2$). Since eddies of scale ℓ are distorted most effectively by motions with scale slightly larger than ℓ, the strain rate $S(\ell)$ acting on eddies of scale ℓ is

$$\breve{S}(\ell) \sim \left[\int_{\ell^{-1}}^{k} k^2 E(k) \, dk \right]^{1/2}. \tag{2.7}$$

The energy of the small scales $u_0^2(\ell)$ is (if $E(k)$ decreases rapidly enough i.e. $2p > 1$) determined by the integral of the energy over all wave numbers greater than ℓ^{-1}, i.e. $u_0^2(\ell) \sim \int_{\ell^{-1}}^{\infty} E(k) \, dk$. Therefore

$$\bar{\epsilon}(\ell) \sim \left[\int_{\ell^{-1}}^{k} k^2 E(k) \, dk \right]^{1/2} \left(\int_{k}^{\infty} E(k) \, dk \right). \tag{2.8}$$

If $E(k) = C^* k^{-2p}$, it follows that $2p = 5/3$, and that the dimensional factor $C^* = c\epsilon^{-2/3}$, and c is a dimensionless coefficient of O(1), as observed. One consequence of (2.7), (2.8) is that the relaxation time is indeed of the order of the time scales for the eddies of scale ℓ, i.e.

$$\tau \sim [S(\ell)]^{-1} \sim \ell/u_0(\ell). \tag{2.9}$$

This result (which is not valid for all kinds of spectra) is used in Section 3 to estimate the spectra of anisotropic turbulence undergoing a steady distortion, such as shear.

Whether the turbulence is at very high, or only moderate, values of Re, from (2.8) it follows that $\bar{\epsilon}(\ell)$ increases with ℓ, to a maximum where $\ell \sim L_X$. Then either $\bar{\epsilon}(\ell)$ is constant for $L_X \gtrsim \ell > \ell_K$ or decreases monotonically (if $Re \lesssim 10^3$ and $2p \gtrsim 2$). In both cases $\bar{\epsilon}$, the mean rate of transfer is equal to ϵ, the mean rate of dissipation at the small scales, and is of order u_0^3/L_X. This implies that the dimensionless ratio $\epsilon/ (u_0^3/L_X)$ is indeed rather insensitive to the form of the spectrum and to the value of Re when it is greater than about 30. Note that the local value of the dissipation rate $\tilde{\epsilon}(\mathbf{x}, t) = \nu (\partial u_i/\partial x_j)^2$ is most intense either in the vortex sheets or at the outer perimeter of vortex tubes. This is why $\tilde{\epsilon}(\mathbf{x}, t)$ is even more intermittent than the vorticity, which itself is more intermittent than the strain because the latter extends several diameters outside the vortex sheets tubes (note that $\epsilon = \bar{\tilde{\epsilon}}$).

Another statistical property of homogeneous turbulence that is insensitive to the form of its spectrum is the skewness of the velocity difference Δu over a distance ℓ i.e. $sk(\ell) = -\overline{\Delta u(\ell)^3}/(\ell)^3 / \left[\left(\overline{\Delta u^2(\ell)^2}/\ell^2 \right) \right]^{3/2}$. Typically $sk(\ell) \rightarrow -0.5$ for $\ell \ll L_x$. A similar analysis of small scale straining as in

Section 2.2 indicates that if the skewness is expressed in terms of the large
and small scale strain, then

$$sk\,(\ell) \simeq \frac{\langle(\partial U/\partial x)\,(\partial u/\partial x)^2\rangle}{\left\langle\left(\frac{\partial u}{\partial x}\right)^{3/2}\right\rangle}.$$

Since $(\partial u/\partial x)^2$ is amplified until it is comparable to the large scale strain S^2,
it follows that

$$sk(\ell) \sim -\left[S\,(\ell)\,/\,(u_0/\ell)\right]^3. \tag{2.10}$$

Since $S\,(\ell) \sim (u_0/\ell)$ for high Reynolds number turbulence, this suggests that
$sk\,(\ell) \sim -1$ independent of ℓ, and that for moderate or low Reynolds number
turbulence, $sk\,(\ell)$ varies slowly with ℓ. This is consistent with Betchov's (1956)
inequality for sk and his conclusion that its value of about -0.5 is not very
sensitive to Re or to the form of the spectrum. In other words the value of
sk is not a very sensitive indicator of the nature of a turbulent flow, or the
state of its development.

2.4.4   Interactions between vortices and extreme value statistics

Observations show that isolated tube-like vortices are continually being generated in turbulent flows; our analysis has shown quantitatively and qualitatively how this can be part of a repeating cycle of events. So far the only
interactions considered have been between those vortices and larger scale
strains and small scale background fluctuations, while neglecting any direct
interaction between the vortical eddies. Some investigations have suggested
that these are highly significant because they may, locally, lead to infinitely
large values of vorticity and even velocity, not only for inviscid but even viscous flows.

A possible classification of such direct interaction might follow the language
of theoretical physics and begin by discriminating between the strengths of
the interactions: weak when the vortices simply move around each other causing only small changes in the flow fields; moderate, when the structures do
not touch, but do significantly change each others' movements and shapes;
and strong, when the vortices impinge on each other, so as to change their
topologies (for example by creating new vortices) and to exchange vorticity
(along with heat and matter) by molecular processes. These classifications
need further clarification when the vortices interact in the presence of background shear, because this, if strong enough, can cause strong interactions
with an isolated vortex (e.g. Moin, Leonard, & Kim 1986, Ohkitani & Gibbon 1999). In two-dimensional flows simple numerical and physical examples
given by Kiya *et al.* (1983) of pairs of line vortex of varying strength with
an isolated line vortex showed that only for narrow ranges of initial orientation angle and relative strengths of the vortices could the interactions be

classified as strong enough for the pair to be broken up and the whole flow pattern changed (Kiya *et al.* 1983). The strong interaction of three-dimensional ring vortices are similarly rare. When they occur their nature differs. This is because now the vorticity can be stretched, distorted and diffused by the interactions, which leads to strong dissipation of energy; after the interaction when the main vortices have moved away in a changed form a residual small scale vorticity field may be 'left behind'. This might be another coherent vortex, or may be a chaotic field of vorticity (e.g. Kida *et al.* 1991). Whatever their strengths, all types of interactions tend to make the vorticity field more isotropic. The strongest interactions contribute most to the development of the energy spectrum, both upscale when unequal vortices merge (e.g. Browand 1966) and downscale when the small scale vorticity is produced as in the above example.

In general, coherent regions of vorticity within turbulent flows are not nicely formed in closed loops, but rather they are local concentrations of vortex lines. Typically they are nearly straight (e.g. Vincent & Meneguzzi 1991); this is because, if they are curved. since the curvature always varies along the vortex, they rapidly deform under their own induction (e.g. Hussain & Husain 1999). What happens when such locally straight vortices interact? Following Moffatt (2000), consider the special situation of a strong interaction when two pairs of antiparallel straight vortices approach each other, the axes of the two pairs are at right angles to each other. The velocity field of each pair stretches and deforms vorticity in the other pair; then because of the symmetry, the impinging path of each pair is not significantly deflected and the interaction results in values of ω and $\mathbf{u}$ that grow without limit, in a finite time, a possibility first proposed theoretically by Leray (1993).

One might conclude that even if strong interactions do not determine the low order statistics, they are the most likely cause of the extreme values of velocity and vorticity anywhere in a turbulent flow field. Therefore they need to be included in any theory or simulation of a turbulent flow if the extreme values of the probability distribution are to be represented. Where nonlinear models do include probability density functions in their predictions (e.g. Kraichnan 1959) these effects do not appear to be included.

3 Interactions between Eddies and Vortical Structures and Large-Scale Motions

3.1 The effects of deformation of eddies

The analysis of the previous section was directed towards showing how large scale strains distorted the eddies and their motions, and also how this affected the statistics of the velocity fluctuations. This can now be applied to the particular types of strain caused by uniform mean shear flows $U(x_3)$ acting on turbulence whose r.m.s. velocity u_0 is small compared with the typical change ΔU of the mean velocity over the scale Λ_U on which the mean velocity gradient varies. It is assumed that the integral scale L_0 is smaller than Λ_U. First the distortions of the eddies are analysed in detail and thence their effects on the mean velocity produced by the Reynolds shear stress $-\overline{u_1 u_3}$ (assuming the density is unity). Such calculations are performed routinely using models for the one-point turbulence statistics $(-\overline{u_i u_j})$, which are based on certain physical assumptions that can now be clarified and in some cases corrected. Also they may depend on estimates of the integral length scales of turbulence L_0, especially that for the normal component, $L_x^{(3)}$, and the integral dissipation scale $L_\epsilon = \epsilon/u_0^3$. In many shear flows, it found that since $L_x^{(3)}$ is smaller than the scales for the other components, it is a good measure for the scales that determine the straining motions of the smaller scale motions. Therefore $L_x^{(3)} \simeq L_\epsilon$ (Hunt 1992).

We have seen in Section 2 how small scale turbulent eddies tends to be distorted cyclically. This is why the average degree of distortion can be estimated by linear processes acting over a distortion period t_D. If $t_D = \hat{t}_D \alpha^{-1}$, where $\alpha = dU/dx_3$, the non-dimensional distortion time $\hat{t}_D$ varies (typically over a range 2 to 10) according to how long fluid particles are sheared after entering the shear flow (as in jets and wakes) and how they pass through regions of high shearing near a rigid surface (Townsend 1976, Savill 1987).

The stages of the linear and nonlinear development of the turbulence from the initiation of the distortion (which we define for this analysis as $t = 0$) can best be understood in terms of the momentum and vorticity equations and the eddy structure of differing length scales ℓ.

The equations for the fluctuating velocity u are

$$\partial u_i/\partial t + U_1 \partial u_i/\partial x_1 + u_3 \partial U_1/\partial x_3 \delta_{i3} + [u_j \partial u_i/\partial x_j] = -\partial p/\partial x_i + \nu \nabla^2 u_i \quad (3.1)$$

$$\partial u_i/\partial x_i = 0, \quad (3.2)$$

and for the fluctuating vorticity ω_i are

$$\partial \omega_i/\partial t + U_1 \partial \omega_i/\partial x_j = (\omega_3 \partial/\partial x_3) U_1 \delta_{1i} - (\Omega_2 \partial/\partial x_2) u_i +$$

$$\left[\left(\omega_j \frac{\partial}{\partial x_j} \right) u_i - (u_j \partial/\partial x_j) \omega_i \right] + \nu \nabla^2 \omega_i, \quad (3.3)$$

where $\boldsymbol{U}(x) = (U_1, 0, 0)$ is the mean shear velocity, with mean vorticity $\boldsymbol{\Omega} = (0, \Omega_2, 0)$. The terms in square brackets are nonlinear. Note that the fluctuating velocity field on a length scale ℓ is related to the local vorticity field $\boldsymbol{\omega}(\mathbf{x}, t)$ over the same length scale. Similarly the fluctuating pressure p is determined by the inertial forces over a distance of order ℓ around the point **x**.

After this hypothetical re-initiation of the turbulence for a certain period between $t = 0$ and $t \sim T_L = L_0/u_0$ the change in the energy containing eddies of the turbulence is largely determined by the linear stretching and rotation of the turbulent vorticity $\boldsymbol{\omega}$ by the shear (i.e. $\boldsymbol{\omega} \cdot \nabla \boldsymbol{U}$). At the same time comparable levels of fluctuating vorticity are produced by the mean vorticity being deformed by the velocity fluctuations through the term $(\boldsymbol{\Omega} \cdot \nabla)\boldsymbol{u}$. For smaller eddies in high Reynolds turbulence whose natural time scale (see Section 2.3) is much shorter being $O(\ell/L_0)^{2/3} T_L$, the linear process is also shorter. This nonlinearity affects the high wave number spectrum but not the magnitudes of the variances of the turbulent velocity (Kevlahan & Hunt 1997).

If the mean shear rate α is much stronger than the strain rates of the larger eddies, i.e. $\alpha L_0/u_0 \gg 1$, the linear process determines the amplification of $\overline{u_1^2}, \overline{u_2^2}$ and the Reynolds shear stress $-\overline{u_1 u_3}$, even for times longer than T_L, as shown by the DNS results of Lee *et al.* (1990). However the linear theory does not generally predict the correct trends, when $t > T_L$, for the normal, or vertical, component; its variance $\overline{u_3^2}$ is reduced by the linear process, whereas nonlinear effects begin to influence this component soon after the distortion begins when $t > T_L$ (Bertoglio 1981, Brasseur & Lee 1989) and eventually the whole structure and the energy of all the components. The time scale depends on the form of the spectra; the higher the Reynolds number the sooner the nonlinear processes operate. Controlled laboratory homogeneous shear experiments by Tavoularis & Karnik (1989) confirm these relative trends, as well as showing that the nonlinear processes are a critical element of the whole flow.

During the linear phase the eddy structure is organised into sloping vortices (Carruthers *et al.* 1990, Rogers & Moin 1987) – not necessarily round in shape – whose vertical velocity components displace fluid elements in the shear flow and thereby induce elongated streamwise streaks of accelerated and decelerated flow. In strongly distorted turbulence these structures persist even when the nonlinear effects are significant (e.g. Lee *et al.* 1990).

At moderate Reynolds number turbulence (typically $Re \leq 10^3$), the velocity field is quite smooth and the energy spectra $E_{11}(k)$ of the turbulence decrease in proportion to k^{-2p} where $2p > 2$. In the presence of shear the velocity gradients of the 'singular' streaky feature of the flow structure are greater than those of the unsheared turbulence and cause $E_{11}(k)$ to decrease in proportion to k^{-2} (as it does with any vortex sheet structure – see Townsend (1951)). For very high Reynolds number turbulence the locally quasi-isotropic eddies ensure that $E_{11}(k)$ retains its usual inertial range form. This analytical result, which has been computationally and experimentally verified (Hunt

& Carruthers 1990) is rather insensitive to the overall structure of the turbulence (provided $2/p > 2$), such as its anisotropy or spectra in the energy containing range, and has been measured even close to a rigid boundary (Lin et al. 1996).

As the mean shear stretches out the vortices, the numerical simulations of Kida & Tanaka (1994) show how the component of the shear that is pure strain motion at 45 degrees to the flow direction (Townsend 1976) flattens the vortices into sheets, which then roll up into new smaller vortices, in the same way as eddies do in irrotational motion. (Kevlahan & Hunt 1997). At the same time the sharp velocity difference across the central part of these stretched sheets also causes an exponentially growing local instability, leading to Kelvin–Helmholtz 'billows', orientated perpendicular to the mean flow direction. Both the induction and instability nonlinear mechanisms contribute to the growth of the normal or vertical velocity component of the turbulence, and at small scales. This assumes that the Reynolds number is high enough for the intense small scale events not to be damped out. Note that although the linear mechanisms continue to damp the large scales, eventually nonlinear effects cause the amplification of $\overline{u_3^2}$ at these scales. So how should these effects be calculated?

Rotta (1951) developed a model for how anisotropic homogeneous turbulence, which had been produced by a planar irrotational distortion, tended to become more isotropic as it decayed. Since then it has been customary to assume that the chaotic nonlinear motions of eddies – perhaps like the motion of gas molecules – would always act in this way. However experimental measurements of turbulence both behind a gauze and a distorted grid show that, in other cases, the degree of anisotropy can remain constant, or even grow as the energy decays (Townsend 1976, Michard et al. 1987). Clearly this difference demonstrates the inadequacy of any kinetic theory analogy which ignores the history of the distortion. A physical explanation for the difference between the two types of behaviour is that in the type of turbulence examined by Rotta the eddies were flattened into vortex sheets which tend to roll up and therefore cause the exchange of kinetic energy in all directions, while in the case of turbulence produced by a gauze the eddies were in the form of rather stable vortex rings which preserve their anisotropy as they decay.

This distinction can be quantified using the approach introduced by Lumley (1978) in which the anisotropy is characterised statistically in terms of the second II and third III 'invariants' (i.e. they are the same whatever the orientation of the coordinate direction) of the anisotropy tensor $b_{ij} = \left(\frac{\overline{u_i u_j}}{(\overline{u_k u_k})} - (1/3)\delta_{ij} \right)$. Since II $= b_{ij} b_{ij}$ and III $= b_{ij} b_{jk} b_{ki}$, it follows that when the eddy vorticity is in the form of sheets where one velocity component is much larger than the others, III > 0 and conversely when vortex rings form, III < 0. This is why the statistical Reynolds stress transport models that are constructed to be as general as possible (Craft & Launder 1996) now model the different dynamics in these two situations by allowing the coefficients in the models (e.g. that define the terms $p\overline{\partial u_i/\partial x_j}$ in terms of the Reynolds stress) to vary as a function of these invariants. A word of caution is necessary about interpreting these invariants: even if they tend to a constant value it does not necessarily imply that all the significant statistical properties are in a steady state; in particular if $\overline{u_1^2} \gg \overline{u_2^2} \gg \overline{u_3^2}$, so that II $\rightarrow 2/3$, but all the ratios of variances, e.g. $\overline{u_1^2}/\overline{u_3^2}$ are continuing to vary (Rogers 1991, Hunt & Carruthers 1990).

The variations in the eddy structure occuring in shear flow are reflected in forms of the spectra, as explained in Section 4.

In uniform shear at the large scales of eddy motion the experimental and numerical results show that the integral scales $L_x^{(i)}$ and the energy steadily increase. This implies that the mean rate of dissipation ϵ is always less than the rate of production P of energy by the mean shear, where $P = -\overline{u_1 u_3}\frac{\partial U_1}{\partial x_3}$. This is because energy is spreading to ever larger scales just as it does in unsheared turbulence; probably this is associated with eddies merging, despite being stretched out by the shear. It implies that the turbulence is not in a steady state, and consequently does not provide a model for shear flows in general, even though the eddy structure itself does have many features in common with all other shear flows (Townsend 1976).

3.2 'Modal' structures in non-uniform flows

In most shear flows the mean velocity gradient $U' = dU_1/dx_3$ is non-uniform and the velocity fluctuation are not only *inhomogeneous* but large scale weak fluctuations or modes span the entire flow, see Figure 6.

These gradients of the mean velocity and of the turbulence kinetic, so that $U'' \equiv d^2U/dx_3^2 \neq 0$ and $du_0^2/dx_3 \neq 0$, affect the eddy structure in these types of turbulent shear flow.

In shear flows where $U_1'' \neq 0$, consider the growth of certain low amplitude fluctuations, defined by $\breve{u}(x, t; \mathbf{k})$, each labelled with a parameter $\mathbf{k}$, (where $|\mathbf{k}|^{-1} \sim \ell$). These tend to form so that all the velocity components at every point in space vary with time in the same way, typically varying with a frequency σ_i and/or exponentially growing or decaying with time scale σ_r. Thus

$$\breve{u}(x, t; \mathbf{k}) \sim \breve{\breve{u}}(x; \mathbf{k}) \exp\left[(i\sigma_i + \sigma_r)t\right]. \tag{3.4a}$$

In most shear flows these 'modal' fluctuations move parallel to the mean flow with velocity $c \sim \sigma_i/k_1$, so that

$$\breve{u}(x, t; \mathbf{k}) \sim \breve{\breve{u}}(x; \mathbf{k}) \exp\left[ik_1(x_1 - ct) + \sigma_r t\right]. \tag{3.4b}$$

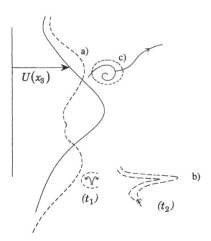

Figure 6. Different types of eddy structure in a mean shear flow:
(a) modal eddies spanning the flow (very strong in free shear
flows); (b) eddies growing algebraically as a result of the mean
flow (typically strong near a wall); (c) coherent (or 'detatched')
eddies moving across the flow (as in the outer part of a boundary
layer).

If they exist these modes are 'eigensolutions' to the linearised form of the
momentum and continuity equations (3.1)–(3.3). The form of the spatial amplitude function $\breve{u}$ and the growth rate for given k_1 depend on the spatial
boundary conditions, but not on any initial, or other temporal conditions.
The relevant hydrodynamic stability theory shows that if the profile has a
point of inflection at $x_3^{(I)}$, i.e. $U''\left(x_3^{(I)}\right) = 0$, then for laminar flows at high
Reynolds number $Re = \Delta U \Lambda/\nu$, the form of $\breve{u}$ and the values of $\sigma(k_1)$ for
the most unstable modes are independent of Re. This why even when such
flows, for example wakes and jets, have become fully turbulent (where the
effective value of Re, based on the eddy viscosity ν_e, $Re_t = \Delta U h/\nu_e$, see
Lessen (1979), Townsend (1976)) the forms of the eddies over the scale Λ
of the mean flow are quite similar to those for transitional flows (Liu 1989).
Similarly two-dimensional vortical structures are observed in both turbulent
and laminar separating boundary layers.

However for velocity profiles, such as zero pressure gradient boundary layers, where there are no inflection points (i.e. $U''(x_3) \neq 0$), the instability of
laminar flows depends on the viscous stresses acting on the disturbances, especially near the 'critical' level $x_3^{(c)}$, where $U(x_3^{(c)}) = c$. Here the perturbed
flow circulates in 'cats eyes' and viscous stresses have time to influence the
mean flow perturbed flow even at high values of Re. This is why on the one

hand the laminar instability mechanism is maximum at finite but large values of Re, and tends to zero as $Re \to \infty$; once the disturbances begin to grow within these recirculating regions, they develop into elongated streamwise vortices, and progressively become more effective in changing the momentum of the mean velocity profile.

However this delicate mechanism is easily disrupted by finite amplitude fluctuations, for example when local turbulent spots are formed, as recent theory, laboratory experiments and numerical simulations have convincingly demonstrated (Smith, Dodson & Bowles 1994, Wu *et al.* 1999). This is why the large eddies in fully turbulent boundary layers are dissimilar to those generated by low amplitude fluctuations in laminar boundary layers at transition, and are more like the linear perturbations of Section 2 that grow 'algebraically' linearly at different rates for each component and in different parts of the flow, e.g. Trefethen *et al.* 1996, Cambon & Scott 1999.

Note that long lived streamwise vortical structures appear in both kinds of layer if they are triggered externally or by an instability such as those generated by wakes of obstacles much smaller than the boundary layer thickness (e.g. Peterka *et al.* 1985). Structures as deep as the boundary layer thickness sustained by the action of normal Reynolds stresses when their widths are about five times as great as the boundary layer thickness (Townsend 1976).

3.3   Movement of eddies

3.3.1   Fluid lumps in uniform shear flows

In the linear analysis of Section 2, the small eddy-like perturbations, as they are being distorted by large scale straining flows, move with the local mean velocity $U(\mathbf{x}, t)$. If the perturbations are low amplitude eigenmodes that span a shear flow, discussed in Section 3.1, they travel with a phase speed c, which is not equal to the local mean velocity U (except at the critical level $x_3^{(c)}$).

These types of disturbances can transport energy and momentum across the flow through the gradients of the pressure fluctuations they generate. However there is another mechanism for such transport through the physical motions of the small packets of waves or lumps of fluid across the mean streamlines, which is generally more effective in turbulent flows and provides the conceptual basis for estimating the length scales and other parameters used in practical models for fluxes of momentum, heat and matter.

In most turbulent flows observations show that the large scale eddies can be identified by coherent patterns of motion, using vorticity filtered on the scale of the eddies as the 'indicator' (Hussain 1986). In general these coherent structures are defined by 'open' surfaces, so that there is some exchange of mass with their surroundings, as well as momentum; they usually move approximately, but not exactly, with the local mean velocity. Prandtl (1925) included sequences of photographs of elliptical eddies moving in an open channel flow in his classic paper where he proposed a model for how three-dimensional eddies transfer momentum across a mean shear flow $U(x_3)$. His basic concept can now be better justified and explained by using recent research, on the deformation and movement of fluid lumps (Prandtl's 'Flussigkeit Ballen') (Magnaudet & Eames 2000, Hunt 1987).

In a shear flow a solid sphere, moving at high Reynolds number with vertical velocity $u_3 (> 0)$ across a shear flow, accelerates as a result of inertial forces. If initially the streamwise perturbation u_1 is zero, then at a time δt later (by which time the particle has moved $u_3 \delta t$ in the x_3 direction)u_1 has increased to $\lambda_1 \sim c_1 \delta t u_3 U'$, where $\lambda_1 = 2/3$. The velocity fluctuation relative to the mean velocity at the new position is $u_1' = (\lambda_1 - 1) \delta t u_3 U'$ and $u_3' = u_3$. Thence the contribution to the Reynolds shear stress (which is determined by the fluctuations) by the fluid inside and outside the lump is

$$-\overline{u_1' u_3'} \gtrsim (1 - \lambda_1) U' \delta t \, \overline{u_3'^2};$$

hereafter we drop the prime superscript for the velocity fluctuation.

This result also applies to a fluid sphere for a short time before it is deformed, and therefore this concept is appropriate at the outer regions of free shear layers. As a fluid lump deforms it tends to roll up (compare the photos of eddies by Head & Bandyophadhay (1980) with those studied by Rottman et al. (1987)), and only slowly loses its momentum defect across the mean flow over a distance ℓ_m. Prandtl's original term 'Bremsweg', or braking distance, was better than his confusing later term 'Mischungsweg' or mixing length (because sometimes this is mistakenly explained as mixing causing momentum transfer). However at larger times when $t_D^* \gg 1$ the lump is stretched out nearly parallel to the flow and then $c_1 \ll 1$, which approximates to its form in shear flows near rigid walls (Landahl 1990). Note that in the presence of a mean scalar gradient $d\bar{\theta}/dx_3$, the contribution to the heat flux by the lump is $-\overline{u_3 \theta'} = u_3 \delta t (d\bar{\theta}/dx_3)$. Thus the lump analysis explains why the eddy Prandtl number or ratio

$$\mathrm{Pr} = K_e / \nu_e \qquad (3.5)$$

of the eddy viscosity ν_e (defined as $-\overline{u_1' u_3'}/(dU/dx_3)$) to the eddy diffusivity K_e (defined as $-\overline{u_3 \theta'}/(d\bar{\theta}/dx_3)$) varies significantly with the period over which the lumps are sheared in different shear flows (Townsend 1976; Prandtl 1956). In both situations of short and long shear time the contribution to K_e from the lump is $(K_e) \sim \delta t \left(\overline{u_3^2}\right)^3$. Therefore Pr, as defined by the lumps, decreases from about 3 to about 1, whereas the linear analysis of isotropic fluctuations shows a decrease from 5/2 to 1; comparing wakes, where the normalised distortion time $\hat{t}_d \sim 2$, and boundary layers, where $\hat{t}_d \sim 10$, shows a decrease in the measured values of Pr from 2 to 1. This is why ℓ_θ

the length scale for estimating K_e differs from ℓ_M. These length scales are determined by the turbulence and mean flow profiles as explained next.

3.3.2 Eddies in non-uniform shear flow

Both small perturbation analyses and experiments of turbulence in uniform shear show that the integral length scale $L_x^{(3)}$ of the normal fluctuations increases without limit. By contrast in flows where the mean shear varies (i.e. $d^2U/dx_3^2 \neq 0$), experiments show that this integral length scale has a finite value related to the local mean $|dU/dx_3|$ shear and to the r.m.s. turbulence $\left(\overline{u_3^2}\right)^{1/2}$ (e.g. Hunt *et al.* 1989).

The eigensolution perturbations (as in (3.4)) which are correlated across the whole shear flow (e.g. boundary layer, wakes etc.) do not determine the fluxes of momentum and shear because their amplitude is too small. Therefore it is necessary to consider how localised disturbances interact with the mean shear, taking into account how they move across the shear flow between levels where the mean velocity gradient differs. This idealised analysis makes use of earlier research by Grosch & Salwen (1968) on perturbations to vortical shear layers and its recent extension and application (Hunt & Durbin, 1999; Jacobs & Durbin 1998).

 Consider low amplitude local (as opposed to large scale) velocity fluctuations moving approximately with the mean flow U and having a length scale ℓ_e (such that $|u_0| \ll \Delta U$) generated in one layer of a mean non-uniform shear flow (say $x_3 > x_3'$,where effectively $\frac{\partial U}{\partial x_3}(x_3) = \Omega_{(+)}$). Because of the very weak $(O(u_0^2))$ pressure fluctuations induced by these eddies they cannot effectively induce (first order) velocity fluctuations at lower layers in the flow, where $x_3 < x_3'$ and $\Omega = \Omega_{(-)}$. Therefore the cross-correlation between velocity fluctuations at different heights is weaker in non-uniform shear flows, compared with their constant velocity or uniform shear flows. For example, in the latter type of flow, the induced velocity u of a vortex ring whose maximum velocity is u_e decays with distance r from it in proportion to $u_e \left(\ell e/overr\right)^3$. But in the former case, if the mean vorticity gradient is strong enough that U decreases significantly below x_3', in relation to u_e, then the component of velocity correlated with that of the vortex ring (denoted by $\tilde{u}$) is

$$\frac{\tilde{u}\left(x_3 < x_3'\right)}{u_e} \sim \frac{|c - U|}{U}. \tag{3.6}$$

Since the eddy tends to travel with the the mean flow, i.e. $|(c - U)/U| \simeq 0$, to first order, the turbulence below x_3' is sheltered from the effects of eddies above that level (Jacobs & Durbin 1998).

However, to second order, a correlated component of velocity is induced, i.e. $\tilde{u}(x_3') \sim u_e^2/|\overline{U(x_3) - U(x_3')}|$. This means that the length scale L_x over which eddies decorrelate, say by a factor $(1/3)$ is such that where $x_3' \sim x_3 - L_x$, the correlated component $\tilde{u}(x_3 - L_x) \sim \frac{u_e^2}{L_X \, dU/dx_3} \sim (1/3) u_e(x_3)$. Since $u_e \sim u_0$, it follows that

$$L_X \sim A_S \frac{u_0}{dU/dx_3},\qquad(3.7)$$

where the shear length scale coefficient A_S is of order unity and typically varies slowly along and across shear flows, and u_0 is the local r.m.s. velocity of the eddies **normal** to the mean flow. Note that this length scale, although proposed by Champagne *et al.* (1970) for uniform shear flows is actually more appropriate for non-uniform shear flows with curvature in their mean velocity profiles, such as wakes, jets etc., because, as explained already, in uniform shear flows the length scale grows without limit. In boundary layers, pipes and channels the formula, which has been tested with Direct Numerical Simulation, has to be modified for the direct blocking effects of the walls on the eddies – see Section 3 – (Hunt *et al.* 1989).

Of course in all these flows the mean shear is strongly coupled to the turbulence. This causes the regions of turbulence to spread. In fact the length L_x and time scales $T_L \sim L_x/u_o$ grow at such a rate that T_L remains of the same order as the travel time (or distortion) T_D of a particle the turbulence was initiated.This can be derived from the results tabulated by Tennekes and Lumley (1971). (Hunt 1992). It explains why these flows always have some dependence on their initial conditions, and why models used in practice are more complex than that given by (3.7)!

3.3.3 Eddy transport or diffusion

Wherever the eddy motion is more energetic in one region compared with another (e.g. at levels above and below $x_3 = x_3'$), one observes that there is a net movement of energetic eddies towards the low energy region, i.e. down the gradient. This is an intermittent and non-Gaussian process. From the momentum equation (and quasi-normal assumption)

$$\overline{u_3^2 \frac{du_3}{dt}} = (1/3) \frac{\overline{du_3^3}}{dt} \sim -\overline{(u_3^2)} \frac{\overline{du_3^2}}{dx_3},$$

whence the magnitude of the third moment can be estimated by using the concept of the eddy time scale T_L, so that

$$\overline{u_3^3} \sim -3T_L \, u_0^2 \, \nabla u_0^2 \sim -3L_x u_0 \nabla u_0^2,\qquad(3.8)$$

(Wyngaard 1979; Hunt, Kaimal & Gaynor 1988). Therefore the rate of transport of energy in slowly developing shear flows from one level to another can be estimated by assuming that on this timescale the turbulent energy diffuses, at a rate given by

$$\frac{\overline{u_3\, \mathrm{d}u_3^2}}{\mathrm{d}x_3} \sim -(2/3)\,\mathrm{d}\frac{\overline{(u_3^3)}}{\mathrm{d}x_3} \sim -\frac{\mathrm{d}}{\mathrm{d}x_3}\left(L_X\, u_0\frac{\mathrm{d}\,(u_0^2)}{\mathrm{d}x_3}\right), \tag{3.9}$$

(cf. Tennekes & Lumley 1971 and Townsend 1976, but this is not consistent with that proposed by George & Castillo 1997). Note that as the large energetic eddies move into low energy regions some of them tend to merge with each other. Although they are also eroded and torn apart, this occurs at a reduced rate because of the weaker small scale eddies surrounding them. As a result, in these regions there is a tendency for the integral scale L_x to increase. From the expression (3.9) it follows that the eddy transport of turbulence at the edge of shear layers where the mean shear is weak is the main source of local kinetic energy. These estimates of the higher moments and diffusion of turbulence can be used to model recirculating mean flows driven by gradients in these normal stresses such as occur in boundary layers where the surface conditions vary across the flow (e.g. Townsend 1976).

3.4   Bounding interface of the turbulent region

The greatest inhomogeneity occurs at the fluctuating interface bounding a region of turbulence, denoted by $[T]$, because across it there is a sudden transition between rotational motions in $[T]$ and irrotational motions in the external region $[E]$ outside it (Figure 10). The surface of the interface S_I cannot be sharply defined because of viscous diffusion across it, but its properties are not sensitive to the precise definition, a typical one being that on S_I the vorticity ω_I is say equal to u_0/L_x (Bissett *et al.* 1999). In the absence of strong stable stratification (McGrath *et al.* 1997), S_I is highly convoluted and even multiply connected with separate pockets of vorticity isolated in $[E]$.

As before, both vorticity dynamics and statistical models are useful for analysing the turbulence near S_I. Consider the turbulence in $[T]$ at time $t = 0$ as a set of self-propagating vortical eddies with velocity u_e and length scale L_e at some level $x_3 = 0$, below S_I. Some of them travel upwards towards S_I. At a time t later a surface could be drawn around those that have travelled furthest, to an average displacement h of order $u_e t$. Because the vortices interact and shed some vorticity as they move (Turner 1973), it follows that the interface is largely defined by the front or propagating faces of the eddies. Surrounding these eddies and in their wakes are the remnants of the shed vorticity. Thus, as many experiments demonstrate, the convolutions on the interface itself and on the coherent structures that move it (e.g. Hussain & Clark 1981) are found to have a length scale L_e, and are largely forced by the large scale dynamics of the shear layer (Gartshore 1966).

An alternative idealised statistical model can be developed based on the assumption that the undulations of the interface are small, for example by considering the interface just after a hypothetical velocity field is initiated, or by analysing the velocity field conditionally in coordinates moving up and down slowly with the interface. Following Phillips (1955), and Carruthers & Hunt (1986) the normal velocity u_3 and fluctuating pressure p in $[T]$ and $[E]$ are calculated so that they match each other across S_I. Linear, inviscid (and therefore approximate) calculations show that, because the vortex lines in $[T]$ when they intersect with S_I have to bend parallel to it, the horizontal velocity fluctuations u_s^2 increase in $[T]$, while the normal fluctuations u_n^2 decrease over a distance L_x from S_I. Across the interface itself, there is a sudden drop in u_s^2 to a lower value in $[E]$, where there are only irrotational velocity fluctuations driven by the normal movements of the interface. These fluctuations decay over a distance of order L_x. These theoretically derived trends are consistent with detailed conditional sampling of the numerically simulated velocity field at the interface of a turbulent wake by Bisset et al. (1999).

In practical turbulence models for the moments of the velocity fluctuations at one point, these differences in the statistics and dynamics either side of the random interface are not accounted for; rather it is assumed that the same type of model applies across the whole shear layer. In the earlier models where it was assumed that the eddy viscosity, was constant at the edge of the layer the computations were straight forward to perform. However, when all the statistical variables in the turbulence model, including the eddy viscosity tend to zero at the outer edge of the layer, the mathematical properties of the model equations are singular. Certain very sensitive computational assumptions have to be made that do not correspond to the real physical situation (e.g. Cazalbau et al. 1994). These interfaces between turbulent and non-turbulent regions also occur **within** turbulent flows at very high Reynolds number, where intense turbulent structure such as elongated vortices or tornadoes exist adjacent to regions of relatively low turbulence. Some research has been done on modelling the intermittency of the interface; but not much on modelling the equations differently each side of the interface and then statistically averaging its position (see Gartshore et al. 1983).

4   Spectra and the Eddy Structure

4.1   Independent eddy fields

It was shown in Sections 2 and 3 how the eddy motions in turbulence, say $u^{(n)}(x)$ for the nth eddy, located near x_n, have characteristic structures determined by the vortex dynamics and, on larger length and time scales, by the initial and boundary conditions. The velocity fields in these structures are significantly correlated over their characteristic length ℓ and time scales ℓ/u_0, but are weakly correlated with motions on much larger or smaller scales, which have been generated far from the structure or at previous times (e.g. by breakup of an earlier eddy). Each of these fields are unique, and are not the same as Fourier waves or any single function in a standard statistical representation (cf. Holmes *et al.* 1997; Kevlahan *et al.* 1993). However any particular velocity profile, e.g. $u^{(n)}(x)$, may be represented as a series of 'reference' functions (e.g. $\exp(i\,k\,x)$) which are usually taken to be orthogonal for greater generality. If the representation is based on localised reference functions such as wavelets (e.g. $f\left(-(x-x_n)^2/\ell_n^2\right)$ where $f_k(\ ) \to 0$ as $(x-x_n)/\ell_n \to \infty$) and if their statistical properties are calculated, e.g. as their spectra and capacity dimension $\tilde{D}_k$, then the significant local features, can be described more precisely, such as self-similarity and the relative effects of coherent and incoherent motions (Kevlahan & Vassilicos 1994; Arneodo 1999).

In general the same 'reference' functions are commonly used for the whole flow and for the eddies, as in Fourier analysis. So that when turbulence is described in terms of the spectra $E(k)$ (i.e. the mean square amplitude of the Fourier coefficient) one cannot tell, without further information, whether $E(k)$ is more a measure of the distribution of the energy of the eddies of scale k^{-1} or of the particular forms of the eddies (e.g. velocity sheets or spiral vortices), over a range of wave numbers where the spectra are self-similar. For previous discussion of this difficult point see Hunt & Vassilicos (1991), Pullin & Saffman (1998).

To study this question the one-dimensional spectra $\phi(k)$ of idealised one-dimensional turbulent flow fields $\mathbf{u}(\mathbf{x},t)$ are calculated where the fields are determined by the statistical distribution $B(n)$ and by the forms of the velocity profile $u^{(n)}(x,t)$ of the **independent** eddy flow fields. Since the vortical field of an eddy induces velocities that distort other vortical fields, in general the eddies cannot be statistically independent. However the relative location and orientation of some eddy structures of comparable scale are such that their motions are approximate independent, as with the vorticity of streamwise elongated roll-vortices in boundary layers (Townsend 1976) or well-separated eddies in two-dimensional turbulence (McWilliams 1990). If the structures have very different length and time scales they can also be independent whatever their relative location and orientation as discussed in Secion 2. Therefore in the analysis of a set of independent eddies, it is necessary to define the range of scales $\Delta\ell$ over which they are independent, or lie within a particular statistical distribution; so the length scale of the nth eddy in the set of N eddies lie in the range

$$\ell_0 < \ell_n < \ell_0 + \Delta\ell.$$

The analysis focuses on the most general types of eddy motion that is self similar over a wide range of scales so that

$$\ell_n = \ell_0 + n\delta\ell, \tag{4.1a}$$

where $\delta\ell = \Delta\ell/N$ and $\Delta\ell \gg \ell_0$. Thence $\ell_n \sim n\,\delta\ell$.

For simplicity a one-dimensional random homogeneous velocity field $u(x)$ in an internal $2X$ with a one-dimensional spectrum $\phi(k)$ is considered (over a finite range of scales). This is related to the set of random eddy velocity fields $u^{(n)}(x)$ that make up the overall flow and their distributions $B(n)$ or to their spectra $\phi^{(e)}(k, n)$. The analysis follows that of random water waves by Belcher & Vassilicos (1997). Thus

$$u(x) = \sum_1^N B(n)\, u^{(n)}(x), \tag{4.1b}$$

where $u(x) = 0$ for $|x| > X$ and where the eddy velocity fields are localised around the centre of each eddy at x_n, so that $|u^{(n)}(x - x_n)| \to 0$ as $(x - x_n)/\ell_n \to \infty$. These fields are assumed to be statistically independent so that, taking an ensemble average denoted by $\langle\ \rangle$,

$$\langle u^{(r)}(x)\, u^{(s)}(x) \rangle = 0 \quad \text{if } r \neq s. \tag{4.1c}$$

Note that the eddy fields are inhomogeneous and therefore their spectra $\phi^{(e)}(k, n)$ are a measure of their structure over the whole scale ℓ_n of the eddy. Also only for high wave numbers do the Fourier components $\tilde{u}(k)$ have the usual orthogonality properties of homogeneous turbulence or stationary random functions. Here

$$\phi^{(e)}(k, n) = |\tilde{u}^{(n)}(k)|^2\, (\pi/X), \tag{4.2a}$$

where $\tilde{u}^{(n)}(k) = \frac{1}{2\pi}\int_{-\infty}^{\infty} u^{(n)} \exp(i\,k\,x)\,\mathrm{d}x$. To estimate the magnitude of $\phi^{(e)}(k, n)$ it is useful to note from (4.2a) that

$$\phi^{(e)}(0, n) = \frac{1}{4\pi X}\left(\int_{-\infty}^{\infty} u^{(e)}(x, n)\,\mathrm{d}x\right)^2.$$

Thence from (4.1b) and (4.1c)

$$\phi(k) = \sum_1^N B^2(n)\, \phi^{(e)}(k, n). \tag{4.2b}$$

As $N \to \infty$ we have

$$\phi(k) = \int_0^{\infty} \hat{B}(n)^2\, \phi^{(e)}(k, n)\,\mathrm{d}n, \tag{4.2c}$$

where

$$B^2(n) = \int_{n-1}^n \hat{B}(n')^2\,\mathrm{d}n'. \tag{4.2d}$$

For sets of self-similar eddies, the fields can be expressed generally in terms of a stretched coordinate $\hat{x} = (x - x_n)/\ell_n$; $\hat{k} = k\ell_n$ so that

$$u^{(n)}(x) = u^{(e)}(\hat{x}, n),\qquad(4.3a)$$

whence

$$\phi^{(e)}(k, n) = \hat{\phi}^{(e)}\left(\hat{k}, n\right) = \left|\tilde{u}^{(n)}\left(\hat{k}\right)\right|^2 (\pi/X).\qquad(4.3b)$$

For the range of length scales over which each eddy velocity field is coherent and defined by $u^{(n)}(x)$, the form of $\phi^{(e)}(k, n)$ when $k \gg \ell_n^{-1}$ depends on the degree of the un-smoothness and/or singularities of the function $u^{(n)}(x)$, and on their spatial distribution.

When one component of a general velocity field $u(x)$ along the x- (or x_1-) axis within an interval of length $2X$ is represented by a Fourier series, then

$$u(x) = \sum_{\infty}^{-\infty} C_n \exp(i\pi n\, x/X), \quad \text{for } |x| < X.\qquad(4.4)$$

The magnitude of the coefficients C_n for large values of n vary in a particular way if there are discontinuities of the derivative (say at the mth order) of the velocity field that are well separated from each other (as with surface streaks in the turbulent boundary layer). Then the coefficients for large values of n decrease in proportion to n^{-p}, where $p = m + 1$ (the case $m = 0$ corresponds to the case of discontinuities in $u(x)$). However if the number, say n_s, of such mth 'singularities' in a small interval $\ell_s(x)$ at a point x, becomes very large as ℓ_s tends to zero, for example at the centre of a spiral vortex where $u(x) \sim \sin\left(x^{-\lambda}\right)$, $(\lambda > 0)$ then the Fourier coefficients decrease with n more slowly, i.e. $p < m + 1$, and p is not in general an integer. However the Fourier coefficients are correlated and except at a finite number of points all the derivatives exist. The spectrum of this velocity field which is proportional to $|C_n|^2$, at high wave numbers is proportional to $n^{(-2p)}$ (Hunt & Vassilicos 1991).

If the characteristic magnitude of the velocity $u_0^{(n)}$ of self similar eddies (defined as $1/\ell_n \int_{-\infty}^{\infty} \left|u^{(n)}\right| dx$) increases with their length scale ℓ_n, in proportion to ℓ_n^γ and therefore n^γ, then

$$u_0^{(n)} \sim u_0 n^\gamma \text{ and } u^{(e)}(\hat{x}, n) = u_0 n^\gamma \hat{u}^{(e)}(\hat{x}), \text{ where } u_0^{(0)} = u_0.$$

Similarly the self-similar spectrum may be rescaled, so that

$$\hat{\phi}^{(e)}\left(\hat{k}, n\right) = n^{(2\gamma+2)} \hat{\phi}_0^{(e)}\left(\hat{k}\right),\qquad(4.5a)$$

where the magnitude of

$$\hat{\phi}_0^{(e)}\left(\hat{k}\right) \sim \hat{\phi}_0^{(e)}(0) \sim u_0^2 \ell_0^2/X.\qquad(4.5b)$$

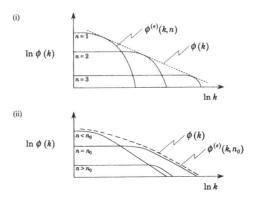

Figure 7. Schematic demonstration of how the full one-dimensional spectrum $\phi(k)$ is related to the spectra of the sequence of independent and self-similar eddies $\phi^{(e)}(k, n)$ for different distributions $B(n)$: (i). $\phi(k)$ is determined by a broad distribution $B(n)$ of eddies, whose spectra $\phi^{(e)}(k, n)$ decrease rapidly as k increases; (ii). Here there is a narrow range of significant eddies, centred on $n = n_0$, so that $\phi(k)$ is determined by the form of the eddy spectrum in the range $\phi^{(e)}(k, n_0)$.

Therefore from (4.5a) and (4.3b), the expression for the total spectrum (4.2c) becomes

$$\phi(k) = \int_0^\infty \hat{\hat{B}}^2(n) \; \hat{\phi}_0^{(e)}\left(\hat{k}\right) dn, \qquad (4.6)$$

where $\hat{\hat{B}}^2(n) = n^{(2+2\gamma)}\hat{B}^2(n)$.

This result (4.6) shows how the spectrum varies depending firstly on the distribution $\hat{B}^2(n)$ of the spatially averaged velocity in the self-similar eddies that make up the flow field, and secondly on how the form of the internal structure, especially the singularities and discontinuities determine the self-similar eddy spectrum $\hat{\phi}_0^{(e)}(k)$; see Figure 7.

In the limit of large scale motion (with finite integral scale), or when vertical motions are considered near a rigid horizontal surface, $\hat{\phi}_0^{(e)}\left(\hat{k} \to 0\right)$ is **constant**; it follows from (4.6) and (4.2a) that $\phi(k \to 0)$ is also a constant proportional to $\int_0^\infty \hat{B}^2(n)\,dn$, as is generally observed (as shown later).

4.2 Narrow band spectra

Consider flows where there is a small variation in the type of eddy motion (which is assumed here) and also a very narrow range of length scales so that the form of the eddy velocity field largely determines the total spectrum $\phi(k)$.

This means that $\hat{B}^2 (n)$ decreases more rapidly as $|n - n_0|$ increases from its maximum value where $n = n_0$ than $\hat{\phi}^{(e)} \left(\hat{k} \right)$ decreases as $\hat{k} \to \infty$. For example if there are singularities in the eddy profile $u^{(e)} (\hat{x})$, e.g. discontinuities in certain levels of derivatives, then,

$$\text{for } \hat{k} \gg 1, \quad \phi^{(e)} \left(\hat{k} \right) \propto \hat{k}^{-2p^{(e)}}. \tag{4.7}$$

Assuming that the variance of the eddy velocity is finite, there is a lower limit on $p^{(e)}$, i.e. $-2p^{(e)} \geq 1$. The criterion for satisfying the 'narrowness' condition is relative to the singularity of the eddy structure; in general it depends on $p^{(e)}$ as well as on $\hat{B}(n)$. If $\hat{B}_{mx} \sim B_{mx}| (n - n_0) |^{-\beta}$, the criterion is that $3 + 2\gamma - 2\beta < 0$. Thence the integral (4.6) and (4.5a) lead to

$$\phi(k) \sim \lambda n_0^{(2\gamma+2)} \hat{\phi}^{(e)} \left(\hat{k} \right) \sim \lambda \phi^{(e)} (k, n_0), \tag{4.8}$$

where

$$\lambda = \int_{-n_0}^{\infty} \hat{B}^2 (\hat{n}) \, \mathrm{d}(\hat{n}) \quad \text{and} \quad \hat{n} = n - n_0.$$

If the distribution is so narrow that $B(n) \sim \delta(n - n_0)$ where $\delta(\ )$ is a Dirac delta function, then irrespective of the value of $p^{(e)}$ the same result (4.8) is obtained. This is a case of highly structured turbulence where the dominant eddies for every value of k lie in a narrow range of scales. The coherent motions in these eddies determine the main properties of the flow field (see Holmes *et al.* 1997). Consequently the high wave number spectrum is determined by the complexity of the structure in the eddies and therefore by the sharpness of the gradients and topological complexity of their flow fields. For example in turbulent shear layers, where the eddies undergo significant straining (i.e. $t_D^* \gg 1$), as in the wall regions of turbulent boundary layers, 'streaks' form with a characteristic spacing ℓ_0 (defined by initial or external conditions) which can be regarded as eddy structures (e.g. Lee *et al.* 1989), as explained by Hunt & Carruthers (1990). Thus $\hat{B}(n)$ has a sharp peak at $n = n_0$. The singularity means that $E^{(e)} \propto k^{-2p}$ because of the step-singularity in the velocity profile in these structures, $p^{(e)} \simeq 1$. Therefore from (4.6) the whole spectrum $E_{11}(k) \propto k^{-2}$ (where k is the three-dimensional wave number). In this model calculation, at $t = 0$, the initial spectrum is determined by a broad distribution of 'smooth' eddies with exponent $p^{(e)}$, so that $E(k) \propto k^{-2p}$ where $2p^{(e)} > 2p > 2$. During the distortion as streaks develop $p^{(e)}$ decreases so that the final spectrum is determined by a narrow distribution and the singular form of the eddies. This interpretation of spectra is consistent with measurements and simulations in homogeneous and wall bounded shear flows (Lin *et al.* 1996). Also Murlis *et al.*'s (1967) measurements in a turbulent boundary layer at a moderate turbulent Reynolds number ($Re \sim 10^2$) show the dependence of p on $\hat{t}_D^*$. They found that $2p$ increased from its lowest value $\simeq 5/3$ near the wall (where $\hat{t}_D^* \sim 4$) to higher values ($2p \leq 3$) in the outer part of the layer (where $\hat{t}_D^* \sim 2$).

4.3   Broad band distribution

Now consider the other limit where there is a wide range of scales of inten-
sities of independent eddies. In this case in (4.6) the distribution function
$\hat{B}^2(n)$ decreases more slowly than $\phi^{(e)}(k,n)$ as $|n-n_0|$ increases; as a consequence the form of $\phi^{(e)}(k,n)$ does not significantly affect the total spectrum
$E(k)$. But note that the normalised eddy spectrum in (4.6) must decrease
fast enough for the integral to converge. As with the narrow distribution the
criterion for this limit depends on the form of both $\hat{B}^2(n)$ and $\hat{E}_0^{(e)}(\hat{k})$ as
$n \to \infty$ and $\hat{k} \to \infty$. If $\hat{B}^2(n) \propto n^{-2\beta}$ and if $\hat{\phi}_0^{(e)}(\hat{k}) \propto \hat{k}^{-2p^{(e)}}$ for $\hat{k} \gg 1$,
when these 'broad' conditions are applied to the integral (4.6) they imply
that, for convergence,

$$2p^{(e)} > 3 + 2\gamma - 2\beta. \tag{4.9a}$$

Then the total spectrum derived from (4.6) is obtained by rescaling n with
the integral as $n\ell_0 k/(k\ell_0) = \hat{k}/k\ell_0$, whence

$$\phi(k) = (k\ell_0)^{-2p} \int_0^\infty \hat{\phi}_0(\hat{k}) \, d\hat{k} \tag{4.9b}$$

$$\sim \left(u_0^2 \ell_0^2/X\right)(k\ell_0)^{-2p}, \tag{4.9c}$$

where

$$2p = 3 + 2\gamma - 2\beta. \tag{4.9d}$$

The exponent 2β of the distribution function $B(n)$ is determined by how
extensively the eddies fill the space, for which a suitable measure is the fractal or capacity dimension of contours of vorticity (Vassilicos & Hunt 1991;
Arneodo 1999).

Following Perry *et al.* (1986) consider elongated boundary layer eddies
whose length scale distribution $B(n)$ is such that there is an equal probability of a point lying within any size of eddy (over the range of these structures).
This corresponds to a physically plausible mechanism for the eddy formation
and breakup, and implies that $B(n) \simeq B_{mx}n^{-1}$ when $n \gg 1$. In the high
Reynolds number boundary layer the variances of each **horizontal** velocity
component $\overline{u_1^{(e)2}}$, $\overline{u_2^{(e)2}}$ are of the order of u_*^2 (the friction velocity) and vary
very slowly with height z above the surface (when $z \leq h/10$), where h is
the boundary layer thickness. This implies that the peak longitudinal and
lateral velocities in the longitudinal self similar eddies of all scales are approximately the same, so that $\gamma = 0$. It follows therefore from (4.9d) that for
this 'broad distribution' for the horizontal component the spectrum exponent
$2p = 1$. There is also a physical or dimensional argument for this result, that
$E_{11}(k)$ should only depend on u_*^2 which leads to $2p = 1$. By contrast for
'blocked' vertical fluctuations where the eddy exponent $2p^{(e)} = 0$, from (4.6),
$E_{33}(k)(\hat{k} \to 0) \sim u_*^2$ and therefore $E_{33}(k) \sim u_*^2 z$ (Hunt & Morrison 2000).

In fully developed two-dimensional turbulence, e.g. McWilliams (1990), the eddies form into a wide range of coherent vortices that are approximately independent. In some cases they are approximately space filling so that in the exponent of $B(n)$, $\beta \lesssim 1$ (e.g. $\beta \simeq 1.3$ in the study of Dritschel (1993)). However these eddies differ significantly from elongated boundary layer eddies in how the magnitude of their peak velocity varies with their scale ℓ_n. Here the modulus of the vorticity of most eddies is approximately constant and is of the order of ω_0, because in the inviscid two-dimensional dynamics, ω is unchanged by interactions between eddies. Thus the typical velocity $u_0^{(n)}$ in each eddy is of order $\ell_n \omega_0 \sim n\, u_0$, so that $\gamma \simeq 1$. Since the vorticity outside the eddies is typically less than ω_0, there is a discontinuity in vorticity or velocity gradient at the edges of these eddies (e.g. Dritschel 1993). Therefore by the singularity analysis following equation (4.4) this determines the form of the exponent $2p^{(e)}$ of the (one-dimensional) eddy spectrum, i.e. when $k\ell_n \gg 1$, $E_{11}^{(e)}(k\ell_n, n) \propto k^{-2p^{(e)}}$ where $3 < 2p^{(e)} \leq 4$. If the eddies have smooth shapes (e.g. circles or ellipses) $2p^{(e)} = 4$. But usually they have smooth spiral shaped accumulation points, in which case $3 < 2p < 4$ (Gilbert 1987; Hunt & Vassilicos 1991). Therefore

$$E_{11}^{(e)}(k\ell_n, n) \sim \left(\omega_0^2 \ell_n^4 / X\right) f(k\ell_n),\tag{4.10}$$

where the dimensionless function $f(k\ell_n) \sim (k\ell_n)^{-2p^{(e)}}$ when $k\ell_n \gg 1$, and $f = 0(1)$ as $k\ell_n \to 0$. Therefore from (4.5), $\hat{E}_{11_0}^{(e)}(\hat{k}) \sim (\omega_0^2 \ell_0^4 / X) f(\hat{k})$ and $\gamma = 1$. Thence for the case $\beta = 1$, the criterion (4.9a) implies $2p^{(e)} > 3 + 2\gamma - 2\beta = 3$. Since this is satisfied, the distribution function $\hat{B}(n)$ determines the spectrum, so that when $k\ell_0 \gg 1$, from (4.9d),

$$E_{11}(k) \sim \left(\frac{\ell_0^2 u_0^2}{X}\right)(\ell_0 k)^{-3}.\tag{4.11}$$

Note that this result depends on the assumption that the turbulence is self-similar. Dritschel (1993) explains how this is sensitive to the initial conditions, especially the distribution $B(n)$ of eddies of a given scale ℓ_n. This is reflected, though not explicitly explained in recent literature.

How does this accord with the usual assumption for three-dimensional turbulence that the spectrum is not determined by any particular features of the velocity fields of the individual eddies (e.g. Tennekes & Lumley (1971); Synge & Lin 1943; McComb 1990)?

Despite the fact that in all types of three-dimensional small scale turbulence at all ranges of Re, qualitative observations (e.g. Brown & Roshko 1974) and measurements show that coherent motions at scale ℓ can be identified over a limited range of length scales, say $\Delta\ell$ or over wave numbers Δk, where $\Delta k \sim \ell^{-1}$. In other words any particular coherent structure may appear in relation to much larger scales as a random element of the ambient turbulence $\mathbf{u}'$ (see sec 2.3). Smaller scale motions where $k \gg \ell^{-1}$ which appear random at the scale ℓ, again have their own distinct coherent structure which is largely uncorrelated with that at larger scales. As a result of the stretching and 'rolling up' of inhomogeneous (or unstable) sheets of vorticity (or its gradients) and accumulation points in the coherent eddy on scale ℓ, the form of its energy spectrum (following the Fourier analysis discussion below (4.4)) for large $k\ell_n$ is given by

$$E^{(e)}\left(k\ell_n, n\right) \propto C^* k^{-2p^{(e)}}, \qquad (4.12a)$$

where

$$1 < 2p^{(e)} < 2. \qquad (4.12b)$$

Applying Townsend's approximate dynamic argument (see Section 2.4), based on that of Kolmogorov (1941), for inertial range coherent eddies leads to $C^* \simeq \bar{\epsilon}^{2/3}$ and $2p^{(e)} = 5/3$, where ϵ is the average rate of transfer of energy to smaller scales in the eddies.

However over ranges of wave number significantly greater than Δk (or k), where the eddy motions are randomly orientated and uncorrelated, the coherent structure analysis is not relevant and the fundamental randomness assumption of the statistical physics models (e.g. Kraichnan 1991, McComb 1999, Lesieur 1990), is justifiable physically. Over shorter length and time scales where the motions are coherent, such models do not describe the rapid response of the turbulence structure or its effects on turbulence processes. Since both types of model (coherent, and statistical, non-coherent) are consistent with the equilibrium Kolmogorov inertial range spectrum

$$E\left(k\right) = C\bar{\epsilon}\, k^{-5/3}, \qquad (4.13)$$

the total spectrum may be determined both by the eddy distribution function and by the characteristic eddy structure.

If the latter explanation was relevant, at high wave numbers, the form of the three-dimensional spectrum $E\left(k\right)$ might largely be attributed to one type and scale of coherent complex eddy motion, as explained in (4.12). In this case the inertial range turbulence corresponds closely to the former (narrow distribution) model. Then from (4.8), n, the exponent of the eddy spectrum, must satisfy the criterion $2p^{(e)} \lesssim 3 + 2\gamma - 2\beta$, or $B\left(n\right) = \delta\left(n - n_0\right)$, and then $p^{(e)} = p$. Certain types of observed eddy structure are consistent with this form, such as 'spiral' eddies (Moffatt, 1984). But this idealisation is probably only valid over a limited wave number range, because of random small scale motions. On the other hand if inertial range turbulence corresponds to the latter (broad distribution) model, the singularities or discontinuities in the eddies should be sufficiently insignificant that

$$2p^{(e)} \geq 3 + 2\gamma - 2\beta. \qquad (4.14a)$$

In that case from (4.9d), it follows that if the exponent of the spectrum $E(k)$ for inertial range turbulence is

$$2p = 3 + 2\gamma - 2\beta = 5/3. \qquad (4.14b)$$

Since from statistical measurements the typical eddy velocity difference over a scale ℓ varies as $\ell^{1/3}$, and if this can be attributed to the typical eddy velocities, then it follows that $\gamma = 1/3$. Therefore from (4.14b) the distribution parameter $\beta \simeq 1$, which is broadly consistent with observations of progressively larger number of small eddies in turbulence; but it would be interesting to test this prediction experimentally.

This idealisation also has its limitations because it cannot be valid over a limited range of length scales of the order of $\Delta\ell < \ell$ (or $\Delta k \lesssim k^{-1}$), because over these scales and on time scales of order $\ell_n/u_0^{(n)}$, the dynamical processes discussed in Section 2.3 ensure that the motions are significantly correlated.

Thus when one is interested in the overall spectrum and any changes to it occuring over long time scales the statistical description of three-dimensional turbulence should be based on the concept of a distribution of independent relatively smooth eddy structures. But if, for certain practical problems, one is interested in changes to turbulence over quite short length and time scales that are typical of the eddy times, then the statistical description in term of coherent structures is necessary for a dynamical analysis of these problems. This is briefly discussed in Section 5.

Note that the dynamics of the flow are equally important in determining the spectra (and other statistics) whether they are more dominated by distribution or by internal eddy structure. Distributions are largely determined by the interactions between eddies and the overall field, including their breakup formation while the internal eddy structure is, as we show in Section 2, more controlled by local dynamics.

5 Final Remarks

There is now increasing experimental and numerical evidence that there are well defined kinematical and dynamical eddy structures occurring in turbulent flows on all scales, and that many of their features are described by quantitative models. Some further developments of modelling theory have been proposed here which all need detailed investigation, namely: quantifying how the shape of an eddy structure changes; the interactions within an eddy structure with local motions on larger and smaller scales and their effects on the overall dynamics; how very intense velocities might be generated within an eddy or interaction between them; and how the spectra for the whole field of turbulence relates to that for the eddy structures, in terms of their individual structure and of their distribution.

What use are these models of eddy structures? Do they help explain turbulence phenomena and turbulence statistics? A useful guideline for deciding between the models is that the better they are the more they should explain.

Although much fundamental research in turbulence is focussed on the higher moments and the intermittency (Frisch 1995) and the extreme values of the velocity statistics, these probably do not provide much insight into the basic small scale eddy structure that has been reviewed and analysed here. It was argued in Section 2.5 that these statistics are likely to be dominated by infrequent (Jimenez *et al.* 1993) strong interaction between vortical structures.

As Lumley has remarked a good test of the 'normal' working of a model of any complex system including turbulence is its response to perturbations of different types and on different scales. Multi-point statistical models (e.g. EDQNM, DIA, ..., reviewed by Lesieur (1990)) and conceptual models based on eddy structure models can approximately describe the long term change to the turbulence spectrum following the sudden injection over a finite time of small scales turbulence whose spectra lie in a narrow band of wave numbers near $k = k^*$. They show how the two point turbulence spectrum tends to be at small scales even if the large scale is anisotropic and inhomogeneous (Mann 1994). In such models the exchange of energy between wave numbers is approximated by diffusion-like terms such as $\left(\propto \frac{d}{dk} \left(D\left(k\right) dE/dk \right) \right)$ which can only be valid on time scales significantly larger than the natural time scale of the turbulence at the relevant wave number i.e. $\tau\left(k^*\right) \sim \left(\int_0^{k^*} k^2 E\left(k\right) dk \right)^{-1/2}$ (cf. Section 2.6). This is why they predict that any delta function input of energy is simply spread to smaller and larger wave numbers. Slow transitions of this kind were measured in the atmospheric boundary layer passing over changes in the surface roughness and were successfully modelled along these lines by Panofsky *et al.* (1982).

However on timescales less than $\tau\left(k^*\right)$, the change of the energy spectrum is quite different; the large scale motions are damped by the increased eddy viscosity at the small scales, resulting from their increased energy (see Section 2.4), while the input motions for $k \gtrsim k^*$ are amplified as a result of straining by the large scales. Consequently the gradients in the spectrum $|\partial E/\partial k|$ near $k \sim k^*$ increase on this time scale, as opposed to decrease as predicted by the diffusion-like models. This explains why, when grids of bars, or obstacles or small inertial particles are introduced into turbulent flows the large scale

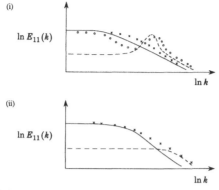

Figure 8. Schematic representation of how an initial one-dimensional spectrum $E_{11}(k)$ (denoted by —) of three-dimensional turbulence changes (i) when it interacts with a perturbation spectrum (- - -) which is confined to scales smaller than ℓ, e.g. produced by a small scale motion in the wakes of particles. Note that initially, when $t \lesssim \tau$ ($k = \ell^{-1}$) in the distorted spectrum, denoted by ooo the larger scales **decrease** while the smaller scales are further amplified; but at yet larger times $t > \tau(\ell^{-1})$ in the distorted spectrum now denoted by $\times \times \times$ the large scales increase and the spectrum relaxes towards the equilibrium state; (ii) when it interacts slowly with an imposed smaller scale turbulence (denoted by - - -). Then the distorted spectrum $\times \times \times$ 'simply relaxes' towards the imposed spectrum on a typical time of the order of $\tau(k)$.

eddy motions and their effects on diffusion are observed to be damped (Hinze 1959 p. 342; Davidson *et al.* 1995; Ghosh *et al.* 1991).

A similar sharp transition is observed near a particular wave number k^* in the energy spectrum when the turbulence is subjected to anisotropic body forces and straining motions, such as a mean shear $(\partial U/\partial x_3)$. Here $k^* \sim \bar{\epsilon}^{-1/2}(\partial U/\partial x_3)^{3/2}$. For $k > k^*$ the effects of the large scale distortion are weak and the normal spectra (e.g. E_{11}), though not the co- (or shear stress E_{13}) spectra, are effectively the same as for isotropic turbulence (Mann 1994). In this case the transition near k^* (which persists over time) is generally sharper than would be predicted using a diffusive energy transfer model. One explanation is that the vortex sheets of the coherent eddy structures that determine the spectrum on the scale k^* act to shelter the inner motions of the structures from the external large scale motions. (Hunt & Durbin 1999; Kevlahan & Farge 1997).

Note that for certain higher order moments, e.g. two-point, third moments $E_{3,33}(k_1)$, there may be direct statistical coupling between large and small sale motions so that the small scale spectra do not have a universal, isotropic form, (as demonstrated in convection by Hunt, Kaimal & Gaynor 1988; Warhaft 2000). A better understanding of the governing mechanisms of eddy structures on these different time and space scales both at moderate and high Reynolds numbers is necessary to improve the prediction of various processes of this turbulence, such as mixing, combustion, two-phase interactions, noise production and of complex flows generated by groups of obstacles and jets, all of which introduce energy at many different scales. As J.C. Wyngaard has commented, it is likely that subgrid scale models for large eddy simulations of such flows will have to be specified for each type of flow, which is another reason for the study of these eddy structures.

Acknowledgements

This work, which was performed while I was holding a Senior Research Fellowship at Trinity College, arose from many conversations during the Isaac Newton Institute Programme on Turbulence held in 1999 and afterwards at Cambridge and Delft. I am especially grateful for the advice and comments of J.C. Vassilicos, P. Carlotti, K. Ohkitani, S. Kida, H.K. Moffatt, F. Hussain, A. Leonard, J. Brasseur, A. Perry, Y. Couder, A. Pumir, A. Tsinober, J. Lumley, Z. Warhaft, F. Nieuwstadt, G. Brethouwer, F.T. Smith.

I am grateful to my colleagues at the Center for Turbulence Research, Stanford University, D.K. Bissett and M.M. Rogers, for their permission to describe at this preliminary stage our joint work on turbulent interfaces.

References

Andreotti, B., Douady, S. & Couder, Y., 1997, 'About the interaction between vorticity and stretching in coherent structures'. In *Proceedings of the Workshop on Turbulence Modeling and Vortex Dynamics*, O. Borafar *et al.* (eds.), Springer.

Angilella, R.J. & Vassilicos, J.C., 1999, 'Time-dependent geometry and energy distribution in a spiral vortex layer', *Phys, Rev. E.* **59**, 5427.

Arneodi, A., Manneville, S., Muzy, J.-F. & Roux, S.G., 1999, 'Revealing a lognormal cascading process in turbulent velocity statistics with wavelet analysis', *Phil. Trans. R. Soc. Lond. A* **357** number 1760 pp. 2415–2438.

Batchelor, G.K., 1953, *The Theory of Homogeneous Turbulence*, Cambridge University Press.

Batchelor, G.K. & Proudman, I., 1954, 'The effects of rapid distortion of a fluid in turbulent motion', *Quart. J. Mech. Appl. Maths.* **7**, 83–103.

Belcher, S. E. & Vassilicos, J.C., 1997, 'Breaking waves and the equilibrium range of wind-wave spectra', *J. Fluid Mech.* **342**, 377–410

Bertoglio, J.P, 1982, 'A model of three-dimensional transfer in non-isotropic homogeneous turbulence'. In *Turbulent Shear Flows* 3, 253–261, L.J.S. Bradbury *et al.* (eds.) Springer.

Betchov, R., 1956, 'An inequality concerning the production of vorticity in isotropic turbulence', *J. Fluid Mech.* 1, 497–504.

Bisset, D.K., Hunt, J.C.R., Cai, X. & Rogers M.M., 1998, 'Interfaces at the outer boundaries of turbulent motions', Annual Research Briefs 1998, 125–135, Center for Turbulence Research, Stanford University.

Brasseur, J.G. & Lee, M.J., 1989, 'Pressure-strain rate events in homogeneous turbulent shear flow'. In *Advances in Turbulence* 2, 306–312, H.A. Fernholz & H.E. Fiedler (eds.), Springer.

Browand, F.K., 1966, 'An experimental investigation of the instability of an incompressible, separated shear layer', *J. Fluid Mech.* 26, 281–307.

Brown, G.L. & Roshko, A., 1974, 'On density effects and large structure in turbulent mixing layers, *J. Fluid Mech.* 64, 775–816.

Cambon, C. & Scott, J.F., 1999, 'Linear and nonlinear models of anisotropic turbulence', *Ann. Rev. Fluid Mech.* 31, 1–53.

Carruthers, D.J., Fung, J.C.H. & Hunt, J.C.R., 1990, 'The emergence of characteristic (coherent?) motion in homogeneous turbulent shear flows'. In *Turbulence and Coherent Structures*, Proceedings of Grenoble Conference 1989, M. Lesieur & O. Métais (eds.) Kluwer, 29–44.

Cazalbau, J.B., Spalart, P.R. & Bradshaw, P., 1994, 'On the behaviour of two-equation models at the edge of a turbulent region', *Phys. Fluids* 6, 1797–1804.

Champagne, F.H., Harris, V.G. & Corrsin, S., 1970, 'Experiments on nearly homogeneous turbulent shear flows', *J. Fluid Mech.* 41, 81–138.

Chertkov, M., Pumir, A. & Shraiman, B.I., 1999, 'Lagrangian tetrad dynamics and the phenomenology of turbulence', *Phys. Fluids* 11, 2394–2410.

Constantin, P., 2000, 'Mathematical theory of turbulence'. In *Proceedings of ICIAM 1999*, J. Ball & J.C.R. Hunt (eds.), Clarendon Press.

Craft, T.J. & Launder, B.E., 1996, 'A Reynolds stress closure designed for complex geometries', *Int. J. Heat & Fluid Flow* 17, 245–254.

Davidson, M.J., Mylne, K.R., Jones, C.D., Phillips, J.C., Perkins, R.J., Fung, J.C.H. & Hunt, J.C.R., 1995, 'Plume dispersion through large groups of obstacles: – a field investigation', *Atmos. Env.* 29, 3245–3256.

Domaradskii, J.A. & Rogallo, R.S., 1990, 'Local energy transfer and nonlocal interactions in homogeneous isotropic turbulence', *Phys. Fluids* 2. 413-426.

Douady, S., Couder, Y. & Brachet, M.E., 1991, 'Direct observation of the intermittency of intense vorticity filaments in turbulence', *Phys. Rev. Lett.* 67, 983–986.

Dritschel, D.G., 1993, 'Vortex properties of two-dimensional turbulence', *Phys. Fluids* 5, 984–997.

Etemadi, N., 1990, 'On curve and surface stretching in isotropic trubulent flow', *J. Fluid Mech.* **221**, 685–692.

Flohr, P. & Vassilicos, J.C., 1997, 'Accelerated scalar dissipation in a vortex', *J. Fluid Mech.* **348**, 295–317.

Frisch, U., 1995, *Turbulence*, Cambridge University Press.

Fung, J.C.H., Hunt, J.C.R., Malik, N.A. & Perkins, R.J., 1992, 'Kinematic simulation of homogeneous turbulent flows generated by unsteady random Fourier modes', *J. Fluid Mech.* **236**, 281–318.

Gartshore, I.S., 1966,'An experimental examination of the large-eddy equilibrium hypothesis', *J. Fluid Mech.* **24**, 89–98.

Gartshore, I.S., Durbin, P.A. & Hunt, J.C.R., 1983, 'The production of turbulent stress in a shear flow by irrotational fluctuations', *J. Fluid Mech* **137**, 307–329.

George, W. K. & Castillo, L., 1997, 'Zero-pressure-gradient turbulent boundary layer', *App. Mech. Reviews* **50**(12), 689–729.

Ghosh, S., Phillips, J.C. & Perkins, R.J., 1990, 'Modelling the flow in droplet-driven sprays'. In *Advances in Turbulence* 3, 405–413, Springer.

Gilbert, A. D., 1988, 'Spiral structures and spectra in two-dimensional turbulence', *J. Fluid Mech* **193**, 475–497.

Grosch, C.E. & Salwen, H., 1968, 'The stability of steady and time-dependent plane Poiseuille flow', *J. Fluid Mech.* **34**, 177–205.

Head, M.R. & Bandyopadhyay, P.R., 1981, 'New aspects of turbulent boundary-layer structure', *J. Fluid Mech.* **107**, 297–338.

Hinze, J.O., 1959, *Turbulence*, McGraw-Hill.

Holmes, P., Lumley, J.L.& Berkooz, G., 1997, *Coherent Structures in Turbulence*, Cambridge University Press.

Hopfinger, E.J., Browand, F.K. & Gagne, Y., 1982, 'Turbulence and waves in a rotating tank', *J. Fluid Mech.* **125**, 505–534.

Hunt J.C.R., 1987, 'Vorticity and vortex dynamics in complex turbulent flows', *Trans. Can. Soc. Mech. Eng.* **11**, 21–35.

Hunt, J.C.R., 1992, 'Developments in computational modelling of turbulent flows'. In *ERCOFTAC Workshop on Numerical Simulation of Unsteady Flows*, O. Pirroneau, W. Rodi, I.L. Rhyming, A.M. Savill & T.V. Truong (eds.), Cambridge University Press, 1–76. (Updated and shortened version 'Theoretical limitations of computational modelling of turbulent flows'. In *Proceedings of the Fifth Symposium on Refined Flow Modelling & Turbulence Measurements*, P.L. Viollet (ed.), Presses Ponts et Chaussees, Paris, 13–38.)

Hunt, J.C.R., Kaimal, J.C. & Gaynor, J.E., 1988, 'Eddy structure in the convective boundary layer – new measurements and new concepts', *Q. J. R. Met. Soc.* **114**, 821–858.

Hunt, J.C.R., Moin, P., Lee M., Moser, R.D., Spalart, P., Mansour, N.N., Kaimal, J.C. & Gaynor, E., 1989, 'Cross correlation and length scales in turbulent flows near surfaces'. In *Advances in Turbulence* **2**, (2nd European Turbulence Conf., Berlin, August 1988), Springer, 128–134.

Hunt, J.C.R., Sandham, N., Vassilicos, J.C., Launder, B.E., Monkewitz, P.A. & Hewitt, G.F., 2000, 'Developments in turbulence research: a review based on the 1999 Programme of the Isaac Newton Institute, Cambridge', submitted to *J. Fluid Mech.*

Hunt, J.C.R. & Carruthers, D.J., 1990, 'Rapid distortion theory and the 'problems' of turbulence', *J. Fluid Mech.* **212**, 497–532.

Hunt, J.C.R. & Durbin, P.A., 1999, 'Perturbed vortical layers and shear sheltering', *Fluid Dyn. Res.* **24**, 375–404

Hunt, J.C.R. & Morrison, J.F., 2000, 'Eddy structure in turbulent boundary layers', *Eur. J. Mech.*, to appear.

Hussain, F., 1986, 'Coherent structures and turbulence', *J. Fluid Mech.* **173**, 303–356.

Hussain, F. & Clark, A.R., 1981, 'On the coherent structure of the axisymmetric mixing layer: a flow-visualisation study', *J. Fluid Mech.* **104**, 263–294.

Husain, H.S. & Hussain, F., 1991, 'Elliptic jets: Part 2. Dynamics of coherent structures: pairing', *J. Fluid Mech.* **233**, 439–482. 1993, 'Elliptic jets: Part 3. Dynamics of preferred mode coherent structure', *J. Fluid Mech.* **248**, 315–361.

Jacobs, R.G. & Durbin, P.A., 1998, 'Shear sheltering and the continuous spectrum of the Orr–Sommerfeld equation', *Phys. Fluids* **105**, 2006–2001.

Jiménez, J., Wray, A.A., Saffman, P.G. & Rogallo, R.S., 1993, 'The structure of intense vorticity in isotropic turbulence', *J. Fluid Mech.* **255**, 65–90.

Kerr, O.S. & Dold, J.W., 1994, 'Periodic steady vortices in a stagnation-point flow', *J. Fluid Mech.* **276**, 307–325.

Kevlahan, N.K.R. & Farge, M., 1997, 'Vorticity filaments in two-dimensional turbulence: creation, stability and effect', *J. Fluid Mech.* **346**, 49–76.

Kevlahan, N.K.R. & Hunt, J.C.R., 1997,'Nonlinear interactions in turbulence with strong irrotational straining', *J. Fluid Mech.* **337**, 333–364.

Kevlahan, N.K.R., Hunt, J.C.R.& Vassilicos, J.C., 1993, 'A comparison of different analytical techniques for identifying structures in turbulence'. In *Free Turbulent Shear Flows*, Proceedings of the IUTAM Conference on Eddy Structure Identification, Poitiers, J.P. Bonnet & M.N. Glauser (eds.), Kluwer, 311–324.

Kevlahan, N.K.R. & Vassilicos, J.C., 1994, 'The space and scale dependencies of the self-similar structure of turbulence', *Proc. R. Soc.* **447**, 341–363.

Kida, S. & Hunt, J.C.R., 1989, 'Interaction between different scales of turbulence over short times', *J. Fluid Mech.* **201**, 411–445.

Kida, S., Takaoka, M. & Hussain, F., 1991, 'Collision of two vortex rings', *J. Fluid Mech.* **230**, 583–646.

Kida, S. & Tanaka, M., 1994, 'Dynamics of vortical structures in a homogeneous shear flow', *J. Fluid Mech.* **274** 43–68.

Kiya, M., Ohyama, M. & Hunt, J.C.R., 1986, 'Vortex pairs and rings interacting with shear layer vortices', *J. Fluid Mech.* **172**, 1–15

Kolmogorov, A.N., 1941, 'Local structure of turbulence in an incompressible fluid at very high Reynolds numbers', *Dokl. Akad. Nauk SSSR* **30**, 299–303.

Kraichnan, R.H., 1959, 'The structure of isotropic turbulence at very high Reynolds numbers', *J. Fluid Mech.* **5**, 497–543.

Landahl, M.T., 1990, 'On sublayer streaks', *J. Fluid Mech.* **212**, 593–614.

Landau, L.E. & Lifschitz, L.M., 1959, *Fluid Mechanics*, Pergamon.

Leonard, A. 2000, 'Evolution of localised packets of vorticity and scalar in turbulence', this volume.

Lee, M.J., Kim, J. & Moin, P., 1990, 'Structure of turbulence at high shear rate', *J. Fluid Mech.* **216**, 561–583.

Leray, J., 1933, 'Etude de diverses équations intègrales nonlinèaires et de quelques problêmes que pose l'hydrodynamique', *J. Math. Pures Appl.* **12**, 1–82.

Lesieur, M., 1990, *Turbulence in Fluids*, Kluwer.

Lessen, M., 1978, 'On the power laws for turbulent jets, wakes and shearing layers and their relationship to the principle of marginal instability', *J. Fluid Mech.* **88**, 535–540.

Leweke, T. & Williamson, C.H.K., 1998, 'Cooperative elliptica instability in a counter-rotating vortex pair', *J. Fluid Mech.* **360**, 85–119.

Lin, Z.-C., Adrian, R.J. & Hanratty, T.J., 1996, 'A study of streaky structures in a turbulent channel flow with particle image velocimetry'. In *Proceedings of the 8th International Symposium on Applications of Laser Technology in Fluid Mechanics*, Lisbon.

Liu, J.T.C., 1989, 'Contributions to the understanding of large scale coherent structures in developing free turbulent shear flows', *Adv. Appl. Mech.* **26**, 183–309.

Lundgren, T.S., 1982, 'Strained spiral vortex model for turbulent fine structure', *Phys. Fluids* **25**, 2193–2203

McComb, W.D., 1990, *The Physics of Fluid Turbulence*, Clarendon Press.

McGrath, J.L., Fernando, H.J.S. & Hunt, J.C.R., 1997, 'Turbulence, waves and mixing at shear-free density interfaces. Part II – laboratory experiments', *J. Fluid Mech.* **347**, 235–261.

McWilliams, J.C., 1990, 'The vortices of two-dimensional turbulence', *J. Fluid Mech.* **219**, 361–385.

Magnaudet, J. & Eames, I., 2000, 'Dynamics of high Reynolds number bubbles in inhomgeneous flows', *Ann.Rev. Fluid Mech.* **32**, 659–708.

Malkus, W.V.R. & Waleffe, F., 1991, 'Transition from Order to disorder in elliptical flow: A direct path to shear flow turbulence'. In *Advances in Turbulence 3*, 197–203, A.V. Johansson & P.H. Alfredsson (eds.), Springer.

Mann, J., 1994, 'The spatial structure of neutral atmospheric surface-layer turbulence', *J. Fluid Mech.* **273**, 141–168.

Maxey, M.R., 1987, 'The gravitational settling of aerosol particles in homogeneous turbulence and random flow fields', *J. Fluid Mech.* **174**, 441–465.

Melander, M.V. & Hussain, F., 1993, 'Coupling between a coherent structure and fine-scale turbulence', *Phys. Rev. E.* **48**, 2669–2689.

Michard, M., Mathieu, J., Morel, R., Alcaraz, E. & Bertoglio, J.P., 1987, 'Grid generated turbulence exhibiting a peak in the spectrum'. In *Advances in Turbulence* 1, 163–169, G. Cante-Bellot & J. Matthieu (eds.), Springer.

Miyazaki, T. & Hunt, J.C.R., 2000, 'Linear and nonlinear interactions between a columnar vortex and external turbulence', *J. Fluid Mech.* **402**, 349–378.

Moffatt, H.K., 1984, 'Simple topological aspects of turbulent vorticity dynamics'. In *Turbulence and Chaotic Phenomena in Fluids*, T. Totsumi (ed.), Elsevier.

Moffatt, H.K., 2000, 'The interaction of skewed vortex pairs: a model for blow-up of the Navier–Stokes equations', *J. Fluid Mech.* **409**, 51–68.

Moffatt, H.K., Kida, S. & Ohkitani, K., 1994, 'Stretched vortices – the sinews of turbulence; large-Reynolds number asymptotics', *J. Fluid Mech.* **259**, 241–264; Corrigendum, 1994, **266**, 311.

Moin, P., Leonard, A. & Kim, J., 1986, 'Evolution of a curved vortex filament into a vortex ring', *Phys. Fluids* **29**, 955–965.

Ohkitani, K., 1999, 'Characterization of nonlocality in turbulence,' Isaac Newton Institute Workshop on Perspectives in the Understanding of Turbulent Systems.

Ohkitani, K. & Gibbon, J.D., 1999, 'Numerical study of singularity formation in a class of Euler and Navier–Stokes flows'. Preprint.

Ooi, A, Martin, J., Soria, J. & Chong, M.S., 1999, 'A study of the evolution and characteristics of the invariants of the velocity-gradient tensor in isotropic turbulence', *J. Fluid Mech.* **381**, 141–174.

Panofsy, H.A., Larks, D., Kipshutz, R., Stone, G., Bradley, E.F., Bowers, A.J. & Hojstrop, X., 1982, 'Spectra of velocity components over complex torsion', *Q. J. R. Met. Soc.* **108**, 215.

Perry, A.E., Herbert, S. & Chang, M.S., 1986, 'A theoretical and experimental study of wall turbulence', *J. Fluid Mech.* **165**, 163–199.

Peterka, J.A., Meroney, R.N. & Kothani, K.M., 1985, 'Windflow patterns about buildings', *J. Wind Energy & Ind. Aero* **21**, 21–38.

Phillips, O.M., 1955, 'The irrotational motion outside a free boundary layer', *Proc. Camb. Phil. Soc.* **51**, 220–229.

Prandtl, L., 1925, 'Bericht uber Untersuchung zur ausgebildeten turbulenz', *Zs. Angew. Math. Mech.* **5**, 136–139.

Prandtl, L., 1956, *Fluid Dynamics*, Blackie.

Pullin, D.I. & Saffman, P.G., 1998, 'Vortex dynamics in turbulence', *Ann. Rev. Fluid Mech.* **30**, 31–51.

Richardson, L.F., 1922, *Weather Prediction by Numerical Processes*, Cambridge University Press.

Rogers, M.M., 1991, 'The structure of a passive field with a uniform mean gradient in rapidly sheared homogeneous turbulent flow', *Phys. Fluids* **3**, 144–154.

Rogers, M.M. & Moin, P. 1987, 'The structure of the vorticity field in homogeneous turbulent flows', *J. Fluid Mech.* **176**, 33–66.

Rotta, J., 1951, 'Statistische theorie nichthomogeneous turbulenz', *Z. Phys.* **129**, 547–572.

Rottman, J.W., Simpson, J.E. & Stansby, P.K., 1987, 'The motion of a cylinder of fluid released from rest in a cross-flow', *J. Fluid Mech.* **177**, 307–337.

Ruetsch, G.R. & Maxey, M.R. 1992, 'The evolution of small-scale structures in homogeneous isotropic turbulence', *Phys. Fluids* **4**, 2747–2760.

Sandham, N.D. & Kleiser, L., 1992, 'The late stages of transition to turbulence on channel flow', *J. Fluid Mech.* **245**, 319–348.

Savill, A.M., 1987, 'Recent development in rapid-distortion theory', *Ann. Rev. Fluid Mech.* **19**, 531–575.

Schwarz, K.W., 1990, 'Evidence for organised small-scale structure in fully developed turbulence', *Phys. Rev. Lett.* **64**, 415.

Sene, K.J., Thomas, N.H. & Hunt, J.C.R., 1993, 'Role of coherent structures in bubble transport by turbulent shear flows', *J. Fluid Mech.* **259**, 219–240.

Smith, F.T., Dodia, B.T. & Bowles, R.G.A., 1994, 'On global and internal dynamics of spots; a theoretical approach', *J. Eng. Maths.* **28**, 73–91.

Spalding, D.B., 1987, 'A turbulence model for buoyant and combusting flows', *Int. J. Methods. Eng.* **24**, 1–23.

Sreenivasan, K.R., 1991, 'Fractals and multifractals in fluid turbulence', *Ann. Rev. Fluid Mech.* **23**, 539–600.

Squires, K. & Eaton, J.K., 1990, 'Particle response and turbulene modification in isotropic turbulence', *Phys. Fluids* **2**, 1191–1203.

Sulem, P.L., She, Z.S., Scholl, H. & Frisch, U., 1989, 'Generation of large-scale structures in three-dimensional flows lacking parity-invariance', *J. Fluid Mech.* **205**, 341–358.

Synge, J.L. & Lin, C.C., 1943, 'On a statistical model of isotropic turbulence', 1943, *Trans. R. Soc. Canada* **37**, 45–79.

Tavoularis, S. & Karnik, U., 1989, 'Further experiments on the evolution of turbulent stresses and scales in uniformly sheared turbulence', *J. Fluid Mech.* **204**, 457–478.

Tennekes, H. & Lumley, J.L., 1971, *A First Course in Turbulence*, MIT Press.

Townsend, A.A., 1951, 'On the fine scale structure of turbulence', *Proc. R. Soc.* **208**, 534–542.

Townsend, A.A., 1976, *Structure of Turbulent Shear Flow*, Cambridge University Press.

Trefethen, L.N., Trefethen, A.E., Reddy, S.C. & Driscoll, T.A., 1993, 'Hydrodynamic stability without eigenvalues', *Science* **262**, 578–584.

Tsinober, A., 1998, 'Is concentrated vorticity that important', *Eur. J. Mech. B Fluids* **17**, 421.

Turner, J.S., 1973, *Buoyancy Effects in Fluids*, Cambridge University Press.

Vassilicos, J.C. & Hunt J.C.R., 1991, 'Fractal dimensions and spectra of interfaces with application to turbulence', *Proc. R. Soc. A* **435**, 505–534.

Vincent, A. & Meneguzzi, M., 1991, 'The spatial structure and statistical properties of homogeneous turbulence', *J. Fluid Mech.* **225**, 1–20.

Vincent, A. & Meneguzzi, M., 1994, 'The dynamics of vorticity tubes in homogeneous turbulence', *J. Fluid Mech.* **258**, 245–254.

Warhaft, Z., 2000, 'The issue of local isotropy of velocity and scalar turbulent fields in vortex dynamics and turbulence', this volume.

Werle H.A., 1973, 'Hydrodynamic flow visualisation', *Ann. Rev. Fluid. Mech.* **5**, 361–382.

Widnall, S., 1975, 'The structure and dynamics of vortex filaments', *Ann. Rev. Fluid. Mech.* **7**, 141–166.

Wu, X., Jacobs, R.G., Hunt, J.C.R. & Durbin, P.A., 1999, 'Simulation of boundary layer transition induced by periodically passing wakes', *J. Fluid Mech.* **398**, 109–153.

Wyngaard, J.C., 1980, 'The atmospheric boundary layer-modeling and measurement'. In *Turbulent Shear Flows* 2, pp 352–365 , L.J.S. Bradbury *et al.* (eds.), Springer.

Stability of Vortex Structures in a Rotating Frame

Claude Cambon

1 Introduction

System rotation does not affect the motion of an incompressible two-dimensional (2D) flow but it alters its stability with respect to three-dimensional (3D) disturbances. When the background flow consists of arrays of vortices, this problem has many applications in geophysical or industrial flows. When considering both cyclonic and anticyclonic vortices in a rotating frame, it is well accepted that moderate anticyclones are preferentially destabilised, but explanations for this and precise ranges of parameters (Rossby number especially) are often not consistent in the literature. This problem, which has been the subject of an abundent literature, – with analyses, physical and numerical experiments – is revisited in this article by looking at the linear stability, *with system rotation*, of simple flows. Special emphasis will be placed on a street of Stuart vortices, an interesting model for the sheared mixing layer with spanwise billows. Results of classic analyses in terms of normal modes are recalled and contrasted with results of an asymptotic analysis (Lifschitz & Hameiri, 1991) for short wave disturbances, which are localised around fluid trajectories.

The article is organised as follows. Linear approaches, which are useful in both turbulence modelling and stability analyses, are recalled in Section 2. They range from 'homogeneous RDT' to WKB theories, as the 'geometrical optics' used by Lifschitz and Hameiri (1991). Three 'background' instabilities are presented in Section 3, they are the centrifugal, elliptic and hyperbolic ones. Simple explanations are given – or recalled – about the simplest academic case where each instability appears individually, and how it is altered by system rotation. Results of the stability analysis of the Stuart vortices in a rotating frame are shown and discussed in Section 4. Section 5 is devoted to a summary of generic findings about the linear stability of 2D vortices in rotating frame. The final part of the paper is devoted to more prospective and/or controversial aspects, with discussion of WKB theories in the stability context and approaches to nonlinear effects in the turbulence context (Section 6). Finally, in Section 7, the problem of *creation* of coherent vortices from 3D turbulence in a rotating frame is touched upon, ending with a discussion about nonlinear interactions of inertial waves, the effects of solid boundaries and the role of forcing (as in the experimental study by Hopfinger *et al.* 1982).

2   From RDT to zonal stability analysis

The velocity and pressure fields (U_i, P) are first split into mean $(\overline{U_i}, \overline{P})$ and fluctuating components (u_i, p) and equations for their time evolution derived from the basic equations of motion of the fluid. Assuming incompressibility, the result is the mean flow equations, which include a Reynolds stress term in addition to background Navier–Stokes equations, and the equations for the fluctuating component (u_i, p). (not recalled here for the sake of brevity, see Cambon & Scott, 1999, for more details on this section.) The 'overbar' denotes a statistical averaging, a Reynolds spatial filter, ...

Neglecting nonlinear and viscous terms in equations for (u_i, p) is the background for Rapid Distortion Theory (or RDT), introduced by Batchelor and Proudman (1954), not to mention older seminal works by Kelvin, Orr, Prandtl, Taylor, . . . (see Savill (1987) and Hunt & Carruthers (1990) for recent reviews). In neglecting nonlinearity entirely, the effects of interaction of turbulence with itself are supposed small compared with those resulting from mean-flow distortion of turbulence. One often has in mind flows such as weak turbulence encountering a sudden contraction in a channel or an aerofoil. Implicit is the idea that the time required for significant distortion by the mean flow is short compared with that for turbulent evolution in the absence of distortion.

Another simplifying assumption which is often made is that the size of turbulent eddies, ℓ, is small compared with the overall length scales of the flow, L, for instance, the size of a body encountering fine-scale free-stream turbulence;see e.g. Hunt (1973), (1978). In that case, one uses a local frame of reference convected with the mean velocity and approximates the mean velocity gradients as uniform, but time-varying. Thus, the mean velocity is approximated by

$$\overline{U}_i = \lambda_{ij}(t)x_j \tag{2.1}$$

in the moving frame of reference.

2.1   RDT for homogeneous turbulence and stability analysis for extensional flows

If (2.1) is assumed to be valid in all the space, it is a necessary condition to preserve statistical homogeneity of the fluctuating field. In turn, the gradient of the Reynolds stress tensor disappears in the equations for the mean, so that there is no feedback of the fluctuating field in the equation governing the mean, and $\overline{U}_i$ has to be a particular solution of the Euler equations. Hence, solving the linearised equations which govern (u_i, p) in the presence of a given mean velocity gradient, is exactly the same problem as that occuring for the linear stability of the flow $(\overline{U}_i, \overline{P})$, with u_i and p small amplitude disturbances to that field.

Using the approximation (2.1) in the equation for the fluctuating velocity, without the nonlinear and viscous terms, the solution is most easily obtained via Fourier synthesis. An elementary Fourier component of the form

$$u_i = a_i(t) \exp[\imath \mathbf{k}(t) \cdot \mathbf{x}] \tag{2.2}$$

yields a solution of the problem if $\mathbf{k}$ and a_i satisfy a linear system of simple ordinary differential equations, referred to as the Townsend equations. The pressure fluctuation, which is a solution of a Poisson equation, is given by an algebraic relationship in terms of a_i. Decomposition of the turbulence into Fourier components, usually referred to as spectral analysis, allows straightforward treatment of the nonlocal dependency of pressure on velocity, which leads to the appearance of spatial integrals if spectral analysis is not used. In this way, the problem of nonlocality is rendered relatively innocuous, which remains true when nonlinearity is allowed for, although in that case the closure problem rears its ugly head.

Time dependency of the wavenumber represents convection of the plane wave $\exp[\imath \mathbf{k}(t) \cdot \mathbf{x}]$ by the mean flow (2.1). Both the direction and magnitude of $\mathbf{k}$ change as wave crests rotate and approach or separate from each other due to mean-velocity gradients. If $\mathbf{k}$ is given at some time t, it can be related to its value at any other time t_0, $\mathbf{k}(t_0) = \mathbf{K}$, by

$$K_i = F_{ji}(t, t_0) k_j, \tag{2.3}$$

where the matrix $\mathbf{F}$ characterises deformation of an imaginary material convected at the mean velocity (2.1) between the times t_0 and t (in fact, in the language of continuum mechanics, it is the Cauchy tensor associated with deformation by the mean flow). Likewise, the solution of the Townsend equations with the above time evolution for $\mathbf{k}$ in (2.3) has the form

$$a_i(t) = G_{ij}(\mathbf{k}, t, t_0) a_j(t_0), \tag{2.4}$$

where G_{ij} is a spectral Green's function, which is a real *deterministic* quantity.

At this stage, it may be noticed that the RDT for homogeneous turbulence in the presence of mean velocity gradients includes two problems:

- A *deterministic problem*, which consists of solving in the more general way the initial value linear system of equations for a_i. This is done by determining the spectral Green's function, which is also the key quantity requested in linear stability analysis.

- A *statistical problem* which is useful for prediction of statistical moments of u_i and p. Interpreting the initial amplitude $a_j(t_0)$ as a random variable with a given dense spectrum, equation (2.4) yields a prediction of the statistical moments by products of the basic Green's function.

Possible applications to statistics are outside the scope of this article. They are touched upon in Section 6. Note that solutions (2.4) are valid even if the nonlinear term is not discarded *a priori*, provided that the perturbation consists of a single mode, since a single Fourier mode cannot interact with itself (Craik & Criminale 1986). Exactly the same deterministic problem as the one of 'homogeneous RDT' was addressed in the context of flow stability (see, for instance, Lagnado *et al.* 1984, Bayly 1986, Craik & Criminale 1986), although the two communities seem to be largely unaware of each other's work. In particular, the stability analysis in terms of time-dependent, distorted, Fourier modes is attributed to Kelvin (1887) by the stability literature. In agreement with the generality of the RDT formulation, which is not restricted to a special case of parallel pure shear flow (as in Kelvin), I propose to refer to (2.2)–(2.3) as 'Lagrangian Fourier modes', which are governed by 'Townsend equations'.

For a mean flow of the form (2.1) to be a particular solution of Euler equations, requires (Craya 1958) that the matrix

$$d\lambda_{ij}/dt + \lambda_{ik}\lambda_{kj} \qquad (2.5)$$

be symmetric and that $\lambda_{ii} = 0$. The case of irrotational mean flows is easy to solve and not discussed here; see Batchelor & Proudman (1954), Cambon (1982) Lagnado *et al.* (1984) for details.

Rotational mean flows are more complicated and only the steady case has received much attention; see Bayly, Holm & Lifschitz (1996) for recent developments in unsteady cases. It can be shown that symmetry of λ^2 and $\lambda_{ii} = 0$ imply that λ_{ij} takes the form

$$\lambda = \begin{pmatrix} 0 & S - \Omega_0 & 0 \\ S + \Omega_0 & 0 & 0 \\ 0 & 0 & 0 \end{pmatrix} \qquad (2.6)$$

when axes are chosen appropriately, where $S, \Omega_0 \geq 0$. This corresponds to steady plane flows, combining vorticity $2\Omega_0$ and irrotational straining S. The special case $S = 0$ yields pure rotation, which is perhaps better treated in the rotating frame of reference and leads to inertial waves and hence oscillating solutions. The general RDT problem with arbitrary S and Ω_0 was analysed by Cambon (1982), while experimental realisations of grid turbulence interacting with the mean flow represented by (2.6) were carried out by Leuchter *et al.* (1992). The above class of steady mean flows is also compatible with homogeneity in a frame of reference rotating about an axis perpendicular to the plane of the flow. Cambon *et al.* (1994) give details of RDT calculations for such flows. The limiting case, $S = \Omega_0$ (Townsend 1956), corresponds to simple shearing and forms the borderline between two distinct regimes, namely those in which the mean flow streamlines are closed and elliptic about

the stagnation point at the origin ($S < \Omega_0$) and those for which they are open
and hyperbolic ($S > \Omega_0$). These two cases will be discussed in Section 3.

2.2   Zonal RDT and local stability analysis

Without making the assumption of fine-scale turbulence, solving the linearised equations for (u_i, p) with $\overline{U}_i$ assumed known is a difficult problem
in general, but becomes somewhat simpler if the mean flow is irrotational, as
in the classical case of high Reynolds number flow past a body, outside the
wake and boundary layer.

Assuming weak inhomogeneity, considerably more progress can be made
without the need for irrotationality of the mean flow, although simplifications
occur in the latter case. Turbulence which is fine-scale compared with the
overall dimensions of the flow can be treated under RDT by following a no-
tional particle moving with the mean velocity. Particles are convected by the
mean velocity field according to

$$\dot{x}_i = \overline{U}_i(\mathbf{x}, t) \tag{2.7}$$

Thus, the results obtained for strictly homogeneous turbulence can be extended to the weakly inhomogeneous case, but with a mean velocity gradient
matrix $\lambda_{ij}(t)$ which reflects the $\partial \overline{U}/\partial x_j$ seen by the moving particle (Hunt
1973, 1978, Durbin & Hunt 1980).

This idea has been formalised in the context of flow stability (see Lifschitz
& Hameiri 1991) using an asymptotic approach based on the classical WKB
method, which is traditionally used to analyse the ray theoretic limit (i.e.
short waves) in wave problems (see, e.g., Lighthill 1978). The solution is
written as

$$u_i(\mathbf{x}, t) = a_i(\mathbf{x}, t) \exp[\imath \Phi(\mathbf{x}, t)/\epsilon] \tag{2.8}$$

with a similar expression for the fluctuating pressure, where Φ is a real phase
function and ϵ is a small parameter expressing the small scale of the 'waves'
represented by (2.8), while $a_i(\mathbf{x}, t)$ is a complex amplitude which is expanded
in powers of ϵ according to the WKB technique. Over distances of $O(\epsilon)$, one
can use a spatial Taylor series representation for Φ, up to the linear term,
and approximate a_i as constant. It is then apparent that (2.8) is locally a
plane-wave Fourier component of wavenumber

$$k_i(\mathbf{x}, t) = \epsilon^{-1} \partial \Phi / \partial x_i \tag{2.9}$$

The amplitude $a_i(\mathbf{x}, t)$ in (2.8) and the corresponding equation for the
fluctuating pressure are expanded as an asymptotic series in powers of ϵ and
the result injected into the linearised equations without viscosity. At leading
order, one finds that

$$\dot{\Phi} = \partial \Phi / \partial t + \overline{U}_j \partial \Phi / \partial x_j = 0, \tag{2.10}$$

i.e. the wave crests of (2.8) are convected by the mean flow. The spatial derivatives of (2.10) yields

$$\dot{k}_i = -\lambda_{ji}(t)k_j \tag{2.11}$$

where, as before, $\lambda_{ij} = \partial \overline{U}_i/\partial x_j$ and the dot represents the mean-flow material derivative $\partial/\partial t + \overline{U}_i \partial/\partial x_i$. At the next order, one obtains

$$\dot{a}_i^{(0)} = -\underbrace{\left(\delta_{ij} - 2k_ik_j/k^2\right)}_{M_{ij}(t)} a_j^{(0)} \tag{2.12}$$

after elimination of the pressure using the leading-order incompressibility condition $k_i a_i^{(0)} = 0$, where $a_i^{(0)}$ is the leading-order term in the expansion of a_i.

Equations (2.11) and (2.12) have exactly the same form as the Townsend equations (for **k** and **a** in (2.2) in homogeneous RDT), which therefore describe the weakly inhomogeneous case at leading order. The only difference is that, rather than being a simple time derivatives, the dots represent mean-flow material derivatives, implying that one should follow convection by the mean flow in weakly inhomogeneous RDT.

3   The three background instabilities

Stability of 2D vortices was the object of an abundant literature. Regarding circular vortices characterised by a radial distribution of vorticity $W(r)$ or circulation $\Gamma(r)$, two instabilities were particularly investigated: the centrifugal instability is found when the circulation admits a maximum, and the so-called 'barotropic instability' is related to an extrmum in the vorticity distribution (see Kloosterziel & van Heijst 1991). Since the latter involves 2D disturbances, it is not directly affected by system rotation and thus is not a good candidate to explain the asymmetry in terms of cyclonic and anticyclonic eddies. In addition, the superimposed straining process generates departure from circularity (elliptic core inside the vortex and hyperbolic stagnation point outside) with related elliptic and hyperbolic instabilities.

3.1   The centrifugal instability in the rotating frame

The classical criterion for a circular vortex of azimuthal velocity $U(r)$, and related vorticity $W(r)$, in a Galilean frame was extended by Kloosterziel & van Heijst (1991) in the rotating frame of angular velocity Ω, by considering the absolute circulation Γ_a. Accordingly, a generalised discriminant is defined as follows:

$$D(r) = \frac{1}{4\pi^2 r^3}\frac{d\Gamma_a^2}{dr} = 2(\Omega + \frac{U}{r})(W + 2\Omega), \quad \Gamma_a(r) = 2\pi r^2(\Omega + U(r)/r), \tag{3.1}$$

and the instability is located around circular trajectories where $D(r)$ becomes negative after a change of sign. If $U(r)$ characterises a standard distribution with $\Gamma(r) = 2\pi r U(r)$ varying monotonically from 0 at the centre (with $\Gamma(r) \sim 2\pi r^2 \Omega_0$) to a finite value Γ_∞ at $r \to \infty$, it is clear that no instability can occur in the fixed frame. This situation is unchanged in a rotating frame with cyclonic system rotation, since a monotonic system circulation is added to the monotonic relative circulation. In contrast, when the basic rotation is anticyclonic with $\Omega < \Omega_0$, the absolute circulation is positive near the centre, and negative at large values of r, choosing $\Omega_0 > 0$ without loss of generality, so that a centrifugal instability may appear. This suggests that vortices with simple radial distribution of relative circulation are destabilised by a centrifugal instability through anticyclonic basic rotation, provided that the local Rossby number at the core $Ro < -1$.

3.2 The elliptic and hyperbolic instabilities in the rotating frame

By contrast with the centrifugal instability, it is not necessary to look at the distribution of vorticity or strain over a large radial distance to understand the basic mechanisms; only the topology of the stagnation point is important, and the simplest model flow is the extensional flow given by (2.1)–(2.6).

The elliptic instability is often characterised as a cooperative instability which results from the additional strain induced by two adjacent vortices, according to works since 1975 (Moore & Saffman 1975, Tsai & Widnal 1976, Pierrehumbert 1986, see also Williamson 1996 for a historical survey and recent works). But it is simpler to get rid of the two vortices problem and to consider a single vortex with elliptic streamlines, as (2.1)–(2.6) with $S < \Omega_0$. At weak ellipticity, $S \ll \Omega_0$, the disturbances consist of unbounded plane inertial waves with the dispersion relation $\sigma = \pm 2\Omega_0 \cos\theta_k$; resonance is found for $\sigma = \Omega_0$, resulting in the selective amplification of oblique modes $\cos\theta_k = \pm 1/2$ by a Floquet mechanism (Bayly 1986); here θ_k is the angle of the wave vector with the mean vorticity vector. In the rotating frame, this angle for destabilised oblique modes and the Floquet coefficient related to their exponential amplification are

$$\cos\theta_k = \pm(1/2)Ro/(1 + Ro) \quad \text{and} \quad (\sigma/S) = (1/16)(3Ro + 2)^2/(Ro + 1)^2 \tag{3.2}$$

outside the range $-2 < Ro < -2/3$ (stable), with $Ro = \Omega_0/\Omega$. The first equation is obtained by replacing the vorticity by the absolute vorticity in the dispersion relation, so that the resonance condition becomes $2(\Omega + \Omega_0)\cos\theta_k = \pm\Omega_0$, whereas the second equation is a recent generalisation by Le Dizès (1999) of the value 9/16 found by Waleffe (1990) in the non-rotating case (or $Ro = \pm\infty$). At significant ellipticity, the role of basic rotation, which shifts the range of oblique modes and modifies the amplification rate, is not significantly different from the one for $E \sim 1$, as illustrated by Cambon *et al.* (1994).

The hyperbolic case is found for $S > \Omega_0$ in (2.6). In the particular case where $\Omega_0 = 0$ (irrotational), vorticity disturbances are governed by a Cauchy equation, which exhibit exponential amplification of vorticity along the axis of stretching. Similar exponential growth also appears when looking at the Fourier modes of the velocity disturbance, especially for pure spanwise modes. In the rotating frame, this exponential amplification for spanwise modes can be cancelled for a sufficiently high value of Ω, but the role of the hyperbolic stretching for explaining the asymmetry in term of the cyclonic and anticyclonic case is not clear.

3.3 Pure spanwise, pressureless modes

The role of 'pressureless' modes was emphasised by Bayly (1988) for centrifugal instability and by Leblanc & Cambon (1997) in connection with generalised criteria. In the latter works, was clarified the fact that the same criterion can be derived from a rigorous stability analysis (Pedley 1969) and from a semi-empirical, apparently 2D and pressureless analysis (Bradshaw 1969, Bidokhti & Tritton 1992). Spanwise modes correspond to $k_1 \sim k_2 \ll k_3$ for perturbations to an extensional 2D flow, under the form (2.1), and to

$$u_i(x_1, x_2, x_3, t) = \tilde{u}_i(x_1, x_2, t)e^{\imath k_3 x_3} \qquad (3.3)$$

with $1/L_h \ll k_3$, L_h being a length scale characteristic of the motion in the plane (x_1, x_2). These modes are not affected by the pressure disturbance for k_3 large enough, and satisfy the incompressibility constraint. The 2D limit, which is recovered in the limit of vanishing k_3, is radically different, spanwise modes have variability with respect to spatial coordinates concentrated in the spanwise direction, whereas 2D modes have variability concentrated in the other two directions. Simplified pressureless dynamics is not correct for the whole velocity field u_i (using, for instance, the 'particle displaced' analysis of Tritton), but it is enough to replace u_i by $\tilde{u}_i$ in (3.3) to recover correct dynamics leading to the *same simplified equations* and thus the same criteria. For 2D extensional flows in rotating frame with parameters S, Ω_0, Ω, the generalisation of the discriminant becomes

$$D = -S^2 + (\Omega_0 + 2\Omega)^2 \qquad (3.4)$$

with amplification as $\exp(\sqrt{-D}t)$ and suggests a maximum destabilisation for a typical Rossby number $Ro = -2$. We refer to $\Omega_0 + 2\Omega$ as the 'tilting vorticity' discussed in Cambon *et al.* (1994), which differ from the absolute vorticity $2\Omega_0 + 2\Omega$.

4 Taylor–Green and Stuart vortices in a rotating frame

4.1 Stability of Taylor–Green vortices. A survey of recent results

The 2D array of Taylor–Green vortices is periodic in both x_1 and x_2 directions. The basic cell is a square and consists of four rotors bounded by squared streamlines and counter-rotating with each other. This is the 'four rollers mill' of Taylor. In addition, ellipticity is possibly introduced by considering a rectangular cell of aspect ratio E, and the flow is seen in a frame rotating in planes (x_1, x_2) with angular velocity Ω (see Figure 1). When $\Omega \neq 0$, two non-adjacent vortices are cyclonic and the two others are anticyclonic. In non-dimensional form, the streamfunction is given by $\Psi = (\sin x_1 \sin x_2)/(1 + E^2)$ and the Rossby number $Ro = W_0/(2\Omega)$, with W_0 the absolute value of the core vorticity.

A preliminar study of the rotating case using LES (Cambon et al. 1994, Section 5) at $E = 1$, showed the preferential destabilisation of anticyclonic vortices at $Ro = -2$ (see their Figure 11), and stabilisation both in the cyclonic case and in the case of zero absolute vorticity at the core $Ro = -1$. The initial data consisted of a 3D disturbance field, as generated for isotropic turbulence, superimposed to the 2D Taylor–Green flow. Both linear stability analysis and DNS were performed by Lundgren & Mansour (1991) in the non-rotating case, but they isolated one vortex by using symmetrised Fourier components for the disturbance field, so that their method is not suitable to study the asymmetry (cyclonic–anticyclonic) induced by system rotation on the whole four-roller mill (Denis Sipp, private communication).

Complete 3D stability analyses in the circular $E = 1$ and elliptic case $E = 2$, were achieved by Sipp & Jacquin (1998) and by Sipp, Lauga & Jacquin (1999) with the effect of system rotation. The latter paper includes a detailed study of the three background instabilities, with a particular emphasis on the centrifugal one, which is activated only in the presence of anticyclonic system rotation. These studies, which were partly motivated by the analysis of Leblanc & Cambon (1998) for the Stuart vortices (see next subsection), offer a quantitative comparison of the local analysis (Lifshitz and Hameiri 1991) for short-wave disturbances localised around mean trajectories, with a more classical nonlocal analysis in terms of normal modes

$$(u_i, p) = e^{\sigma t} e^{i k_3 x_3} f(x_1, x_2) \tag{4.1}$$

with $f(x_1, x_2) = \sum \tilde{f}(k_1, k_2) e^{i k_1 x_1} e^{i E k_2 x_2}$, and k_1, k_2 integers. The main striking result is the precise identification by both methods of a centrifugal mode of instability, as illustrated by Figure 2.

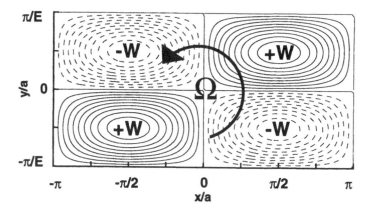

Figure 1. The Taylor–Green flow: iso-values of the vorticity. Case $E = 2$ (courtesy Sipp *et al.* 1999).

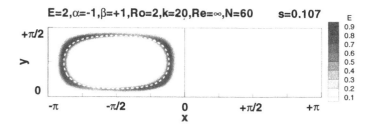

Figure 2. Energy repartition of a typical centrifugal unstable normal mode obtained by the matrix eigenvalue method. The dashed white line sketches the streamline $\tilde{\psi} = -0.3558$ in the vicinity of which local asymptotic Bayly's method provides the centrifugal unstable normal mode. This mode is odd both with respect to the origin and with respect to the centre of the anticyclonic vortex (courtesy Sipp *et al.* 1999).

4.2 Methodology and typical results for the Stuart vortices

The array of Stuart vortices is periodic in the streamwise direction x_1 only, and the vorticity of the eddies has the same sign. The streamfunction is given

by
$$\Psi = \ln(\cosh x_2 - \rho \cos x_1) \qquad (4.2)$$

where ρ, $0 < \rho < 1$, characterises the vorticity distribution inside the vortices. This is a good model of the first instability of the plane mixing layer. The limiting case $\rho = 0$ gives the classic parallel flow with tangent-hyperbolic profile, with no dependency on the streamwise coordinate and no concentrated eddies. The other limit $\rho = 1$ corresponds to an array of concentrated point vortices. In the general case, ρ gives both the ellipticity and the vorticity in the core of the eddies (see Figure 3 for $\rho = 1/3$). The Rossby number is defined as the ratio $Ro = W_0/(2\Omega)$ of core vorticity to system vorticity, with $W_0 = -(1+\rho)/(1-\rho)$.

The linear stability approach of Leblanc & Cambon (1998) included both a nonlocal method of normal modes with $f(x_1, x_2) = e^{\mu x_1} \tilde{f}(x_1, x_2)$ in (4.1), and a particular application of the local analysis, limited to stagnation points. For high values of the spanwise wave number k_3, both analyses were shown to coincide well with identification of the role of elliptic and hyperbolic points. Especially, the coefficient of amplification $Re(\sigma)$ of the normal mode of type (4.1) was shown to coincide with the one given by the temporal Floquet analysis (see below) around the elliptic stagnation point at the core. Nevertheless, no clear evidence of a centrifugal mode of instability was given by the nonlocal method, probably due to the somewhat low values of ρ investigated, and perhaps due to a lack of numerical resolution. The local method, applied here only to stagnation points, was, of course, not relevant for identifying such a mode. Other numerical results for nonlocal stability in the same case (Stuart vortices in a rotating frame) have suggested that a centrifugal mode does exist in the anticyclonic cases (Potylitsin & Peltier 1999). These results, and the relevance of local analysis proved by Sipp *et al.* (1999) for identifying the centrifugal mode, have suggested extending the local analysis to streamlines of Stuart vortices other than the stagnation points. This work is in progress (Cambon, Godeferd & Leblanc 1999); its method and typical results are summarised in the following.

Recall that the Townsend equations, recovered by the WKB analysis of Lifshitz & Hameiri (1991), have to be numerically solved in the rotating frame following a given trajectory. The system of linear equations becomes

$$\dot{x}_i = \overline{U}_i \qquad (4.3)$$
$$\dot{k}_i = -\overline{U}_{j,i} k_j \qquad (4.4)$$
$$\dot{a}_i = -\left[\left(\delta_{ij} - 2\frac{k_i k_j}{k^2} \right) \overline{U}_{jl} + 2\Omega \left(\delta_{ij} - \frac{k_i k_j}{k^2} \right) \epsilon_{j3l} \right] a_l. \qquad (4.5)$$

They are the trajectory equation, the eikonal equation, and the amplitude equation, respectively. The overdot denotes a Lagrangian (or substantial) derivative following trajectories. In the above system of ODE, the velocity components $\overline{U}_i$ and the velocity gradient matrix $\overline{U}_{i,j}$ are analytically expressed at any point using (4.2). These equations are solved given initial data, denoted by capital letters, so that $\mathbf{X}$ is the initial position on the trajectory (Lagrangian coordinate), $\mathbf{K}$ is the Lagrangian wavevector, and $\mathbf{A}$ is the initial amplitude.

Only closed trajectories, identified by the abcissa $0 < x_0 < \pi$, with $\mathbf{X} = (x_0, 0, 0)$ are considered here. The Lagrangian, initial, wave vector is chosen normal to the initial velocity, so that $\mathbf{K} = (\sin\theta_k, 0, \cos\theta_k)$, $0 \leq \theta_k \leq \pi/2$. The value $\theta_k = 0$ characterises *pure spanwise, pressureless* modes, whereas $\theta_k = \pi/2$ characterises 2D modes.

The Green's function of the linear system of equations (4.3)–(4.5), as G_{ij} in (2.4) is numerically computed, after a period T, which corresponds to the time to run a complete loop,

$$a_i(\mathbf{X}, \mathbf{K}, T) = G_{ij}(\mathbf{X}, \mathbf{K}, T, 0)A_j \tag{4.6}$$

choosing $A_j = \delta_{j1}$, $A_j = \delta_{j2}$, $A_j = \delta_{j3}$, successively. Then the Floquet parameter $Re\,[\sigma(\mathbf{X}, \mathbf{K}, T)]$ related to the maximum eigenvalue s of the Floquet matrix though $\sigma = \ln(s)/T$, is identified for each trajectory and each initial wavevector.

The distribution of the Floquet parameter is shown for different closed trajectories and different positions of the wave vector in Figures 4, 5 and 6. The thick line curve on the bottom plane represents the absolute circulation Γ_a plotted versus x_0; this is computed numerically for the different non-circular trajectories. The case $\rho = 1/3$ is considered in Figure 4, whithout rotation, and in Figure 5 ($Ro = -2$ at the core). This case was addressed by Leblanc and Cambon (1998). The values of vorticity and ellipticity at the core are $W_0 = -2$ and $E = 1.732$, respectively. The case $\rho = 3/4$, with larger core vorticity ($W_0 = -7$) and smaller core ellipticity ($E = 1.155$) is considered in Figure 7. This case was addressed by Potylitsin and Peltier (1999). These figures enable us to identify the three typical modes and their alteration by system rotation in a clear way, using an inexpensive 'local' (trajectory by trajectory) computation.

5 General conclusions about linear stability of 2D vortex flows in a rotating frame

5.1 Results on the three background instabilities

Results of classical, nonlocal, stability analyses in terms of normal modes, were contrasted with results of the local analysis (Lifshitz & Hameiri 1991).

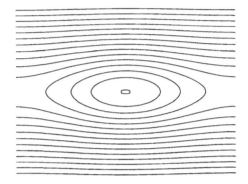

Figure 3. The Stuart flow. Isovalues of the streamfunction (4.2).
Case $\rho = 1/3$.

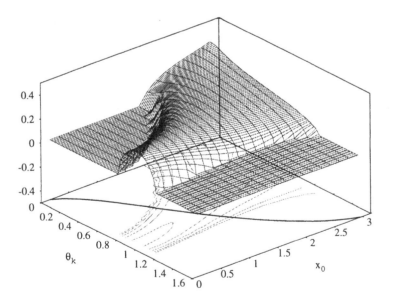

Figure 4. Floquet parameter distribution for different closed trajectories, given by (4.2) and initialised at $\mathbf{X} = (x_0, 0, 0)$, and for
different angles θ_k. Case $\rho = 1/3$, $\Omega = 0$.

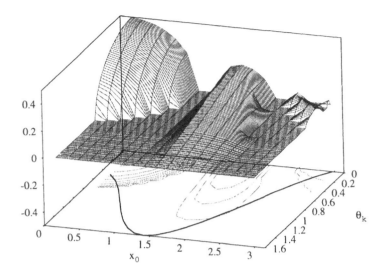

Figure 5. Legend as in Figure 4. Case $\rho = 1/3$, $Ro = -2$.

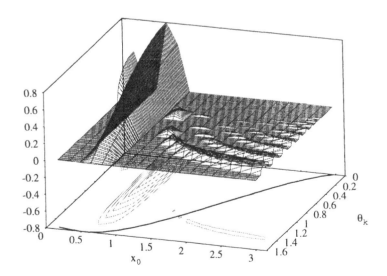

Figure 6. Legend as in Figure 4. Case $\rho = 3/4$, $Ro = -2$.

The latter approach allows to apply zonal RDT equations along individual streamlines for short-wavelength disturbances. Some simple mechanisms identified for simplified extensional flow given by (2.1) and (2.6), in rotating frame, were recovered in more complex flows, such as the Taylor–Green and the Stuart vortices. Local analysis is invaluable for substituting the *informative* taxonomy of unstable modes, *hyperbolic, elliptic, centrifugal*, to the *zoological* one, *braid, core, edge*, used by Peltier and coworkers since 1994.

As shown in Figures 4, 5 and 6 for the Stuart vortices, the elliptic mode is captured as an oblique mode whose angular location θ_k and amplification rate σ is found for individual streamlines near the core, in a way consistent with previous RDT and stability analyses for extensional flows (Cambon 1982, Bayly 1986, Cambon *et al.* 1994). The relevance of equations (3.2) is confirmed for predicting the shift of the angular location θ_k and modification of the amplification rate by system rotation. In particular, the elliptic mode, which is located at $\theta_k \sim \pi/3$ with no rotation, is shifted towards a spanwise mode $\theta_k \sim 0$ and more amplified in the anticyclonic case, in agreement with a maximum amplification for $Ro = -2$. These results are also consistent with the ones of Sipp and coworkers for the flattened Taylor–Green vortices.

The identification of the centrifugal mode in anticyclonic cases ($Ro < 0$) is also clear and accurate using the local analysis along intermediate streamlines between the core and the periphery. This mode is confirmed to be essentially spanwise ($\theta_k = 0$), and located nearly outside the streamline where absolute circulation reaches a maximum. This characteristic streamline moves towards the periphery when the anticyclonic system rotation is smaller and smaller, so that the centrifugal and the hyperbolic modes can eventually merge.

It is confirmed that the instable hyperbolic mode is essentially spanwise ($\theta_k = 0$), located near peripheral streamlines, and cancelled by large enough rotation rate, without a net distinction between cyclonic and anticyclonic cases.

The most important result illustrated by figures 4, 5 and 6 (see also Cambon, Godeferd & Leblanc 1999 for more details) is the competition between centrifugal and elliptic instability in the anticyclonic case. For values of the Rossby number around $Ro = -2$, where both types of modes are important, the elliptic instability is shown to be dominant for the lowest value of ρ (Figure 5). Of course, centrifugal instability is dominant for the cases with weaker core ellipticity. Finally, the centrifugal instability explains the asymmetry of the effect of system rotation and destabilisation of anticyclonic vortices for quasi circular vortices, whereas this explanation is provided by the elliptic instability if the core of the vortex is elliptic enough. A quantitative study of the domain (ρ, ψ, Ro) in which each instability is relevant and delineating the limit of the influence of the centrifugal instability, was recently proposed by Sipp & Jacquin (1999), in close connection with local criteria for spanwise modes of disturbances (see below).

5.2 Validity of criteria from simplified 'pressureless' analyses

Even if the basic elliptical instability involves oblique modes at $\theta_k \sim \pi/3$, so that spanwise modes ($\theta_k = 0$) are stable without rotation, maximum amplification in the rotating case, which is found at $Ro = -2$, corresponds again to a spanwise mode. Hence the special case of spanwise modes, which are naturally unaffected by pressure disturbances, requires a particular attention. Short-wave dynamics is simplified for these modes since the $k_i k_j / k^2$ factors disappear in equation (4.5), as they do in equation (2.12). Leblanc & Cambon (1997) proposed generalising the inertial discriminant, which reduces to (3.4) in the case of an extensional flow, under the form

$$D = 2(\Omega + U/\mathcal{R})(W + 2\Omega) - (\partial_s U)^2, \tag{5.1}$$

where D is the determinant of the inertial matrix in curvilinear coordinates (s, Ψ), with s the curvilinear abcissa, $\mathcal{R}$ the local curvature radius, $U = ds/dt$ the velocity, and $W = -\nabla^2 \Psi$ the vorticity at the given streamline. In the particular case of centrifugal instabilities, Sipp & Jacquin (1999) showed that the correct criterion is

$$D = 2(\Omega + U/\mathcal{R})(W + 2\Omega) \tag{5.2}$$

instead, generalising the demonstration of Bayly (1988), so that $D < 0$ on a whole trajectory is a sufficient condition for having a centrifugal instability. Introducing the intrinsic shear rate (or shear vorticity) $\mathcal{S}$ defined by $W = -\mathcal{S} + 2U/\mathcal{R}$, (5.2) becomes $D = (W + 2\Omega + \mathcal{S})(W + 2\Omega)$, or, equivalently $D = -(\mathcal{S}/2)^2 + (W + 2\Omega + \mathcal{S}/2)^2$. The criterion (5.1) differs from (5.2) through the diagonal contribution $\partial_s U$ to the strain rate in the inertial tensor. Note that in the general case in which the matrix of the linear system of equations is time dependent over a streamline, the stability is not necessarily determined by the determinant of this matrix. In the analysis of Sipp and Jacquin, the 'pressureless' system (4.5) is projected onto a special basis, so that its matrix becomes triangular and its determinant does characterise the stability; this matrix differs from the inertial tensor from the term $\partial_s U$. Accordingly, a suitable generalisation of (3.4) to (5.2) is found by replacing the strain rate S by half the shear rate $\mathcal{S}/2$, discarding the diagonal contribution to S in the inertial tensor, and replacing the 'tilting vorticity' $\Omega_0 + 2\Omega$ by $W + \mathcal{S}/2 + 2\Omega$. Of course, the particular forms of the discriminant D for parallel flows ($\mathcal{S} = -W$, Pedley-Bradshaw criterion) and circular flows (Kloosterziel criterion (3.1)) are immediately recovered, as (3.4) for the extensional flow.

A significant difference can be found when comparing the *local* Rossby numbers which give the maximum destabilisation for elliptic and centrifugal instability. When the elliptic instability is involved, the maximum destabilisation is found near the Rossby number $Ro = -2$, and a significant range of Rossby numbers around $Ro = -1$, the case of zero absolute vorticity, yields stability. This is in agreement with the analysis at vanishing ellipticity, which predicts stability for $-2/3 > Ro > -2$, and unstability with amplification rate (3.2) outside this domain. In the elliptic case, the choice of the vorticity at the core for defining the Rossby number is convenient, with no alternative. The situation is not the same when the centrifugal instability is involved, since it may appear near an intermediate streamline, not necessarily close to the core, so that the local vorticity of this streamline can be different (smaller in absolute value) from the core vorticity. The results in Figures 5 and 6 show that the centrifugal instability takes place for streamlines nearly outside the streamline where the absolute circulation has a maximum. It is easy to show that this limiting streamline corresponds to a change of sign of the absolute vorticity, in agreement with the generalisation of the Kloosterziel criterion to non-circular streamlines, and with the analysis of Sipp & Jacquin (1999). Accordingly, the centrifugal instability occurs over an annulus of streamlines, whose local Rossby number Ro_Ψ is in the range $0 < Ro_\Psi < -1$. Hence the domain of maximum amplification of elliptic instability by system rotation and the domain of activation (by rotation too) of centrifugal instability, involve different ranges of Rossby numbers, if they are defined locally.

6   Open issues and perspectives in stability analysis and turbulence modelling

The validity of linearised approaches close to zonal RDT is strongly connected to the existence of real modes, possibly obtained in linearly combining individual wave packets of form (2.8). Lifschitz and coworkers have proposed a sophisticated evaluation of upper bounds for higher order terms in the ϵ-expansion, so that they demonstrate that the unbounded growth in time of $a_i^{(0)}$ given by (2.12) along a trajectory means actual instability. The problem of rebuilding the disturbance flow in physical space, however, is not simple, even in homogeneous RDT, as illustrated by the well-known case of pure plane shear (or unbounded Couette flow). In this case, the Green's function (2.4) was found analytically (Townsend 1956), but the calculation of the resulting history of the kinetic energy of the disturbance, which involves integration of $G_{ij}G_{ij}$ over all directions of $\mathbf{k}$, is complex and requires asymptotic analysis (Rogers 1991). Existence of normal modes of instability given by (2.12) near the pressureless limit (k_3 large but not infinite) was proven by Bayly (1988) for centrifugal instability (see also Sipp and Jacquin 1999). These localised unstable modes in physical space consist of Hermite polynomials with Gaussian envelope in the plane of the background flow. Note that the recent semi-empirical 'WKB RDT' approach of Nazarenko *et al.* (1999) involves similar localised disturbances, which are obtained by introducing a Gaussian filter in the Fourier synthesis (a Gabor transform). As a bonus, these approaches allow one to link the small parameter ϵ in (2.8) to the spatial extension of the actual mode of disturbance in physical space.

Other problems posed, or perspectives offered, by the mathematical background of Lifshitz and Hameiri (1991) are touched upon below.

- Viscous cut-off. The possible short-wave instability can be cancelled by viscous effects, which can be easily accounted for in the equations (as done in homogeneous RDT for a long time). For a mode characterised by an amplification rate σ, the characteristic viscous cut-off would be $k_c = \sqrt{\sigma/\nu}$. Both conditions, k large enough (short-wave) and $k < k_c$ may be a strong constraint.

- Connection with Lagrangian chaos. The phase Φ is purely advected by the background flow in (2.10), and it is known that complex trajectories can result from a very simple velocity field.

- Open spiral-type trajectories. Another challenge, for the stability problem following trajectories.

- Connection with spatial growth and convective-absolute instability concepts. Disturbances with slowly-varying phase $\frac{1}{\epsilon}(\int \mathbf{k} \cdot d\mathbf{x} - \sigma t)$ are also considered, and WKB approximations used, in this context, but $\mathbf{k}$ is complex, not time-dependent, and the disturbances are essentially 2D. Is it relevant to consider Φ as complex in (2.8)?

Briefly looking at turbulence modelling, it is necessary to distinguish the nonlocal problem, which is mainly due to the relationship between pressure and velocity fluctuation, and the nonlinear problem. Although purely linear theory closes the equations without further ado and simplifies mathematical analysis, it is rather limited in its domain of applicability, ignoring as it does all interactions of turbulence with itself, including the physically important cascade process. Multi-point turbulence models which account for nonlinearity via closure lead to moment equations with a well-defined linear operator and nonlinear source terms; see Cambon & Scott (1999) for a review.

Growth of the fluctuating velocity according to RDT hastens the demise of the linear approximation, which is, in any case, suspect over long periods of time, as, for example, in the case of rapidly rotating turbulence for which no instability arises. An analysis of the importance of nonlinearity in an unstable mean-flow, namely pure plane straining, was recently given by Kevlahan &

Hunt (1997). One may nonetheless conjecture that linear effects continue to
be important in structuring the turbulence, although linear theory itself no
longer gives a precise description. It is not possible to discuss in a general
way the nonlinear problem, which is very complex and multiform, particularly
when the turbulence is subjected to effects of mean gradients and/or body
forces. This problem will be only illustrated by the effects of pure rotation
in the last section. A rotating fluid leads to inertial waves (Greenspan 1968)
whose linearised properties are well-understood and whose amplitudes provide
a set of variables for which the linear behaviour is particularly simple. Linear
theory describes rotating turbulence as a combination of inertial waves, which
interact when nonlinearity is allowed for via selective resonant transfers whose
origin is best understood with linear theory as a background.

7 Creation of vortices by rotation from unstructured turbulence

In the absence of mean gradients in the rotating frame, the vorticity is governed by

$$\frac{\partial \omega_i}{\partial t} + 2\Omega_j \frac{\partial u_i}{\partial x_j} = \frac{\partial u_i}{\partial x_j}\omega_j - u_j\frac{\partial \omega_i}{\partial x_j} + \nu\frac{\partial^2 \omega_i}{\partial x_j \partial x_j} \qquad (7.1)$$

Only the linear second term in the left-hand side explicitly involves the angular velocity of the rotating frame of reference. Nonlinear and viscous terms are gathered on the right-hand side.

In agreement with the Proudman theorem, a 2D state, or $\Omega_j \partial u_i/\partial x_j = 0$, is found in the limit of low Rossby number, high Reynolds number, and slow motions. The first two conditions yield neglecting right-hand side terms, whereas the last one amounts to neglect the temporal derivative. It is important to point out that the Proudman theorem says only that the *slow manifold* of the linear regime necessarily is the *2D manifold*, but it cannot predict the transition from 3D to 2D structure, which is a typically nonlinear and unsteady process; see Cambon *et al.* (1997) for a survey.

For unforced, unbounded, turbulent field, the linear regime consists of superposition of inertial waves (for u_i, ω_i, p, ...), of the form $\sum A_i \exp[\imath(\mathbf{k} \cdot \mathbf{x} - \sigma t)]$, in which the dispersion law is given by

$$\sigma = \pm 2\Omega \cos\theta_k = \pm 2\Omega\frac{k_3}{k} \qquad (7.2)$$

and the slow manifold is recovered as the wave plane orthogonal to the rotation axis, for $k_3 = 0$ which corresponds to $\partial/\partial x_3 = 0$ in physical space. But the linear regime conserves the spectral density of energy, or $\hat{u}_i^*\hat{u}_i$, so that the transition from 3D to 2D turbulence must be interpreted as an angular drain of energy from oblique wavevectors towards the waveplane $k_3 = 0$, energy drain which is mediated by nonlinear interactions.

This tendency, which is a partial two-dimensionalisation induced by *nonlinear* interactions, has been extensively studied in homogeneous turbulence, using experimental, theoretical and DNS/LES high resolution results. Looking at the distribution of spectral energy, or $e(k, cos\theta_k = k_3/k, t) \sim \overline{\hat{u}_i^* \hat{u}_i}$, it is directly linked to the concentration of energy, as shown in Figure 7 from Cambon *et al.* (1997). In physical space, the partial two-dimensionalisation is reflected by the rise of the anisotropic relationship for the integral length scales, with dominant increase of axially separated integral length scales. The Reynolds stress tensor anisotropy remains weak, with a significant dominance of the axial component, or $\overline{u_3^2} > \overline{u_1^2} = \overline{u_2^2}$. This illustrates the essential difference between the 2D limit, or $\partial/\partial x_3 = 0$, and the two-component limit, or $u_3 = 0$. In addition, anisotropic features linked to the two-dimensionalisation process were shown to be triggered at a *macro-Rossby number* $Ro^L = \dfrac{u'}{(2\Omega L)}$ close to one. Studies of temporal evolution of initially isotropic turbulence at high Rossby number ($Ro_L > 1$) display an *intermediate* range of Rossby numbers, which is delineated by $Ro_L < 1$ and $Ro^\omega = \omega'/(2\Omega) \sim u'/(2\Omega\lambda) > 1$, for anisotropisation effects, the last Rossby number being a *micro-Rossby* number (see also Jacquin *et al.*, 1990, for first – experimental – evidence).

Although they confirm the significant anisotropisation linked to partial two-dimensionalisation, high resolution DNS and LES do not really show creation of coherent quasi-2D vortices, as far as the conditions for reproducing homogeneity are fulfilled. One of these conditions is to stop the computation when the most amplified integral length scale becomes of the same order of magnitude as the length of the computational box. A good compromise to reach higher elapsed times without developing spurious anisotropy was obtained by Mansour with an elongated computational box with length in the axial direction four times the length in the other two directions (corresponding to $512 \times 128 \times 128$ LES used in Cambon *et al.*, 1997). Apparently more complete two-dimensionalisation with creation of strong axial rotors was shown in the low resolution 64^3 LES of Bartello *et al.* (1994), but this seems to be a numerical artifact due to the blocking of the integral length scales when the computation is performed for too large elapsed times. Another difference of the latter study with DNS and LES studies, in which homogeneity is fulfilled (Bardina *et al.*, Cambon *et al.*), was the rise of a two-component limit for the Reynolds stress tensor ($\overline{u_3^2} \ll \overline{u_1^2} \sim \overline{u_2^2}$), in close connection with interference with periodic boundaries.

The last numerical study has suggested that boundary effects are important for reinforcing the rise of coherent axial vortices. Hence a numerical simulation of rotating turbulence between two solid parallel walls has been performed by Godeferd & Lollini (1999) by means of a pseudo-spectral code.

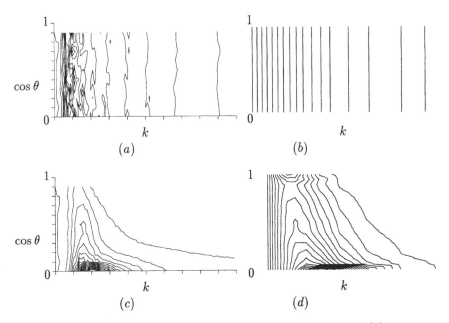

Figure 7. Isolines of kinetic energy for LES computations (a) at $\Omega = 0$ at time $t/\tau = 427$, (b) EDQNM2 with $\Omega = 0$; (c) LES with $\Omega = 1$ at $t/\tau = 575$; and (d) EDQNM2 calculation with $\Omega = 1$ at time $t/\tau = 148$. The vertical axis bears $\cos\theta_k$ (from 0 to 1 upwards) and the horizontal one the wave number k (see Cambon et al. 1997).

Another, more important, motivation for the DNS was to try to reproduce the essential results of the experiment by Hopfinger et al. (1982) in which confinement and local forcing are additional, essentially inhomogeneous, effects with respect to the Coriolis force. Typical DNS results are briefly presented and discussed as follows.

A transition is shown to occur between the region close to the forcing and an outer region in which coherent vortices appear, the number of which depends on the Reynolds and Rossby numbers. Identification of vortices is shown in Figure 8 using both iso-vorticities (noisy spots) and a specific criterion (Normalised Angular Momentum), which was suggested by experimentalists in PIV for obtaining smooth isovalues.

Asymmetry in terms of cyclones-anticyclones is mainly induced by the Ekman pumping near the solid boundaries, yielding helical trajectories. This is illustrated in Figure 9, in which a pair cyclone-anticyclone is isolated. Even if the Ekman pumping generates a three-component motion, the presence of the horizontal walls, and the presence of the forcing in a horizontal plane

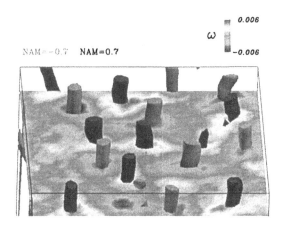

Figure 8. Vortex structures identified by NAM iso-values (tubes) and vorticity iso-values (courtesy Godeferd & Lollini 1999.)

between them, are essential for enforcing coherent vortices.

Nevertheless, and in contrast with the experimental results, the asymmetry cyclones-anticyclones was not obtained in term of number and intensity, but for a case. In the same way, the typical distance between adjacent vortices is of the same order of magnitude as their diameter, and the Rossby number in their core is close to one. It was expected that for a given symmetric distribution of more intense and concentrated vortices (higher Rossby number), the centrifugal instability could act in destabilizing the anticyclones, so that dominant cyclones could emerge, as in the physical experiment. It seems that the insufficiently high Reynolds number is responsible for the lack of intensity and concentration. Hence, the conditions of stability discussed throughout this paper are no relevant for strongly favouring the cyclonic eddies, but they could do for higher Reynolds DNS or LES.

Acknowledgments

The author is grateful to F. S. Godeferd, D. Sipp, J. C. Vassilicos, and A. Leonard for helpful discussion, and assistance (FSG).

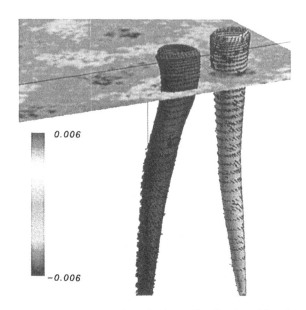

Figure 9. Selected pair of cyclonic-anticyclonic eddy structures (courtesy Godeferd & Lollini 1999.)

References

Bardina, J., Ferziger, J.M., Rogallo, R.S. (1985) 'Effect of rotation on isotropic turbulence: computation and modelling', *J. Fluid Mech.* **154**, 321–326.

Bartello, P., Métais, O., Lesieur, M. (1994) 'Coherent structure in rotating three-dimensional turbulence', *J. Fluid Mech.* **273**, 1–29.

Batchelor, G.K. (1953) *The Theory of Homogeneous Turbulence*, Cambridge University Press.

Batchelor, G.K., Proudman I. (1954) 'The effect of rapid distortion in a fluid in turbulent motion', *Q. J. Mech. Appl. Maths*, **7**, 83.

Bayly, B.J. (1986) 'Three-dimensional instability of elliptical flow', *Phys. Rev. Lett.*, **57**, 2160.

Bayly, B.J. (1988) 'Three-dimensional centrifugal-type instabilities in inviscid two-dimensional flows', *Phys. Fluids* **31**, (1), 56–64.

Bayly, B.J., Holm, D.D., Lifschitz, A. (1996) 'Three-dimensional stability of elliptical vortex columns in external strain flows', *Phil. Trans. R. Soc. Lond.* A. **354**, 895–926.

Bidokhti, A.A., Tritton, D.J. (1992) 'The structure of a turbulent free shear layer in a rotating fluid', *J. Fluid Mech.* **241**, 469.

Cambon, C. (1982) 'Etude spectrale d'un champ turbulent incompressible soumis à des effets couplés de déformation et de rotation imposés extérieurement', *Thèse de Doctorat d'État*, Université de Lyon, France.

Cambon C., Benoit J.P., Shao L., Jacquin L. (1994) 'Stability analysis and large eddy simulation of rotating turbulence with organized eddies', *J. Fluid Mech.* **278**, 175–200.

Cambon, C., Godeferd, F.S., Leblanc (1999) 'Stability of Stuart vortices in a rotating frame', (LMFA preprint).

Cambon C., Mansour N.N., Godeferd F.S. (1997) 'Energy transfer in rotating turbulence', *J. Fluid Mech* **337**, 303–332.

Cambon C., Scott, J. (1999) 'Linear and nonlinear models for anisotropic turbulence', *Ann. Rev. J. Fluid Mech* **31**, 1–53.

Craik, A.D.D., Criminale, W.O. (1986) 'Evolution of wavelike disturbances in shear flows: a class of exact solutions of the Navier–Stokes equations', *Proc. R. Soc. Lond. A* **406**, 13–26.

Craya A. (1958) 'Contribution à l'analyse de la turbulence associée à des vitesses moyennes', *P.S.T. n⁰ 345*, Ministère de l'Air, France.

Durbin, P.A., Hunt J.C.R. (1980) 'On surface pressure fluctuations beneath turbulent flow round bluff bodies', *J. Fluid Mech.* **100**, 161–164.

Godeferd, F.S., Lollini, L. (1999) 'Direct numerical simulations of turbulence with confinment and rotation', *J. Fluid Mech.* **393**, 1–51.

Greenspan, H.P. (1968) *The Theory of Rotating Fluids*, Cambridge University Press.

Hopfinger, E.J., Browand, F.K., Gagne, Y. (1982) 'Turbulence and waves in a rotating tank', *J. Fluid Mech.* **125**, 505.

Hunt, J.C.R. (1973) 'A theory of turbulent flow around two-dimensional bluff bodies', *J. Fluid Mech.* **61**, 625–706.

Hunt J.C.R. (1978) 'A review of the theory of rapidly distorted turbulent flows and its application', *Fluid Dyn. Trans.* **9**, 121–152.

Hunt J.C.R., Carruthers D.J. (1990) 'Rapid distortion theory and the 'problems' of turbulence', *J. Fluid Mech.* **212**, 497–532.

Jacquin L., Leuchter O., Cambon C., Mathieu J. (1990) 'Homogeneous turbulence in the presence of rotation', *J. Fluid Mech.* **220**, 1–52.

Kelvin, Lord (1887) 'Stability of fluid motion – Rectilineal motion of viscous fluid between two parallel planes', *Phil. Mag.* **24**, 188–196.

Kevlahan N.K.R., Hunt J.C.R. (1997) 'Nonlinear interactions in turbulence with strong irrotational straining', *J. Fluid Mech.*, **337**, 333–364.

Kloosterziel, R.C., van Heijst, G.J.F. (1991) 'An experimental study of unstable barotropic vortices in a rotating fluid', *J. Fluid Mech.* **223**, 1–24.

Lagnado R.R., Phan-Thien N., Leal L.G. (1984) 'The stability of two-dimensional linear flows', *Phys. Fluids* **27**, 1094–1101.

Leblanc S., Cambon C. (1997) 'On the three-dimensional instabilities of plane flows subjected to Coriolis force', *Phys. Fluids* **9** (5), 1307–1316.

Leblanc S., Cambon C. (1998) 'The effect of the Coriolis force on the stability of the Stuart vortices', *J. Fluid Mech.* **357**, 353–379.

Le Dizès S. (1999) 'On 3D instabilities to multipolar vortices', IRPHE Preprint. (private communication)

Leuchter O., Benoit J.P., Cambon C. (1992) 'Homogeneous turbulence subjected to rotation-dominated plane distortion', *Fourth Turbulent Shear Flows*, Delft University of Technology.

Lifschitz A., Hameiri E. (1991) 'Local stability conditions in fluid dynamics', *Phys. Fluids* A 3, 2644–2641.

Lighthill M.J. (1978) *Waves in Fluids*, Cambridge University Press.

Lundgren T.S., Mansour N.N. (1996) 'Transition to turbulence in an elliptic vortex', *J. Fluid Mech.* **307**, 43–62.

Moore, D.W., Saffman, P.G. (1975) 'The instability of a straight vortex filament in a strain field', *Proc. R. Soc. Lond.* A **346**, 413–425.

Nazarenko, S., Kevlahan, N.K.R., Dubrulle, B. (1999) 'WKB theory for rapid distortion of inhomogeneous turbulence', *J. Fluid Mech.* to appear.

Pierrehumbert R.T. (1986) 'Universal short-wave instability of two-dimensional eddies in an inviscid fluid', *Phys. Rev. Lett.* **57**, 2157–2159.

Potylitsin, P.G. & Peltier, W.R. (1999) 'Three-dimensional destabilisation of Stuart vortices: the influence of rotation and ellipticity', *J. Fluid Mech.*, to appear.

Rogers M.M. (1991) 'The structure of a passive scalar field with a uniform gradient in rapidly sheared homogeneous turbulent flow', *Phys. Fluids* A 3, 144–154.

Savill A.M. (1987) 'Recent developments in rapid-distortion theory', *Ann. Rev. Fluid Mech.* **19**, 531–575.

Sipp, D., Jacquin, L. (1998) 'Elliptic instabilities in two-dimensional Taylor–Green flattened vortices', *Phys. Fluids* **10**, 839–849.

Sipp, D., Jacquin, L. (1999) 'Three-dimensional centrifugal-type instabilities of two-dimensional flows in rotating systems', *Phys. Fluids*, submitted.

Sipp, D., Lauga, E., Jacquin, L. (1999) 'Vortices in rotating systems: centrifugal, elliptic and hyperbolic type instabilities', *Phys. Fluids*, **11**, 12, 3716–3728.

Townsend, A.A. (1956) *The Structure of Turbulent Shear Flow*, revised version 1976. Cambridge University Press

Tsai, C.Y., Widnall, S.E. (1976) 'The stability of short waves on a straight vortex filament in a weak externally imposed strain field', *J. Fluid Mech.* **73**, 721–733.

Waleffe, F. (1990) 'On the three-dimensional instability of strained vortices' *Phys. Fluid* A 2, 76–80.

Williamson C.H.K. (1996) 'Three-dimensional wake transition', *J. Fluid Mech.* **328**, 345–407.

LES and Vortex Topology in Shear and Rotating Flows

Marcel Lesieur, Pierre Comte and Olivier Métais

Abstract

The scope of this article is to show that large-eddy simulations (LES) of turbulence may predit good statistics, and are also an excellent tool to investigate the vortex topology in high Reynolds number flows, even at quite small scales.

We start with a brief review of various subgrid-models developed in Grenoble on the basis of spectral eddy-viscosity and eddy-diffusivity coefficients. They are expressed thanks to a two-point spectral closure of isotropic turbulence, the EDQNM theory. The spectral dynamic model in particular is applied to the infrared dynamics of three-dimensional decaying isotropic turbulence, where we confirm the existence of a k^4 backscatter in the kinetic-energy spectrum, and predict a new k^2 law for the pressure spectrum in this range. Applied to the channel flow at $h^+ = 390$, the same model provides first and second-order statistics in very good agreement with DNS, with a reduction of the computational cost by a factor of about 100.

Then we show how LES lets us capture the intense thin longitudinal vortices stretched in temporal and spatial free-shear layers, as well as the dislocations of the large-scale vortex fields.

Afterwards, we study transitions in a boundary layer developing on a flat plate. We show how low- and high-speed streaks are well correlated with longitudinal Λ vortices close to the wall during the transition stage. On the other hand, in the developed-turbulence region, the low-speed streaks are about three times longer than the hairpin vortices ejected above. This questions the popular interpretation of the streaks as induced by the longitudinal vortices.

Finally, with the aid of DNS and LES we study shear layers (free- or wall-bounded) rotating about an axis of spanwise direction. We show for developed turbulence (in regions where the initial Rossby number is strictly smaller than -1) a universal character of the mean velocity profile which becomes linear, with a Rossby equal now to -1. This is interpreted in terms of absolute-vortex dynamics.

1 LES: physical vs spectral space

In large-eddy simulations in physical space (assuming that density ρ_0 is uniform), one considers a low-pass filter $G_{\Delta x}$, where Δx characterizes the grid size (see e.g. Lesieur & Métais, 1996, for a review). If Δx is uniform, the filtered Navier–Stokes equations become

$$\frac{\partial \bar{u}_i}{\partial t} + \frac{\partial}{\partial x_j}(\bar{u}_i \bar{u}_j) = -\frac{1}{\rho_0}\frac{\partial \bar{p}}{\partial x_i} + \frac{\partial}{\partial x_j}(2\nu \bar{S}_{ij} + T_{ij}), \qquad (1.1)$$

where

$$S_{ij} = \frac{1}{2}\left(\frac{\partial u_i}{\partial x_j} + \frac{\partial u_j}{\partial x_i}\right) \qquad (1.2)$$

is the filtered deformation tensor, and

$$T_{ij} = \bar{u}_i \bar{u}_j - \overline{u_i u_j} \qquad (1.3)$$

is the subgrid-stresses tensor. The filtered continuity equation writes

$$\frac{\partial \bar{u}_j}{\partial x_j} = 0. \qquad (1.4)$$

If an eddy-viscosity assumption is made, we have:

$$T_{ij} = 2\nu_t(\vec{x}, t)\,\bar{S}_{ij} + \frac{1}{3}T_{ll}\,\delta_{ij}, \qquad (1.5)$$

where the eddy viscosity ν_t has to be specified. Then the LES momentum equations become

$$\frac{\partial \bar{u}_i}{\partial t} + \frac{\partial}{\partial x_j}(\bar{u}_i \bar{u}_j) = -\frac{1}{\rho_0}\frac{\partial \bar{P}}{\partial x_i} + \frac{\partial}{\partial x_j}[2(\nu + \nu_t)\bar{S}_{ij}] \qquad (1.6)$$

where the pressure has been modified as

$$\bar{P} = \bar{p} - \frac{1}{3}\rho_0 T_{ll}. \qquad (1.7)$$

Among the eddy viscosities that are classically used, let us mention Smagorinsky's (1963), the family of structure-function models (Métais & Lesieur, 1992) and the dynamic Smagorinsky model (Germano *et al.*, 1991).

The only justification of any eddy-viscosity model is an analogy with molecular dissipation: molecular viscosity ν characterizes for a 'macroscopic' fluid parcel the momentum exchanges with the surrounding fluid due to molecular diffusion across its interface. Here, one assumes a wide separation between macroscopic and microscopic scales. No such scale separation exists in the LES problem, where one observes in general a distribution of energy continuously decreasing from the energetic to the smallest dissipative scales, even in inflexional shear flows with vigorous coherent vortices. We believe that the lack of spectral gap is the major drawback of the eddy-viscosity assumption, responsible for the fact that numerous *a priori* tests (numerical or experimental) invalidate relation (1.5).

In this respect, the spectral eddy-viscosity concept presented now is much preferable. We assume that Navier–Stokes is written in Fourier space. This requires statistical homogeneity in the three spatial directions, but we will see below how to handle flows with one direction of inhomogeneity. Let

$$\hat{u}_i(\vec{k}, t) = \left(\frac{1}{2\pi}\right)^3 \int e^{-i\vec{k}.\vec{x}} \; \vec{u}(\vec{x}, t) d\vec{x} \tag{1.8}$$

be the spatial Fourier transform of the velocity field, and $k_C = \pi \Delta x^{-1}$ a cutoff wavenumber. The filter is a sharp filter which eliminates Fourier modes larger than k_C. We write Navier–Stokes in Fourier space as

$$\frac{\partial}{\partial t}\hat{u}_i(\vec{k}, t) + [\nu + \nu_t(k|k_C)]k^2\hat{u}_i(\vec{k}, t)$$
$$= -ik_m \left(\delta_{ij} - \frac{k_i k_j}{k^2}\right) \int_{|\vec{p}|,|\vec{q}|<k_C}^{\vec{p}+\vec{q}=\vec{k}} \hat{u}_j(\vec{p}, t)\hat{u}_m(\vec{q}, t)d\vec{p} \tag{1.9}$$

where $\nu_t(k|k_C)$ is defined by

$$\nu_t(k|k_C)k^2\hat{u}_i(\vec{k}, t)$$
$$= ik_m \left(\delta_{ij} - \frac{k_i k_j}{k^2}\right) \int_{|\vec{p}|\text{or}|\vec{q}|>k_C}^{\vec{p}+\vec{q}=\vec{k}} \hat{u}_j(\vec{p}, t)\hat{u}_m(\vec{q}, t)d\vec{p} \; . \tag{1.10}$$

is the spectral eddy viscosity. A spectral eddy-diffusivity for a passive scalar may be defined in the same way, by writing the passive-scalar equation in Fourier space. One finds

$$\kappa_t(k|k_C)k^2\hat{T}(\vec{k}, t) = ik_j \int_{|\vec{p}|\text{or}|\vec{q}|>k_C}^{\vec{p}+\vec{q}=\vec{k}} \hat{u}_j(\vec{p}, t)\hat{T}(\vec{q}, t)d\vec{p} \; . \tag{1.11}$$

1.1   Plateau-peak eddy viscosity

Spectral eddy coefficients thus defined by equations (1.10) and (1.11) involve unknown subgrid-scale quantities. We will therefore determine these eddy coefficients at an energetic level, writing the evolution equations for the kinetic-energy and passive-scalar spectra given by a two-point closure of isotropic turbulence, the EDQNM[1] theory (see Lesieur, 1997, for a review). Splitting the transfers across k_C in the same way as was done in (1.9), and dividing by $-2k^2$ times respectively the kinetic-energy and passive-scalar spectra, one can calculate the eddy-coefficients. For three-dimensional isotropic turbulence, and if k_C lies within a long $k^{-5/3}$ range, it is found that

$$\nu_t(k|k_C) = 0.441 \; C_K^{-3/2} X(k/k_C) \left[\frac{E(k_C)}{k_C}\right]^{1/2} \tag{1.12}$$

where C_K is the Kolmogorov constant, $E(k_C)$ the kinetic-energy spectrum at k_C, and $X(k/k_C)$ a non-dimensional function equal to 1 up to about

---

[1]Eddy-Damped Quasi-Normal Markovian

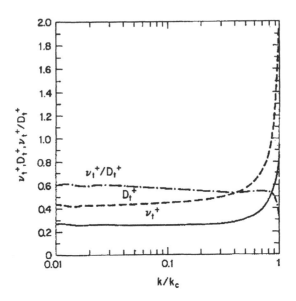

Figure 1: EDQNM spectral eddy coefficients in a 3D Kolmogorov cascade.

$k/k_C = 1/3$, and sharply rising above ('plateau-peak' behaviour, see Chollet & Lesieur, 1981). An analogous study using the Test-Field Model (TFM)[2] had previously been done by Kraichnan (1976). However, Kraichnan did not point out the scaling of the eddy viscosity against $[E(k_C)/k_C]^{1/2}$, which turns out to be essential for LES purposes. The function $0.441\ C_K^{-3/2}X(k/k_C)$ of (1.12) is represented in Figure 1, taken from Chollet & Lesieur (1982). The plateau part can be obtained analytically through leading-order expansions with respect to the small parameter k/k_C. It does in fact correspond to a regular eddy viscosity in physical space, as if there were a spectral gap. As for the 'peak' (Kraichnan called it 'cusp'), it is mostly due to semi-local interactions across k_C (such that $p \ll k \sim q \sim k_C$). Figure 1 presents also the EDQNM eddy diffusivity and the corresponding turbulent Prandtl number, introduced in Chollet & Lesieur (1982). The eddy diffusivity has also the plateau-peak behaviour, and the turbulent Prandtl number is approximately constant (≈ 0.6) in Fourier space. In fact, such a value is the highest one permitted by adjustments of the constants arising in the passive-scalar spectrum EDQNM equation (see Lesieur, 1997, pp. 259–260, and also Herring et al., 1982).

We have carried out LES of decaying isotropic turbulence (Lesieur & Rogallo, 1989, Metais & Lesieur, 1992) using the EDQNM plateau-peak eddy coefficients defined above. Examples of these calculations are shown on Figure 2. Initial spectra are $\sim k^8$ in the infrared region ($k \to 0$). One sees an initial

---

[2]This model is in fact equivalent to the EDQNM model in a Kolmogorov inertial range.

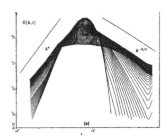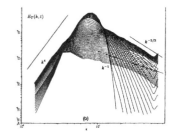

Figure 2: LES of 3D isotropic decaying turbulence. From Lesieur & Rogallo (1989).

infrared k^4 backscatter, well predicted by the EDQNM theory (see Lesieur, 1997, p. 245). We see also the ultraviolet kinetic-energy cascade which builds up: first, a $k^{-5/3}$ slope forms at the cutoff, then it steepens slightly and evolves towards a k^{-2} slope during the self-similar decay. On the other hand, the passive scalar has a very short Corrsin–Oboukhov $k^{-5/3}$ range close to k_C, and a large anomalous range shallower than k^{-1} in the energetic scales. This range is interpreted by Métais & Lesieur (1992) as being due to the rapid stirring of the scalar fluctuations by the coherent vortices of such a turbulence. The latter consist of long, thin tubes of high vorticity and low pressure, which have been characterized numerically by various groups since the work of Siggia (1981). On the figure, the cutoff is $k_C \approx 60$. We have considered a fictitious cutoff $k'_C = k_C/2$, and performed a double filtering in Fourier space across k'_C. Let us decompose the kinetic-energy transfers across k'_C as

$$T_{>k'_C}(k) = T_{>k_C}(k) + T_{k'_C<;<k_C}(k) \tag{1.13}$$

where $T_{k'_C<;<k_C}(k)$ involves triad integrals such that

$$|\vec{p}|_{\mathrm{or}}|\vec{q}| > k'_C \ , \quad |\vec{p}|_{\mathrm{and}}|\vec{q}| < k_C \ .$$

Equation (1.13) is is the exact energetic equivalent of Germano's identity (Germano *et al.*, 1991) in spectral space, stressing that the subgrid transfers across k'_C are equal to the subgrid transfers across k_C plus 'resolved' subgrid transfers across k'_C. Figure 3 shows the eddy viscosity $-(1/2)k^{-2}E(k)^{-1}T_{k'_C<;<k_C}(k)$ derived from the resolved subgrid transfers in the calculations of Métais & Lesieur (1992), and the equivalent eddy diffusivity for a passive scalar. Both are normalized by $[E(k'_C)/k'_C]^{1/2}$.

It confirms that the plateau-peak behaviour does exist in reality for the eddy viscosity, but is questionable for the eddy diffusivity. This turbulent-mixing anomaly is certainly related to the existence of the large-scale k^{-1} scalar spectrum found above. It is shown in Métais & Lesieur (1992) that the anomaly disappears when the temperature is no longer passive and coupled

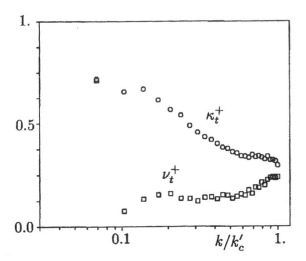

Figure 3: LES of 3D isotropic decaying turbulence, resolved eddy-viscosity and diffusivity calculated trough a double filtering (from Métais & Lesieur (1992).

with the velocity within the frame of Boussinesq approximation (stable stratification). It is possible that the same holds for compressible turbulence, which would legitimate the use of the plateau-peak eddy diffusivity in this case.

1.2 Spectral-dynamic model

The plateau-peak model considered above assumed a $k^{-5/3}$ kinetic-energy spectrum at the cutoff. Let us now consider spectra $\propto k^{-m}$ for $k > k_C$, with $m \neq 5/3$. We re-calculate analytically the plateau with the same EDQNM leading-order expansions in k/k_C as above. We retain the peak shape through $X(k/k_C)$ in order to be consistant with the Kolmogorov spectrum expression of the eddy viscosity. The spectral eddy viscosity becomes now

$$\nu_t(k|k_C) = 0.31 C_K^{-3/2} \sqrt{3-m} \frac{5-m}{m+1} X(k/k_C) \left[\frac{E(k_C)}{k_C}\right]^{1/2} , \qquad (1.14)$$

for $m \leq 3$ (see Métais & Lesieur, 1992). For $m > 3$, the scaling is no longer valid, and the eddy viscosity will be set equal to zero, which is not unreasonable since the spectrum decays rapidly enough for the subgrid effects not to be very important. In the spectral-dynamic model, the exponent m is determined through the LES with the aid of least-squares fits of the kinetic-energy spectrum close to the cutoff. Within this approximation, the turbulent Prandtl number is

$$P_r^t = 0.18 \, (5 - m) \qquad (1.15)$$

(see Lesieur, 1997). This is an advantage with respect to the plain plateau-peak model where the turbulent Prandtl number was constant.

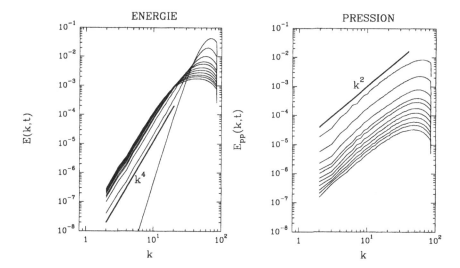

Figure 4: Infrared kinetic-energy and pressure spectra in 3D isotropic decaying turbulence, from Lesieur, Ossia & Métais (1999, courtesy S. Ossia).

We now show recent results of the spectral-dynamic model applied to the infrared velocity and pressure in decaying isotropic turbulence (Lesieur, Ossia & Métais, 1999). Here, the spectrum that allow us to determine the exponent m is calculated by a spatial average in the computational box. In order to have the widest infrared range available, the initial spectral peak is close to the cutoff, and the calculation is de-aliased because of the important amount of kinetic energy at the cutoff. The simulation on Figure 4 concerns an initial k^8 infrared kinetic-energy spectrum, with a Gaussian velocity field. The calculation is run here up to 20 initial large-eddy turnover times. One sees the kinetic-energy spectrum which picks up immediately a k^4 behaviour with a positive transfer, which confirms the existence of the k^4 backscatter. On the other hand, the infrared pressure spectrum immediately follows a k^2 law, and decays rapidly (no pressure backscatter). We have explained this law using non-local expansions of the Quasi-Normal (or EDQNM[3]) pressure-spectrum equation. It persists for very large times (several hundreds of initial turnover times), and is in fact independant of the infrared kinetic-energy spectrum behaviour. In these simulations, the pressure variance decays with time approximately as the squared kinetic energy (see Lesieur, Ossia & Métais, 1999, for details).

---

[3]Both theories are equivalent for the pressure (Larchevêque, 1990).

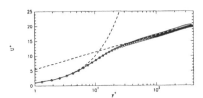 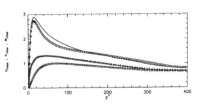

Figure 5: Turbulent channel flow, comparisons of the spectral-dynamic model (straight lines, $h^+ = 389$) with the DNS of Antonia *et al.* (1992, symbols, $h^+ = 395$); mean velocity (left), rms velocity components (right) (courtesy E. Lamballais).

We now present spectral-dynamic model results for a periodic channel. We use a mixed spectral-compact code, the compact scheme being employed in the transverse direction, while pseudo-spectral methods are used in the longitudinal and spanwise directions which are periodic. Calculations start with a parabolic Poiseuille velocity profile, on which a small 3D white-noise perturbation is superposed, and run up to complete statistical stationarity, assuming a constant flow rate of average velocity U_m across the section. The channel has a width $2h$. The kinetic-energy spectrum allowing one to determine the eddy-viscosity is calculated at each time step by averaging in planes parallels to the walls, so that m is a function of (y, t). The code has been validated with a DNS at $h^+ = 160$, and compares very well with purely spectral DNS at same Reynolds, as far as the first and second-order statistics are concerned. At $h^+ = 204$, these statistics still compare well with the LES of Piomelli *et al.* (1993) using the classical dynamic model. We present here one LES at $R_e = 2hU_m/\nu = 14000$ ($h^+ = 389$), taken from Lamballais *et al.* (1998). There is a grid refinement close to the wall, in order to simulate accurately the viscous sublayer (first point at $y^+ = 1$). We have compared the calculation with a DNS at $h^+ = 395$ carried out in Antonia *et al.* (1992). Figure 5 shows the mean velocity and the rms velocity components. The agreement is very good, which is a severe challenge for the model. Notice that the LES allows one to reduce the computational cost by a factor of the order of 100, which is huge.

1.3 Structure-function models

For many practical applications, the geometry of the domain is too complex to allow for pseudo-spectral, or even spectral, methods to be used. In this case, numerical schemes are formulated in physical space, and the LES subgrid model has to be expressed in physical space also.

In fact, the cusp part of the plateau-peak eddy viscosity can be formulated in physical space in the form of a hyperviscosity (Lesieur & Métais 1996), and this is certainly an option to develop in the future. Here, we will be concerned by models of the structure-function family, which we will briefly describe (see Lesieur & Métais, 1996, and Lesieur, 1997, for more details).

The idea of the structure-function model proposed in Métais & Lesieur (1992) is to erase the peak by subgrid-energy conservation, which yields a k-independent eddy viscosity allowing one to go back to physical space. One then takes

$$\nu_{SF}(\vec{x}, t) = \frac{2}{3} C_K^{-3/2} \left[\frac{E(k_C, \vec{x}, t))}{k_C} \right]^{1/2}, \tag{1.16}$$

with $k_C = \pi/\Delta x$, where $E(k_C, \vec{x}, t)$ is a local kinetic-energy spectrum, calculated in terms of the local second-order velocity structure function of the filtered field $F_2(\vec{x}, \Delta x)$ as if the turbulence is three-dimensionally isotropic. This yields for a Kolmogorov spectrum

$$\nu_t^{SF}(\vec{x}, \Delta x) = 0.105 \, C_K^{-3/2} \, \Delta x \, [F_2(\vec{x}, \Delta x)]^{1/2}. \tag{1.17}$$

We calculate F_2 with a local statistical average of square-velocity differences between $\vec{x}$ and the six closest points surrounding $\vec{x}$ on the computational grid. In some cases, the average may be taken over four points parallel to a given plane. Notice also that if the computational grid is not regular (but still orthogonal), interpolations of (1.17) based upon Kolmogorov's 2/3 law for the second-order structure function have been proposed by Lesieur & Métais (1996). Such a structure-function model (SF) works very well for decaying isotropic turbulence, where it yields a fairly good Kolmogorov spectrum (Métais & Lesieur, 1992), better than Smagorinsky's model (with $C_S = 0.2$) and the plateau-peak models (simple or dynamic). It is of interest to compare the structure function and Smagorinsky's models. This may be done by approximating the spatial derivatives in S_{ij} by first-order finite differences. One finds

$$\nu_t^{SF} = 0.777 \, (C_S \Delta x)^2 \sqrt{2 \bar{S}_{ij} \bar{S}_{ij} + \bar{\omega}_i \bar{\omega}_i} + O(\Delta x)^2), \tag{1.18}$$

where $\bar{\omega}$ is the vorticity of the filtered field, whereas C_S is the regular Smagorinsky constant within Lilly's (1987) evaluation in a Kolmogorov cascade. Let us consider for instance the stagnation regions between large vortices, with initial vorticity much smaller than the strain rate: here, the SF model is about 20% less dissipative than Smagorinsky's. This will favour the eventual longitudinal stretching of hairpin vortices. At any rate, these two models are too dissipative in free-shear flows, and other models must be employed. As for wall flows, the SF model does not work for transition in an incompressible boundary layer where, like Smagorinsky's, it is too dissipative and prevents secondary instabilities of TS waves from developing. To overcome this difficulty, two improved versions of the SF model have been developed: the selective structure-function model (SSF), and the filtered structure-function model (FSF).

In the SSF, we switch off the eddy-viscosity when the flow is not three-dimensional enough. The three-dimensionalization criterion is the following:

Figure 6: Vorticity field obtained in the LES of a temporal mixing layer forced quasi two-dimensionally (left) and three-dimensionally (right).

one measures the angle between the vorticity at a given grid point and the average vorticity at the six closest neighbouring points (or the four closest points in the four-point formulation). If this angle exceeds 20°, the most probable value according to simulations of isotropic turbulence at a resolution of $32^3 \sim 64^3$, the eddy-viscosity is turned on. Otherwise, only molecular dissipation is acting.

In the FSF model (Ducros et $al.$, 1996), the filtered field $\bar{u}_i$ is submitted to a high-pass filter in order to get rid of low-frequency oscillations which affect $E(k_C, \vec{x}, t)$ in the SF model. The high-pass filter is a Laplacian discretized by second-order centred finite differences and iterated three times. We find that

$$\nu_t^{FSF}(\vec{x}, \Delta x) = 0.0014 \, C_K^{-3/2} \, \Delta x \, [\tilde{F}_2(\vec{x}, \Delta x)]^{1/2} \ . \tag{1.19}$$

2 Coherent vortices

2.1 Definition

Since we will discuss the ability for LES to predict coherent-vortex dynamics, it is of interest to clarify what we call coherent vortices, and discuss briefly how we can identify them.

Coherent vortices in turbulence are defined by Lesieur (1997, pp. 6–7) as regions of the flow satisfying three conditions:

(i) the vorticity modulus concentration ω should be high enough so that a local roll up of the surrounding fluid is possible,

(ii) they should keep approximately their shape during a time T_c longer enough in front of the local turnover time ω^{-1},

(iii) they should be unpredictable.

In this context, high vorticity modulus ω is a possible candidate for coherent-vortex identification.

With such a definition, the core of the coherent vortices should be pressure lows. Indeed, a fluid parcel winding around the vortex will be (in a frame moving with the parcel) in approximate balance between centrifugal and pressure-gradient effects. We are talking here of the static pressure p. The reasoning may be made more quantitative by considering the Euler equation in the form

$$\frac{\partial \vec{u}}{\partial t} + \vec{\omega} \times \vec{u} = -\frac{1}{\rho}\vec{\nabla}P \qquad (2.1)$$

where $P = p + \rho\vec{u}^2/2$ is now the dynamic pressure. In a frame moving with the coherent vortex and supposed locally Galilean, the ratio of the second to the first term in the l.h.s. of (2.1) is of the order of $T_c\omega$. Then the equation reduces to the cyclostrophic balance

$$\vec{\omega} \times \vec{u} = -\frac{1}{\rho}\vec{\nabla}P \qquad (2.2)$$

if condition (ii) above is fulfilled. It implies that the vortex tube is a low for the dynamic pressure.

2.2 The Q-criterion

If one decomposes the velocity-gradient tensor as $S_{ij} + \Omega_{ij}$, where S_{ij} and Ω_{ij} are respectively the deformation and rotational components, Q is the second invariant of the velocity gradient

$$Q = \frac{1}{2}(\Omega_{ij}\Omega_{ij} - S_{ij}S_{ij}) = \frac{1}{4}(\vec{\omega}^2 - 2S_{ij}S_{ij}) \ . \qquad (2.3)$$

It is well-known that $\nabla^2 p = 2Q$. One can also very easily show using the divergence theorem that $Q > 0$ within thin isobaric low-pressure tubes of convex cross section. In fact, the criterion $Q > 0$ was proposed by Hunt *et al.* (1998) for vortex identification, as an approximation of the discriminant allowing the prediction of the existence of complex eigenvalues for the velocity-gradient tensor, and hence a local rotation of the vorticity (or passive-scalar gradient) vector. Such a criterion is exact in two dimensions and does correspond to 'elliptic regions' in the framework of the so-called Weiss (1981) criterion. In fact, simulations of various types of flows that we have carried out (isotropic turbulence, mixing layers, backstep, channel, boundary layers) confirm that coherent vortices are very well characterized by isosurfaces of Q at a given positive threshold.

3   Mixing layers

3.1   Temporal case

We consider a basic hyperbolic-tangent velocity profile on which is superposed a weak initial random perturbation. In the case of an initial 3D white-noise perturbation, DNS give rise to a helical-pairing type interaction between the big vortices (Comte et al., 1992), where vortex filaments oscillate out-of-phase in the spanwise direction, and reconnect, yielding a vortex-lattice structure. At the same low Reynolds (80, based on the initial vorticity thickness) and with a quasi two-dimensional perturbation, quasi two-dimensional Kelvin–Helmholtz vortices form, but without stretching of the intense longitudinal vortices within the braids. On the other hand, LES using the plateau-peak model can generate these thin longitudinal vortices observed in the experiments (Konrad, 1976, Bernal & Roshko 1986, Huang & Ho 1990). Such a pattern is shown on Figure 6 (left), taken from Silvestrini et al. (1998), and presenting a map of the vorticity modulus. It proves that LES can handle small-scale vortices, which would otherwise be dissipated by viscosity in DNS. We have recovered the same dislocated pattern as in the DNS in our LES with the three-dimensional white-noise initial forcing. Figure 6 (right) shows the vorticity modulus thus obtained.

3.2   Spatial case

We present FSF-based LES of spatially-growing incompressible mixing layers initiated upstream by a hyperbolic-tangent velocity profile superposed on the average flow, plus a weak random forcing regenerated at each time step. The numerical code is the same as for the channel. With an upstream forcing consisting in a quasi two-dimensional random perturbation, intense longitudinal vortices stretched between quasi 2D Kelvin-Helmholtz vortices are found, as in the temporal case (Figure 7). They may undergo a pairing accompanying the pairing of fundamental spanwise vortices, as shown in Comte et al. (1998). When the forcing is a 3D random white noise, helical pairing occurs upstream, as indicated by the low-pressure maps of Figure 8. But none of these simulations has reached self-similarity, and they are still dependant of the computational domain size in the three space directions. Thus calculations in bigger domains are necessary. Another solution is to impose upstream deterministic flows corresponding to two turbulent boundary layers on each side of the splitter plate (Piomelli, 1999, private communication).

4   Boundary layer above a flat plate

We study LES of transition to turbulence in a boundary layer above an adiabatic flat plate at $M_\infty = 0.3$ (perfect gas). The flow is weakly compressible, and the problem is very close to incompressibility. We have developed a compressible LES code where we solve the equations using fourth-order accurate spatial finite-difference schemes of the McCormack type for nonlinear terms.

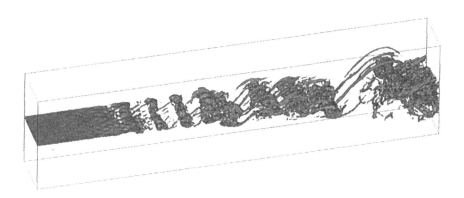

Figure 7: LES of an incompressible mixing layer forced upstream by a quasi two-dimensional random perturbation; vorticity modulus at a threshold $(2/3)\omega_i$ (courtesy G. Silvestrini).

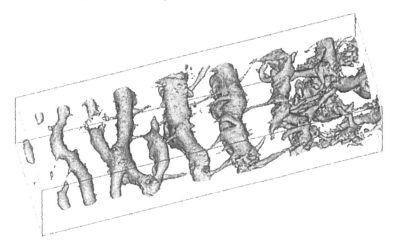

Figure 8: Same as Figure 7, but with a three-dimensional upstream white-noise forcing, low-pressure field (courtesy G. Silvestrini).

The subgrid model used is the FSF model. The resolution at the wall is $y^+ = 1$ or 2, and the upstream conditions are obtained with the aid of nonlinear parabolized stability expansion (PSE) calculations (Bertolotti & Herbert, 1991, Airiau, 1994). To this upstream state (with $R_{\delta_1} = 1000$), one superposes a 3D white-noise of amplitude 0.2 the amplitude of the PSE perturbation. In the 'Klebanov case' (K), one sees in the transitional region formation of big longitudinal Λ-shaped vortices lying on the wall, and in phase with respect to the flow direction (see Figure 9, top). In the 'Herbert case' (H), the vortices are staggered (see Figure 9, bottom).

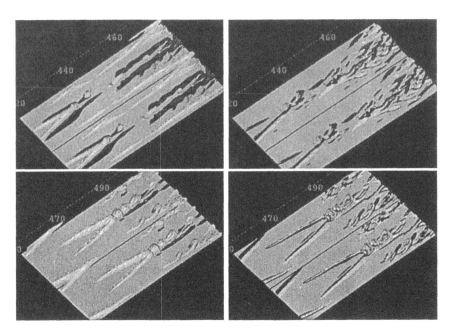

Figure 9: LES of a spatial boundary layer at Mach 0.3; top and bottom, K- and H-transition respectively; the l.h.s. and r.h.s. correspond respectively to velocity and vorticity fluctuation components (positive, dark; negative, light grey). Grey represents an isosurface of positive Q (courtesy E. Briand).

The figures show at the end of transition the longitudinal components of velocity and vorticity (positive in dark, negative in light grey) and also positive Q (in grey). One sees that the Λ vortices are very well correlated with a system of induced high and low-speed streaks[4]. Downstream of $\approx$ 440 δ_i, the streaks become purely longitudinal, with no evidence of associated longitudinal vortices which would drive them. Furthermore, Q isosurfaces show the shedding of secondary arch vortices at the tip of the Λs.

Figure 10 shows for the K-transition the downstream evolution of the friction coefficient at the wall, with comparison against the theoretical predictions of Van Driest[5] and Barenblatt (1993). One sees a good agreement of the LES with these predictions, a resolution of $y^+ = 1$ improving the result. It is even better in the H-case. The peak in the friction coefficient is at 490 δ_i, much further than the change of regime of the velocity streaks, and might be associated to an event such as the localized creation of a big hairpin vortex observed in the simulations of Ducros et al. (1996).

If one looks at developed turbulence further downstream, various plots of

---

[4]The fact that high-speed streaks are not visible in the K-case is due to an ill-chosen threshold.

[5]discussed in Cousteix (1989)

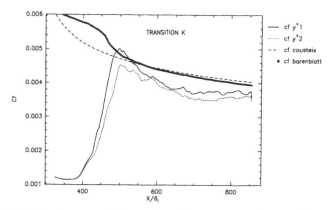

Figure 10: LES of a spatial boundary layer at Mach 0.3: friction coefficients against downstream distance, compared with theoretical predictions of Cousteix and Barenblatt (courtesy E. Briand).

vorticity components and pressure, as well as Q, accompanied by animations, show the very long longitudinal velocity streaks (about 1000 wall units for the low speeds) travelling downstream. Above these streaks are ejected hairpins through a secondary Kelvin–Helmholtz instability which occurs at a height of about $30 \approx 40$ wall units. The length of the hairpins is about 300 wall units, so that there are several hairpins (about 3) above a single low-speed streak. In this sense, we no longer have the perfect correlation vortices-streaks which we observed during the transitional stage. It is therefore difficult to associate in the developed region the streaks with a system of purely longitudinal alternate vortices at the wall.

5   Universality of rotating shear flows

We come back to purely incompressible flows, and study now the influence of a solid-body rotation upon free- or wall-bounded shear flows. We consider a parallel basic (or mean) velocity $\bar{u}(y)$, and assume that the axis of rotation is parallel to the spanwise direction. We work in a relative rotating frame of angular-rotation vector $\vec{\Omega}$. We add a coriolis force to the Navier–Stokes equations, while centrifugal forces are incorporated in the pressure gradient. Let

$$R_o(y, t) = -\frac{1}{2\Omega} \frac{d\bar{u}}{dy} \tag{5.1}$$

be the local Rossby number. Regions with a positive (resp. negative) local Rossby will be called cyclonic (resp. anticyclonic). We recall also that the absolute vorticity vector is $\vec{\omega}_a = \vec{\omega} + f\vec{z}$ (with $f = 2\Omega$), and satisfies Helmholtz' theorem in its conditions of applicability, within which absolute-vortex elements are material.

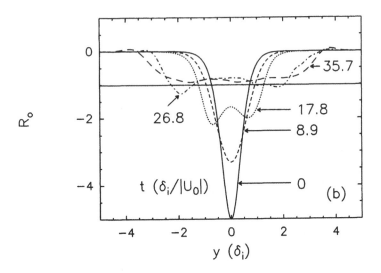

Figure 11: DNS of a rotating periodic mixing layer: time evolution of the local Rossby number.

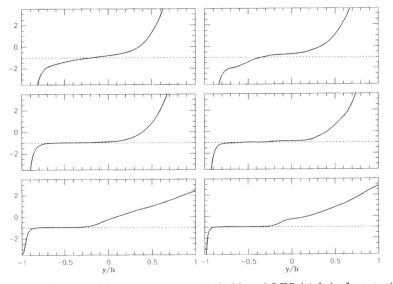

Figure 12: Final local Rossby in the DNS (left) and LES (right) of a rotating channel. From top to bottom, initial Rossby at the wall: 18, 6 and 2.

We first synthesize results concerning free-shear layers coming both from 3D linear-stability studies (Yanase *et al.*, 1993), and DNS or LES (Métais *et al.*, 1995). As in instability studies, we start with a basic parallel velocity profile, weakly perturbed. There is a critical local Rossby number of -1 such that:

(i) In regions where initially $R_o(y) \geq -1$, the shear layer is two-dimensionalized. In a mixing layer, for instance, 3D perturbations are damped, and straight Kelvin–Helmholtz billows form[6]. This result agrees in particular with the Proudman–Taylor theorem when the Rossby number modulus is small.

(ii) For $R_o^{\min} < R_o(y) < -1$ ('weak' anticyclonic rotation), where $R_o^{\min} = -10 \sim -20$ decreases as the Reynolds increases, the flow is highly three-dimensionalized, with production of intense Görtler-like longitudinal rolls. Examination of the vorticity fields shows that they correspond in fact to the condensation of absolute-vortex lines into very long hairpins which are oriented in a purely longitudinal plane. As a result, their spanwise vorticity component is zero, which implies that the mean velocity profile becomes constant and equal to f, so that the local Rossby number uniformizes to the value -1. This is clear for the mixing layer DNS of Figure 11 (Métais *et al.*, 1995), showing a DNS of periodic mixing layer, whose initial minimum Rossby number at the inflexion point of the hyperbolic tangent velocity profile is -5. One sees how the linear velocity profile of Rossby number -1 develops over a large part of the domain. The same results are found in the anticyclonic region of a wake. There is in fact universality of this result for shear flows (free or wall bounded), since this law is also found for the channel, as we have checked (see Figure 12, taken from Lamballais *et al.*, 1998). In the latter case, it is interesting to see that the important physical result corresponding to the establishment of the $R_o(y) = -1$ plateau is well reproduced by the LES, but not by the one-point closure models.

It is not easy to understand why the absolute vorticity is reoriented longitudinally. Indeed, the 3D linear-stability analysis shows for the mixing layer the existence of a purely longitudinal (x-independant) mode (shear-Coriolis instability) which dominates inflexional instabilities at $R_o^{\min} < R_o(y) < -1$, with a maximum effect at an initial Rossby number(at the inflexion point) of -2.5. It is at this value that the linear stage develops. We believe that the longitudinal reorientation of the absolute vorticity is a nonlinear mechanism, which occurs when the growth of the linear mode has concentrated enough longitudinal vorticity in the $y - z$ plane. If these longitudinal vorticity concentrations are strong enough to form vortices, whose core is 'elliptic' in the sense of the Weiss–Hunt criterion seen above, the absolute vorticity will rotate about $\vec{x}$ in such a way that the local Rossby number will increase. We have here for the absolute-vorticity vector an interesting mechanism of nonlinear longitudinal self-reorientation.

---

[6]without stretching of longitudinal vortices nor helical pairing

6 Conclusion

We have presented the general framework of large-eddy simulations (LES) carried out in spectral space, with the plateau-peak type eddy-viscosity, derived from two-point closures of turbulence, and which lets us go beyond the scale-separation assumption inherent to the classical eddy-viscosity concept in physical space. We have verified for isotropic turbulence and with the aid of a double-filtering in spectral space that the plateau-peak does exist. We have proposed a modification of the plateau-peak to account for kinetic-energy spectra decaying differently from Kolmogorov at the cutoff. This spectral-dynamic model has been applied first to the decay of isotropic turbulence, where we have confirmed the k^4 backscatter for the infrared kinetic-energy spectrum, and discovered a new k^2 law for the pressure. Applied to the turbulent channel flow at high Reynolds number, the spectral-dynamic model gives very good results with respect to direct-numerical simulations. Compared with the latter, the LES reduces the computational cost by a factor of the order of 100. Back to physical space, we have reinterpreted these models in terms of velocity-structure functions.

We have also provided a precise definition of coherent vortices, and related them to high-vorticity modulus, low static and dynamic pressure, and positive Q. The latter has been associated to the Weiss–Hunt criterion, which turns out to be an excellent way of vortex identification in isotropic or shear turbulence.

We have developed LES of temporal (resp. spatial) mixing layers, and shown the possibility of controlling the flow topology, depending upon the nature (quasi 2D or 3D) of the initial (resp. upstream) perturbation. Thus, the system may bifurcate from quasi 2D Kelvin–Helmholtz vortices stretching intense thin longitudinal hairpins to a lattice of big vortices undergoing helical pairing.

We have simulated the complete transition to turbulence of a boundary layer developing above a flat plate at Mach 0.3, with upstream conditions corresponding to harmonic (K) or subharmonic (H) situations. In transition, there is an excellent correlation between Λ vortices and a system of low- and high-speed streaks at the wall. But this is no more true in the developed region, with several hairpin vortices ejected above the same low-speed streaks. It is therefore difficult to explain the streaks as induced by the presence of long alternate longitudinal vortices close to the wall.

Finally, we have shown through DNS and LES a universal behaviour of anticyclonic shear layers rotating at a moderate angular velocity Ω (in terms of Rossby numbers) about an axis parallel to the spanwise direction. The mean-velocity profile becomes linear, of gradient 2Ω, and the absolute vorticity condenses into purely longitudinal coherent hairpins.

Aknowledgements

We thank P. Begou, E. Briand, F. Delcayre, Y. Dubief, E. Lamballais, S. Ossia and G. Silvestrini for their contribution to the work presented here. This work was done at LEGI-Grenoble and was suported by Institut National Polytechnique de Grenoble, Université Joseph Fourier, CNRS, Direction Générale pour l'Armement, and Institut Universitaire de France.

References

Airiau, C. (1994) 'Stabilité linéaire et faiblement non linéaire d'une couche limite laminaire incompressible par un systeme d'équations parabolisé (PSE)', PhD Thesis, Toulouse University.

Antonia, R.A., Teitel, M., Kim, J. and Browne, L.W.B (1992) 'Low Reynolds number effects in a fully-developed turbulent channel flow', *J. Fluid Mech.*, **236**, 579–605.

Barenblatt, G. (1993), *J. Fluid Mech.*, **248**, 513–529.

Bernal, L.P. and Roshko, A. (1986) 'Streamwise vortex structure in plane mixing layer', *J. Fluid Mech.*, **170**, 499–525.

Bertolotti, P. and Herbert, T. (1991) *Theoret. and Comp. Fluids Dynamics*, **3**, 117–124.

Chollet, J.P. and Lesieur, M. (1981) 'Parameterization of small scales of three-dimensional isotropic turbulence utilizing spectral closures', *J. Atmos. Sci.*, **38**, 2747–2757.

Chollet, J.P. and Lesieur, M. (1982) 'Modélisation sous maille des flux de quantité de mouvement et de chaleur en turbulence tridimensionnelle', *La Météorologie*, **29**, 183–191.

Comte, P., Lesieur, M. and Lamballais, E. (1992) 'Large and small-scale stirring of vorticity and a passive scalar in 3D temporal mixing layer', *Phys. Fluids A.*, **4**, 2761–2778.

Comte, P., Silvestrini, J.H. and Begou, P. (1998) 'Streamwise Vortices in Large-Eddy Simulations of Mixing Layers', *Eur. J. Mech. B/Fluids*, **17**, 615–637.

Cousteix, J. (1989) *Turbulence et couche limite*, CEPADUES.

Ducros, F., Comte, P. and Lesieur, M. (1996) 'Large-eddy simulation of transition to turbulence in a boundary-layer developing spatially over a flat plate', *J. Fluid Mech.*, **326**, 1–36.

Germano, M., Piomelli, U., Moin, P. and Cabot, W. (1991) 'A dynamic subgrid-scale eddy-viscosity model', *Phys. Fluids A.*, **3**, 1760–1765.

Herring, J.R., Schertzer, D., Lesieur, M., Newman, G.R., Chollet, J.P. and Larchevêque, M. (1982) 'A comparative assessment of spectral closures as applied to passive scalar diffusion', *J. Fluid Mech.*, **124**, 411–437.

Huang, L.S. and Ho, C.M. (1990) 'Small-scale transition in a plane mixing layer', *J. Fluid Mech.*, **210**, 475–500.

Hunt, J., Wray, A. and Moin, P. (1988) *CTR Rep. S-88*, 193.

Konrad, J.H. (1976), 'An experimental investigation of mixing in two-dimensional turbulent shear flows with applications to diffusion-limited chemical reactions', Ph.D. Thesis, California Institute of Technology.

Kraichnan, R.H. (1976) 'Eddy viscosity in two and three dimensions', *J. Atmos. Sci.*, **33**, 1521–1536.

Lamballais, E., Métais, O. and Lesieur, M. (1998) 'Spectral-dynamic model for large-eddy simulations of turbulent rotating channel flow', *Theor. Comp. Fluid Dyn.*, **12**, 149–177.

Larchevêque, M. (1990) 'Pressure fluctuations and Lagrangian accelerations in two-dimensional incompressible isotropic turbulence', *Eur. J. Mech. B.*, **9**, 109–128.

Lesieur, M. and Rogallo, R. (1989) 'Large-eddy simulation of passive-scalar diffusion in isotropic turbulence', *Phys. Fluids A.*, **1**, 718–722.

Lesieur, M. and Métais, O. (1996) 'New trends in large-eddy simulations of turbulence', *Ann. Rev. Fluid Mech.*, **28**, 45–82.

Lesieur, M. (1997) *Turbulence in Fluids*, 3rd edition, Kluwer Academic Publishers.

Lesieur, M., Ossia, S. and Métais, O. (1999) 'Infrared pressure spectra in 3D and 2D isotropic incompressible turbulence', *Physics of Fluids*, in press.

Lilly, D.K. (1987) in *Lecture Notes on Turbulence*, J.R. Herring and J.C. McWilliams (eds.), World Scientific, 171–218.

Métais, O. and Lesieur, M. (1992) 'Spectral large-eddy simulations of isotropic and stably-stratified turbulence', *J. Fluid Mech.*, **239**, 157–194.

Métais, O., Flores, C., Yanase, S., Riley, J. and Lesieur, M. (1995) 'Rotating free shear flows. Part 2: numerical simulations', *J. Fluid Mech.*, **293**, 41–80.

Piomelli, U. (1993) 'High Reynolds number calculations using the dynamic subgrid-scale stress model', *Phys. Fluids A.*, **5**, 1484–1490.

Siggia, E.D. (1981) 'Numerical study of small-scale intermittency in three-dimensional turbulence', *J. Fluid Mech.*, **107**, 375–406.

Silvestrini, J. H., Lamballais, E. and Lesieur, M. (1998) 'Spectral-dynamic model for LES of free and wall shear flows', *Int. J. Heat Fluid Flow*, **19**, 492–504.

Smagorinsky, J. (1963) 'General circulation experiments with the primitive equations', *Mon. Weath. Rev.*, **91**, 99–164.

Weiss, J. (1981) 'The dynamics of enstrophy transfers in two-dimensional hydrodynamics', La Jolla Institute preprint LJI-TN-121ss, see also (1991) *Physica D*, **48**, 273.

Yanase, S., Flores, C., Métais, O. and Riley, J. (1993) 'Rotating free-shear flows. I. Linear stability analysis', *Phys. Fluids*, **5**, 2725–2737.

Conditional Mode Elimination with Asymptotic Freedom for Isotropic Turbulence at Large Reynolds Numbers

David McComb & Craig Johnston

Summary

The numerical simulation of turbulence, in common with other areas of computational physics, requires the resolution of very large numbers of degrees of freedom. In some of these other areas, such as critical phenomena or quantum chromodynamics, the Renormalization Group (RG) has been of great utility, both in facilitating calculation and in helping to improve our understanding of the underlying physics. In RG the basic strategy is first to average out the fluctuations associated with the smallest scales. The system is then rescaled in order to make it look the same as it did to begin with, thus revealing the presence of scale invariance under the RG transformation. As scale invariance has long been thought to be a characteristic of fluid turbulence, it is natural to suppose that RG could also be applied in this case, and many attempts have been made to do this over the last two decades (see, for instance, [1, 2, 3, 4]). Unfortunately, the first stage in the RG transformation – which is relatively easy to perform for microscopic Ising models – cannot be implemented for fluid turbulence, which is a macroscopic phenomenon and in principle fully deterministic. In this article we deal with the bedrock problem of averaging out degrees of freedom corresponding to the largest wavenumbers in our description of fluid turbulence, while holding the remaining modes constant. We shall argue that in classical deterministic systems such an operation can only be an approximation, but that observations of turbulent structure as an interplay between coherence and chaos could justify the introduction of an assumption of asymptotic freedom in wavenumber. On this basis, we show how small scale modes, and hence the part of the non-linear term in the Navier–Stokes equation representing coupling to them, may be averaged out by introducing, as a hypothetical limiting case, a weak conditional average with asymptotic freedom. A residual deterministic part, while important for individual realizations, makes a negligible contribution to the renormalization of the dissipation rate. This is because the full ensemble average, needed to establish the energy balance, releases the constraint on the conditional average. We conclude with a discussion of the present limitations of the method, along with some possible lines of future development.

1   Introduction

The application of renormalization group to dynamical problems in microscopic physics requires an average over small scales in which large scales are held fixed [5]. Unfortunately, the corresponding procedure for classical nonlinear systems, such as Navier–Stokes turbulence, is impossible *in principle*, because of the deterministic nature of such systems. Recently it has been proposed that the chaotic nature of turbulence may justify the use of an approximate conditional average [6]. In this article we argue that the conditional elimination of a band of high wavenumber modes may be accomplished in terms of a deterministic part, which has a coherent phase relation with the retained modes, and a chaotic part, which is asymptotically free, in a sense that we shall discuss in Section 5, and may be averaged out with the introduction of an effective viscosity. The reduction of the number of modes takes place at a constant rate of energy dissipation, and it is further argued that the renormalization of this quantity can be adequately represented by the incoherent part only. This is because the full ensemble average, needed for the spectral energy balance, tends to 'lift' the constraint on the conditional average.

2   The basic equations

We consider incompressible fluid turbulence, as governed by the solenoidal Navier–Stokes equation

$$\left(\frac{\partial}{\partial t} + \nu_0 k^2\right) u_\alpha(\mathbf{k}, t) = M_{\alpha\beta\gamma}(\mathbf{k}) \int d^3 j u_\beta(\mathbf{j}, t) u_\gamma(\mathbf{k} - \mathbf{j}, t), \qquad (2.1)$$

where ν_0 is the kinematic viscosity of the fluid,

$$M_{\alpha\beta\gamma}(\mathbf{k}) = \frac{1}{2i}[k_\beta D_{\alpha\gamma}(\mathbf{k}) + k_\gamma D_{\alpha\beta}(\mathbf{k})], \qquad (2.2)$$

and the projection operator $D_{\alpha\beta}(\mathbf{k})$ is expressed in terms of the Kronecker delta $\delta_{\alpha\beta}$ as

$$D_{\alpha\beta}(\mathbf{k}) = \delta_{\alpha\beta} - \frac{k_\alpha k_\beta}{k^2}. \qquad (2.3)$$

In order to pose a specific problem, we restrict our attention to stationary, isotropic, homogeneous turbulence, with dissipation rate ε and zero mean velocity. We also introduce an upper cutoff wavenumber K_{max}, which is defined through the dissipation integral

$$\varepsilon = \int_0^\infty 2\nu_0 k^2 E(k) dk \simeq \int_0^{K_{max}} 2\nu_0 k^2 E(k) dk, \qquad (2.4)$$

where $E(k)$ is the energy spectrum, so ensuring that K_{max} is of the same order as the Kolmogorov dissipation wavenumber, $k_d = (\varepsilon/\nu^3)^{1/4}$.

3 The filtered equations of motion

Having defined our equations of motion, we first divide up the velocity field at $k = K_c$, where $0 < K_c < K_{max}$, according to

$$u_\alpha(\mathbf{k}, t) = \begin{cases} u_\alpha^-(\mathbf{k}, t) \text{ for } 0 < k < K_c \\ u_\alpha^+(\mathbf{k}, t) \text{ for } K_c < k < K_{max}. \end{cases} \qquad (3.1)$$

The Navier–Stokes equation may then be decomposed using (3.1), to give

$$(\partial_t + \nu_0 k^2) u_k^- = M_k^- (u_j^- u_{k-j}^- + 2 u_j^- u_{k-j}^+ + u_j^+ u_{k-j}^+), \qquad (3.2)$$

$$(\partial_t + \nu_0 k^2) u_k^+ = M_k^+ (u_j^- u_{k-j}^- + 2 u_j^- u_{k-j}^+ + u_j^+ u_{k-j}^+), \qquad (3.3)$$

where, for simplicity, we have contracted all vector indices and independent variables into a single subscript. Given these expressions our aim is then simply to average out the effect of the high wavenumber modes in (3.2) upon the remaining low wavenumber modes.

4 Mode elimination by conditional average

In order to obtain an expression for the average effect of the high wavenumber modes upon a particular low wavenumber mode, we need to average out the u^+ whilst holding the u^- constant. This requires a *conditional average*, which we denote by $\langle \cdots \rangle_c$, satisfying the condition

$$\langle u_\alpha^-(\mathbf{k}, t) \rangle_c = u_\alpha^-(\mathbf{k}, t). \qquad (4.1)$$

This is the *only* rigorous property we can attribute to the conditional average, and it should also be noted that it is vital to distinguish between this operation and that of a filtered ensemble average, in which we essentially perform no average for wavenumbers below K_c and an ordinary ensemble average for wavenumbers above the cutoff.

In establishing the statistical properties of $u_\alpha(\mathbf{k}, t)$ we consider an ensemble $\mathcal{W}$ consisting of the set of M time-independent realizations $\{W_\alpha^{(i)}(\mathbf{k})\}$, each realization being labelled by an integer i. Subject to certain weak conditions, the ensemble average is simply

$$\langle u_\alpha(\mathbf{k},t) \rangle = \lim_{M \to \infty} \frac{1}{M} \sum_{i=1}^{M} W_\alpha^{(i)}(\mathbf{k}) = \bar{U}_\alpha(\mathbf{k},t), \qquad (4.2)$$

where $\bar{U}(\mathbf{k},t)$ is the time average of $u_\alpha(\mathbf{k},t)$. This procedure can then be extended to any well behaved functional, $F[u_\alpha(\mathbf{k},t)]$, thus:

$$\langle F[u_\alpha(\mathbf{k},t)] \rangle = \lim_{M \to \infty} \frac{1}{M} \sum_{i=1}^{M} F[W_\alpha^{(i)}(\mathbf{k})]. \qquad (4.3)$$

This ensemble average is illustrated schematically in Figure 1, where we show how the energy spectrum is obtained by averaging over all the individual members of the ensemble. For comparison, in Figure 2 we also provide an illustration of the filtered ensemble average in which the average is only performed for wavenumbers above K_c, the low wavenumber region not being subject to any averaging procedure.

Now we consider how to perform a *conditional* average. To do this, we first select a subensemble, $\mathcal{Y} \equiv \left\{ Y_\alpha^{(i)}(\mathbf{k}) \right\} \subset \mathcal{W}$, and choose the members of this *biased* (or conditional) subensemble to be those N $(N \le M)$ members of $\mathcal{W}$ satisfying the criterion

$$\lim_{\delta \to 0} \left(\max|\theta^-(k)W_\alpha^{(i)}(\mathbf{k}) - u_\alpha^-(\mathbf{k},t_1)| \right) \le \delta, \qquad (4.4)$$

where t_1 is some fixed time, $u_\alpha^-(\mathbf{k},t_1)$ is the low wavenumber part of the velocity field at time t_1 and $\theta^-(k) = 1$ for $0 \le k < K_c$, and zero otherwise. The conditional average is then obtained by generalizing (4.2) and (4.3) to the biased subensemble, viz

$$\langle u_\alpha(\mathbf{k},t) \rangle_c = \lim_{N \to \infty} \frac{1}{N} \sum_{i=1}^{N} Y_\alpha^{(i)}(\mathbf{k}), \qquad (4.5)$$

and

$$\langle F[u_\alpha(\mathbf{k},t)] \rangle_c = \lim_{N \to \infty} \frac{1}{N} \sum_{i=1}^{N} F[Y_\alpha^{(i)}(\mathbf{k})]. \qquad (4.6)$$

It follows by construction that (4.1) holds, since from (4.4) and (4.5)

$$\langle u_\alpha^-(\mathbf{k},t) \rangle_c = \lim_{N \to \infty} \frac{1}{N} [N u_\alpha^-(\mathbf{k},t)] = u_\alpha^-(\mathbf{k},t). \qquad (4.7)$$

The operation of selecting a subensemble to perform the conditional average is illustrated schematically in Figure 3. Here we illustrate how the members of the biased subensemble are selected from the full ensemble, and

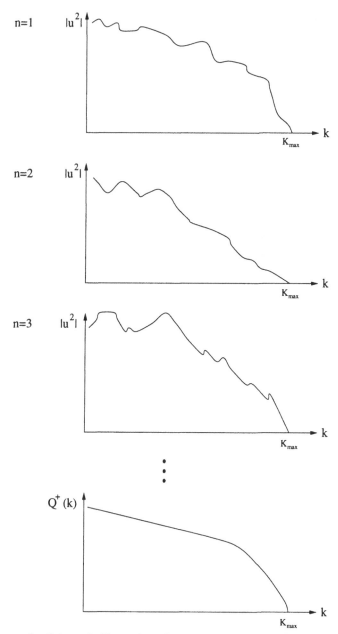

Figure 1: Schematic illustration of the way in which the full ensemble average is performed by summing over all the individual members of the ensemble.

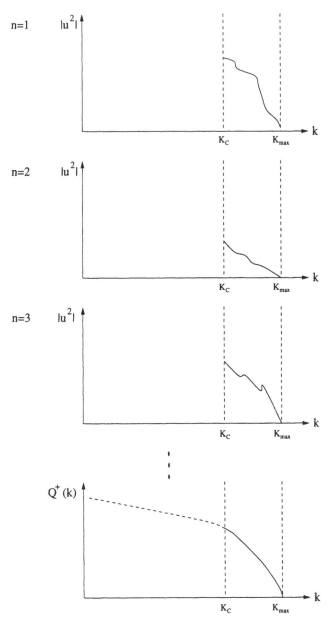

Figure 2: Schematic illustration of the operation involved in performing a filtered ensemble average. For wavenumbers above K_c, the usual ensemble averaging procedure is performed, whilst for wavenumbers below there is no averaging.

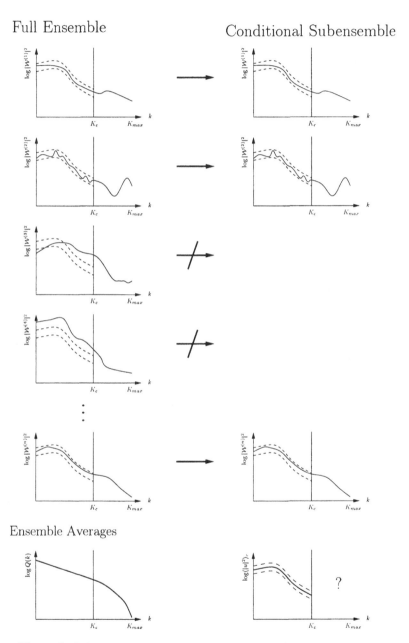

Figure 3: Schematic illustration of how members of the biased (or *conditional*) sub-ensemble $\mathcal{Y}$ (RHS) are selected from the full ensemble $\mathcal{W}$ (LHS), and the respective averages obtained from these ensembles.

the results we would expect for averages performed over the two different ensembles. This figure also serves to illustrate the problem now facing us, which arises from the fact that the Navier–Stokes equation displays deterministic chaos. The physical reasoning behind the conditional average is that since the Navier–Stokes equation displays chaotic behaviour we may constrain the velocity modes at low wavenumbers and yet still view the high wavenumber modes as being unconstrained. However, we cannot be sure of the exact behaviour of the u_k^+ modes under the conditional average. This can be seen if we consider two extreme scenarios for their behaviour.

Firstly, if we assume that subensemble is strictly deterministic, then in this instance, u_k^+ is fully determined by prescribing u_k^-. Equation (4.1) then implies that $\langle u_j^- u_{k-j}^- \rangle_c = u_j^- u_{k-j}^-$, $\langle u_j^- u_{k-j}^+ \rangle_c = u_j^- u_{k-j}^+$ and $\langle u_j^+ u_{k-j}^- \rangle_c = u_j^+ u_{k-j}^-$. Thus, in this scenario, the low-pass filtered Navier–Stokes equation, equation (3.2), reduces back to itself under the conditional average, and no modes are averaged out.

Secondly, if we assume that the subensemble is purely random, then in this case it follows that u_k^+ is independent of u_k^-. Hence, applying the conditional average to the low-pass filtered Navier–Stokes equation we find

$$(\partial_t + \nu_0 k^2)u_k^- = M_k^- u_j^- u_{k-j}^-,$$

the $u_j^- u_{k-j}^+$ term being zero since the ensemble average of u^+ is zero, whilst the $u_j^+ u_{k-j}^+$ is zero due to homogeneity. Thus in this scenario it appears that there is *no* effect from nonlinear coupling.

5 Conditional average with asymptotic freedom

In reality we are faced with a situation somewhere between these two extremes, and so we replace our criterion for members of the biased subensemble, equation (4.4), which is equivalent to the first of these situations if $\delta = 0$, by a less precise criterion

$$\max \left| \theta^-(k) W_\alpha^{(i)}(\mathbf{k}) - u_\alpha^-(\mathbf{k}, t_1) \right| \leq \xi, \tag{5.1}$$

where, in general, ξ is of the order of the turbulent velocities involved.

To obtain a non-trivial conditional average we must now identify those circumstances in which ξ may be neglected as being, in some sense, small. A measure of the 'smallness of ξ' can be identified by constructing the subensemble as

$$W_\alpha^{(i)}(\mathbf{k}) = u_\alpha^-(\mathbf{k}, t_1) + \phi_\alpha^{(i)}(\mathbf{k}, t_1), \tag{5.2}$$

where i is any label satisfying (5.1). If we then further restrict the subensemble to be such that the set $\{\phi_\alpha^{(i)}(\mathbf{k}, t_1)\}$ satisfies (4.1), we find that

$$\langle u_j^- u_{k-j}^- \rangle_c = u_j^- u_{k-j}^- + \langle \phi_j \phi_{k-j} \rangle_c. \tag{5.3}$$

Thus in order to maintain form invariance of the Navier–Stokes equation under conditional averaging, we require

$$\langle \phi_j \phi_{k-j} \rangle_c \to 0 \tag{5.4}$$

in some limit. This is our criterion for the smallness of ξ.

If we further suppose that chaos and unpredictability are local[1] characteristics of turbulence, and there is support for such a view [7, 8], then if K_c and $K_{\max}$ are sufficiently far apart we might expect, due to the development of unpredictability as we move away from K_c, that the effect of the constraint given in equation (5.1) would die away, such that

$$\lim_{k \to K_{\max}} \langle u_\alpha^+(\mathbf{k}, t) \rangle_c \to \langle u_\alpha^+(\mathbf{K}_{\max}, t) \rangle. \tag{5.5}$$

We refer to this property as *asymptotic freedom*. In order extend this concept to higher-order moments, we introduce the following *Hypothesis of Local Chaos*:

For sufficiently large Reynolds' number and corresponding $K_{\max}$, there exists a cut-off wavenumber $K_c < K_{\max}$ such that a mixed conditional moment involving p low wavenumber and r high wavenumber modes takes the limiting form

$$\lim_{\xi \to 0} \langle u_\alpha^-(\mathbf{k_1}, t) u_\beta^-(\mathbf{k_2}, t) \dots u_\gamma^-(\mathbf{k_p}, t)$$
$$\times u_\delta^+(\mathbf{k_{p+1}}, t) u_\epsilon^+(\mathbf{k_{p+2}}, t) \dots u_\sigma^+(\mathbf{k_{p+r}}, t) \rangle_c \to$$
$$u_\alpha^-(\mathbf{k_1}, t) u_\beta^-(\mathbf{k_2}, t) \dots u_\gamma^-(\mathbf{k_p}, t)$$
$$\times \lim_{\{\cdot\} \to K_{\max}} \langle u_\delta^+(\mathbf{k_{p+1}}, t) u_\epsilon^+(\mathbf{k_{p+2}}, t) \dots u_\sigma^+(\mathbf{k_{p+r}}, t) \rangle, \tag{5.6}$$

where $\lim_{\{\cdot\} \to K_{\max}}$ means take the limit for all wavevector arguments of the u^+ modes within the Dirac brackets, with the condition of equation (5.4) satisfied as a corollary.

This provides our definition of an asymptotic conditional average and may be used to evaluate all terms involving mixed products of u^- with u^+. For example

$$\lim_{\xi \to 0} \langle u_j^- u_{k-j}^+ \rangle_c = u_j^- \lim_{\{\cdot\} \to K_{\max}} \langle u_{k-j}^+ \rangle = 0, \tag{5.7}$$

since $\langle u_\alpha(\mathbf{k}, t) \rangle = 0$. Note here also that the hypothesis as defined is more general than is necessary, since we shall only need to consider products containing at most two u^- modes.

---

[1]Note that, as is usual in theoretical work on turbulence, we take 'local' to mean local with respect to wavenumber.

6 Conditionally-averaged equation of motion for the explicit scales

If we then take the conditional average of the low-pass filtered Navier–Stokes equation, equation (3.2), we obtain

$$(\partial_t + \nu_0 k^2) u_k^- = M_k^- \{\langle u_j^- u_{k-j}^- \rangle_c + 2\langle u_j^- u_{k-j}^+ \rangle_c + \langle u_j^+ u_{k-j}^+ \rangle_c\}, \qquad (6.1)$$

where the conditional average of u_k^- on the left hand side has been evaluated using (4.1). This expression may be further re-written as

$$(\partial_t + \nu_0 k^2) u_k^- = M_k^- u_j^- u_{k-j}^- + S^-(k|K_c) + M_k^- \lim_{\xi \to 0} \langle u_j^+ u_{k-j}^+ \rangle_c, \qquad (6.2)$$

where

$$S^-(k|K_c) = M_k^- \{\langle \phi_j^- \phi_{k-j}^- \rangle_c + 2\langle u_j^- u_{k-j}^+ \rangle_c + \langle u_j^+ u_{k-j}^+ \rangle_c - \lim_{\xi \to 0} \langle u_j^+ u_{k-j}^+ \rangle_c\}.$$
$$(6.3)$$

It should also be noted that the hypothesis *must* hold for $K_c \to 0$, as in this instance the equations just reduce to those of the Reynolds average, with $u_\alpha(\mathbf{k}, t) \to \bar{U}_\alpha(\mathbf{k}, t)$ as given by (4.2).

7 Iterative solution for the high-k modes

Our hypothesis does not tell us how evaluate the conditional average in (6.2), which involves a non-trivial projection of a product of u^+ modes in the Hilbert space of the u^- modes, but we may use the high-pass filtered Navier–Stokes equation, equation (3.3), to obtain a governing equation for this quantity. To do this we use (3.3) to write equations for u_j^+ and u_{k-j}^+, multiply these equations by u_{k-j}^+ and u_j^+ respectively, add the resulting equations together, and then take the conditional average. After some rearrangement of dummy variables, this gives

$$\lim_{\xi \to 0} (\partial_t + \nu_0 j^2 + \nu_0 |\mathbf{k} - \mathbf{j}|^2) \langle u_j^+ u_{k-j}^+ \rangle_c$$
$$= \lim_{\xi \to 0} 2M_j^+ \times \{\langle u_p^- u_{j-p}^- u_{k-j}^+ \rangle_c + 2\langle u_p^- u_{j-p}^+ u_{k-j}^+ \rangle_c + \langle u_p^+ u_{j-p}^+ u_{k-j}^+ \rangle_c\} \quad (7.1)$$

Applying the hypothesis, it is easily seen that the first term on the right hand side of (7.1) is zero (since in the limit it involves the ensemble average of u_k^+), and that the second term gives rise to a term linear in u_k^-. The third term may be evaluated by iterating the above procedure to form a dynamical equation for $\langle u_p^+ u_{j-p}^+ u_{k-j}^+ \rangle_c$, which in turn gives rise to higher-order moments.

In general, we can show that a similar pattern occurs for all higher order moments involving only products of u^+. That is, each such moment gives rise to a moment involving two u^- modes, which in general has to be zero for consistency in its wavevector arguments, a term linear in u_k^-, and a moment involving only u^+ modes of next higher order. Hence we may write the general result

$$M_{\alpha\beta\gamma}^-(\mathbf{k}) \lim_{\xi \to 0} \langle u_\beta^+(\mathbf{j}, t) u_\gamma^+(\mathbf{k} - \mathbf{j}, t) \rangle_c = \int_{-\infty}^t ds A(\mathbf{k}, t - s) u_\alpha^-(\mathbf{k}, s), \qquad (7.2)$$

where $A(\mathbf{k}, t - s)$ has the form

$$\begin{aligned}
A(\mathbf{k}, t - s) = &\exp[-(\nu_0 j^2 + \nu_0 |\mathbf{k} - \mathbf{j}|^2)(t - s)] \\
&\times \{ 4 M_k^- M_j^+ \lim_{\{\cdot\} \to K_{\max}} \langle u_{j-p}^+ u_{k-j}^+ \rangle \\
&+ 24 M_k^- M_j^+ L_{03}^{-1} M_p^+ \lim_{\{\cdot\} \to K_{\max}} \langle u_{p-q}^+ u_{j-p}^+ u_{k-j}^+ \rangle + \ldots \}
\end{aligned} (7.3)$$

$L_{03} \equiv \partial_t + \nu_0 p^2 + \nu_0 |\mathbf{j} - \mathbf{p}|^2 + \nu_0 |\mathbf{k} - \mathbf{j}|^2$, and where higher order terms are easily found by induction. Thus, in all, equation (6.2) for the low wavenumber modes may be written as

$$(\partial_t + \nu_0 k^2) u_k^- - \int_{-\infty}^t ds A(\mathbf{k}, t - s) u_\alpha(\mathbf{k}, s) = M_k^- u_j^- u_{k-j}^- + S^-(k|K_c). \quad (7.4)$$

It should perhaps be emphasised that at this stage we have yet to make an approximation. We have merely asserted an hypothesis of asymptotic freedom and, on that basis, solved iteratively for the mean effect of the high-wavenumber modes to (in principle) any order.

8 Two approximations

In order to test our hypothesis, we make two approximations. First, we truncate the expansion of $A(\mathbf{k}, t)$ at lowest non-trivial order. This can be justified by the introduction of a *local* Reynolds number based on a length scale K_c^{-1}, the moment expansion being re-expressed as a power series in this parameter.

The local Reynolds number may be expressed in terms of the energy spectrum as

$$R(K_c) = \frac{1}{\nu_0} \left(\frac{E(K_c)}{2\pi K_c} \right)^{1/2}. \qquad (8.1)$$

In order to evaluate this, we substitute the energy spectrum obtained from a direct numerical simulation, and hence obtain an estimate of its magnitude for any choice of cutoff K_c. The results from such a substitution are illustrated in Figure 4.

As can be clearly seen here, this calculation yields the result that $R(K_c)$ is less than unity for any value of K_c greater than $\sim 0.45 k_d$, taking a value of approximately 0.1 as K_c tends to k_d. Given these results, it would thus

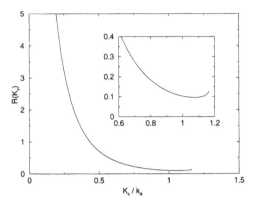

Figure 4: Prediction for the local Reynolds number $R(K_c)$ as found using the time averaged energy spectrum from a 256^3 direct numerical simulation with $\varepsilon = 0.149$, $\nu_0 = 10^{-3}$ and $R_\lambda = 190$. The inset illustrates the high wavenumber region of the graph.

seem a reasonable approximation to neglect higher order terms in the moment expansion. Doing this leaves us with the expression

$$\lim_{\xi \to 0} M_{\alpha\beta\gamma}^-(\mathbf{k})\langle u_\beta^+(\mathbf{j}, t)u_\gamma^+(\mathbf{k} - \mathbf{j}, t)\rangle_c = \int_{-\infty}^t ds\, \exp[-(\nu_0 j^2 + \nu_0|\mathbf{k} - \mathbf{j}|^2)(t - s)]$$

$$\times \{4M_{\alpha\beta\gamma}^-(\mathbf{k})M_{\beta\delta\epsilon}^+(\mathbf{j})\int d^3p \lim_{\{\cdot\} \to K_{\max}} \langle u_\epsilon^+(\mathbf{j} - \mathbf{p}, s)u_\gamma^+(\mathbf{k} - \mathbf{j}, s)\rangle u_\delta^-(\mathbf{k}, s).$$

$$(8.2)$$

Second, there is the question of how to perform the time integral. This reduces to

$$\int_{-\infty}^t ds\, \exp[-(\nu_0 j^2 + \nu_0|\mathbf{k} - \mathbf{j}|^2)(t - s)]u_\delta^-(\mathbf{k}, s), \qquad (8.3)$$

since for stationary, homogeneous, and isotropic turbulence we may write

$$\langle u_\epsilon^+(\mathbf{j} - \mathbf{p}, s)u_\gamma^+(\mathbf{k} - \mathbf{j}, s)\rangle = Q(|\mathbf{k} - \mathbf{j}|)D_{\epsilon\gamma}(\mathbf{k} - \mathbf{j})\delta(\mathbf{k} - \mathbf{p}), \qquad (8.4)$$

where $Q(k)$ is the spectral density and δ is the Dirac delta function, hence meaning that these terms have no bearing on the time integral in (8.2). To perform the integral in (8.3) we change the variable of integration from s to $T = t - s$, expand the resultant $u_\delta^-(\mathbf{k}, t - T)$ as a Taylor series about $T = 0$, and then truncate the expansion at zero order, this approach being based upon the physical idea that the u^- modes are slowly evolving on time scales defined by the inverse of $\nu_0 j^2 + \nu_0|\mathbf{k} - \mathbf{j}|^2$.

As with the truncation of the moment expansion, we have also investigated the validity of this second approximation using results from direct numerical simulations performed on a 256^3 grid, with Taylor-Reynolds number $R_\lambda =$ 190. Although these results are not conclusive, since at this resolution the simulations have a very limited inertial range (see, for instance, [9, 10]), they again indicate that there is a range of K_c (K_c greater than $\sim 0.6k_d$) where the approximation gives rise to an error term of less than unity, and that the magnitude of the error will decrease as we increase R_λ to the large values where we may reasonably expect our hypothesis to hold.

9 Asymptotic eddy viscosity

With these approximations, the right hand side of (7.2) is simple to evaluate, and we are left with the final expression for the conditional average on the right hand side of (6.2) for the low-wavenumber modes

$$M_{\alpha\beta\gamma}^-(\mathbf{k})\lim_{\xi\to 0}\langle u_\beta^+(\mathbf{j},t)u_\gamma^+(\mathbf{k}-\mathbf{j},t)\rangle_c =$$

$$= 4M_{\alpha\beta\gamma}^-(\mathbf{k})M_{\beta\delta\epsilon}^+(\mathbf{j})\lim_{|\mathbf{k}-\mathbf{j}|\to K_{\max}}\frac{Q(|\mathbf{k}-\mathbf{j}|)D_{\epsilon\gamma}(\mathbf{k}-\mathbf{j})}{\nu_0 j^2 + \nu_0|\mathbf{k}-\mathbf{j}|^2}u_\delta^-(\mathbf{k},t),\quad (9.1)$$

which is linear in u_k^- and hence means that the term may be interpreted as an increment to the viscosity. Substituting (9.1) into equation (7.4) we are thus left with

$$(\partial_t + \nu_0 k^2)u_k^- - 4M_k^- M_j^+\lim_{|\mathbf{k}-\mathbf{j}|\to K_{max}}\frac{Q(|\mathbf{k}-\mathbf{j}|)D_{k-j}}{\nu_0 j^2 + \nu_0|\mathbf{k}-\mathbf{j}|^2}u_k^-$$

$$= M_k^- u_j^- u_{k-j}^- + S^-(k|K_c).\quad (9.2)$$

In order to evaluate the limit on the wavevectors, we make a first-order truncation of a Taylor series expansion of $Q^+(|\mathbf{k}-\mathbf{j}|)$ about $K_{\max}$. Doing this we re-obtain the results previously obtained using the two-field theory of McComb and Watt [4]. As they showed, a Renormalization Group calculation based on these equations reaches a fixed point, in terms of the scaled effective viscosity

$$\nu_n^*(k') = \nu_n(k)/(\alpha^{1/2}\varepsilon^{2/3}k_n^{-4/3}),\quad (9.3)$$

where $k' = k/k_n$ and $k_n = (K_c/K_{max})^n K_{max}$, for a range of starting values for the molecular viscosity, see Figure 5, and also gives a prediction for the Kolmogorov constant of 1.60 ± 0.01, in good agreement with experiment, for $0.55K_{\max} \leq K_c \leq 0.75K_{\max}$, as illustrated in Figure 6.

In addition, this calculation also obtains an effective viscosity, as illustrated in Figure 7, where we show how its value evolves as we eliminate successive shells of wavenumber modes within the Renormalization Group calculation. For completeness, we also include Figure 8, which shows how the effective viscosity obtained at the fixed point varies with the choice of K_c/K_d. This calculation obtained the Kolmogorov exponent and pre-factor by assuming

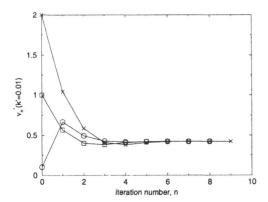

Figure 5: Convergence of the scaled effective viscosity to a fixed point for three different values of the initial viscosity. (Evaluated at $K_c/K_{max} = 0.6$.)

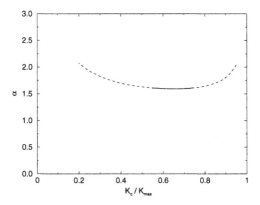

Figure 6: Prediction for the value of the Kolmogorov constant α as obtained from a numerical calculation using the two-field theory of McComb and Watt. The dashed lines illustrate the regions where we expect the theory to break down.

that the effective viscosity and its increment scale in the same way (which is true at the fixed point) and that the rate of energy transfer is renormalized. This latter assumption amounted, in our present terms, to the neglect of the term $S^-(k|K_c)$ in equation (6.2).

A basis for the justification of this step can now be offered as follows. The equation for the energy spectrum is obtained by multiplying the dynamical equation for $u_\alpha^-(\mathbf{k}, t)$ by $u_\alpha^-(-\mathbf{k}, t)$ and then performing an average over the

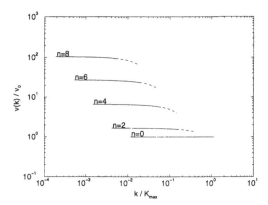

Figure 7: Development of the total effective viscosity Renormalization Group calculation. The fixed point is reached after 8 iterations. (Evaluated at $K_c/K_{max} = 0.6$.)

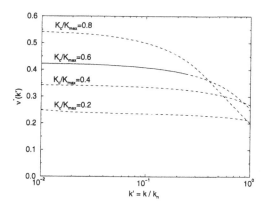

Figure 8: Dependence of the scaled effective viscosity upon wavenumber, showing the effect of varying the bandwidth parameter K_c/K_{max}. The dashed lines illustrate the regions of wavenumber and the bandwidths where we expect the theory to break down.

full ensemble. Thus the effect of the term $S^-(k|K_c)$ is just

$$\langle S^-(k|K_c)u_\alpha^-(-\mathbf{k}, t)\rangle, \tag{9.4}$$

and if we consider the form of $S^-(k|K_c)$, as given in equation (6.3), we see that each of its terms involves a conditional average. In evaluating such terms within (9.4) we perform a double summation, firstly summing over all members with low wavenumber modes close to a particular member of the

ensemble, and then repeating this summation for every member of the ensemble. Now, the initial ensemble was constructed according to the principle
of equal *a priori* probabilities, but this is no longer necessarily true of the
composite ensemble which we are now considering. If it were true, then the
terms in $S^-(k|K_c)$ would vanish identically for all K_c, and although unlikely
to be strictly correct, given the results of the Renormalization Group calcu-
lations [4], it seems likely that the contribution from $S^-(k|K_c)$ is small for
K_c in the range $0.55K_{max} \leq K_c \leq 0.75K_{max}$.

10 Conclusions

To sum up, we began with equation (3.2) for the low-wavenumber modes.
and conditionally-averaged it to obtain (6.1). This equation was immediately
rewritten as (6.2), in which the mean effect of the high-wavenumber modes
was represented according to our hypothesis of asymptotic freedom, along
with a correction term $S^-(k|K_c)$.

In the immediately preceding section we argued that the contribution from
this correction term may be small in the equation for the dissipation rate but
it should be noted that this work does not suggest that $S^-(k|K_c)$ can be
neglected in equation (6.2), which is the governing equation for a single realization. However it does suggest that, having averaged out the chaotic part
to yield an effective viscosity, one should consider modelling the relationship
of $S^-(k|K_c)$ to the u_k^- modes as predominantly deterministic.

It may also be the case that the hypothesis of local chaos, as stated here, is
somewhat over-complicated. As we have previously noted, within our calculation we need to consider mixed moments containing at most two u_k^- modes
and we are currently considering whether it is possible to rewrite these terms
in such a way that they only involve a single u_k^- mode. As is shown in Appendix A, such a form for the hypothesis is given support if we consider the
problem in terms of the probability functional of the velocity field, and if this
is possible then we will be able to redefine the hypothesis in a simpler form.
Determining whether or not this is possible is our next immediate aim.

Acknowledgements

Both authors acknowledge the support and facilities provided by the Isaac
Newton Institute. We also wish to thank Alistair Young for providing results
from numerical simulations. C.J. acknowledges the financial support of the
Engineering and Physical Sciences Research Council.

Appendix: The conditional projector in function space

The reasoning behind the hypothesis of local chaos can be clarified if we consider the Hilbert space projection of products of the u^+ modes onto a u^- mode. We start by introducing the exact probability functional $P[\mathbf{u}(\mathbf{k},t)]$ such that the expectation value of any well behaved functional F is given by

$$\langle F[\mathbf{u}(\mathbf{k},t)]\rangle = \int \mathcal{D}\mathbf{u}(\mathbf{k},t)P[\mathbf{u}(\mathbf{k},t)]F[\mathbf{u}(\mathbf{k},t)],$$

where the functional integration is indicated symbolically. This operation is unaffected by our filtering the modes, and hence

$$\langle F[\mathbf{u}^\pm(\mathbf{k},t)]\rangle = \int \mathcal{D}\mathbf{u}(\mathbf{k},t)P[\mathbf{u}(\mathbf{k},t)]F[\mathbf{u}^\pm(\mathbf{k},t)].$$

In order to extract a conditional projection on the u^- modes, we then construct a projection operator $\mathcal{P}_c^-$, such that its action on an arbitrary functional is given by

$$\mathcal{P}_c^- F[\mathbf{u}(\mathbf{k},t)] = \int d^3j \int ds\, \mathbf{u}^-(\mathbf{j},s) \int \mathcal{D}\mathbf{u}(\mathbf{k},t)P[\mathbf{u}(\mathbf{k},t)]\frac{\delta F[\mathbf{u}(\mathbf{k},t)]}{\delta \mathbf{u}^-(\mathbf{j},s)},$$

where $\delta/\delta\mathbf{u}^-$ denotes functional differentiation. If we assume the proper normalization

$$\int \mathcal{D}\mathbf{u}(\mathbf{k},t)P[\mathbf{u}(\mathbf{k},t)] = 1,$$

and take $F[\mathbf{u}(\mathbf{k},t)] = \mathbf{u}^-(\mathbf{k},t)$, then we can easily see that

$$\begin{aligned}
\mathcal{P}_c^- \mathbf{u}^-(\mathbf{k},t) &= \int d^3j \int ds\, \mathbf{u}^-(\mathbf{j},s)\delta(\mathbf{k}-\mathbf{j})\delta(t-s) \\
&= \mathbf{u}^-(\mathbf{k},t)
\end{aligned}$$

as required.

In general, of course, we wish to project out products of u^- with functionals of the u^+ modes, that is $F[\mathbf{u}(\mathbf{k},t)] = \mathbf{u}^-(\mathbf{k},t)f[\mathbf{u}^+(\mathbf{k},t)]$. Hence, we require

$$\mathcal{P}_c^- \mathbf{u}^-(\mathbf{k},t)f[\mathbf{u}^+(\mathbf{k},t)] = \int d^3j \int ds\, \mathbf{u}^-(\mathbf{j},s) \int \mathcal{D}\mathbf{u}(\mathbf{k},t)P[\mathbf{u}(\mathbf{k},t)]$$

$$\times \frac{\delta \mathbf{u}^-(\mathbf{k},t)f[\mathbf{u}^+(\mathbf{k},t)]}{\delta \mathbf{u}^-(\mathbf{j},s)}$$

$$= \int d^3j \int ds\, \mathbf{u}^-(\mathbf{j},s) \int \mathcal{D}\mathbf{u}(\mathbf{k},t)P[\mathbf{u}(\mathbf{k},t)]$$

$$\times \left\{ \delta(\mathbf{k}-\mathbf{j})\delta(t-s)f[\mathbf{u}^+(\mathbf{k},t)] + \mathbf{u}^-(\mathbf{k},s)\frac{\delta f[\mathbf{u}^+(\mathbf{k},t)]}{\delta \mathbf{u}^-(\mathbf{j},s)} \right\}.$$

The second term in the curly brackets is intractable, and is simply another way of expressing the turbulence problem, but if we are able to find a limit in which u^+ becomes independent of u^-, then this term vanishes and we are left with

$$\mathcal{P}_c^- \mathbf{u}^-(\mathbf{k}, t) f[\mathbf{u}^+(\mathbf{k}, t)] = \mathbf{u}^-(\mathbf{k}, t) \langle f[\mathbf{u}^+(\mathbf{k}, t)] \rangle.$$

In making the hypothesis of local chaos, we postulate that such a limit exists, under the conditions defined.

References

[1] D. Forster, D.R. Nelson, and M.J. Stephen, *Phys. Rev. A* **16**, 732 (1977).

[2] V. Yakhot and S.A. Orszag, *J. Sci. Comp.* **1**, 3 (1986).

[3] Y. Zhou and G. Vahala, *Phys. Rev. E* **47**, 2503 (1993).

[4] W.D. McComb and A.G. Watt, *Phys. Rev. A* **46**, 4797 (1992).

[5] K.G. Wilson, *Rev. Mod. Phys.* **47**, 773 (1975).

[6] W.D. McComb, W. Roberts, and A.G. Watt, *Phys. Rev. A* **45**, 3507 (1992).

[7] R. G. Deissler, *Phys. Fluids* **29**, 1453 (1986).

[8] L. Machiels, *Phys. Rev. Lett.* **79**, 3411 (1997).

[9] J. Brasseur and C.H. Wei, *Phys. Fluids* **6**, 842 (1994).

[10] P.K. Yeung and Y. Zhou, *Phys. Rev. E* **56**, 1746 (1997).